Thermodynamics and Regulation of Biological Processes

Thermodynamics and Regulation of Biological Processes

Editors
I. Lamprecht · A. I. Zotin

W DE G

Walter de Gruyter · Berlin · New York 1985

Editors
Prof. Dr. Ingolf Lamprecht
Institute of Biophysics
Free University Berlin
D-1000 Berlin 33

Prof. Dr. A. I. Zotin
Institute of Developmental Biology
Academy of Science of the U.S.S.R.
Moscow

Library of Congress Cataloging in Publication Data

Main entry under title:
Thermodynamics and regulation of biological processes.
Translated from the Russian.
Bibliography: p.
Includes index.
1. Biophysics. 2. Thermodynamics. 3. Biological control systems. 4. Information theory in biology. 5. Evolution. 6. Ontogeny. I. Lamprecht, Ingolf, 1933- . II. Zotin, A. I. (Aleksandr Il'ich), 1926- .
QH505.T394 1985 574.19'1 84-23302
ISBN 0-89925-007-6 (U.S.)

CIP-Kurztitelaufnahme der Deutschen Bibliothek

Thermodynamics and regulation of biological processes /
ed. Ingolf Lamprecht ; A. I. Zotin. – Berlin ; New York :
de Gruyter, 1985.
ISBN 3-11-009789-3

NE: Lamprecht, Ingolf [Hrsg.]

 Printing: Gerike GmbH, Berlin. – Binding: Dieter Mikolai, Berlin. Printed in Germany.

This monograph deals with thermodynamic aspects of control and regulation in biological systems; relationships between thermodynamics of irreversible processes, information and control theory; extremal principles and regulation in ecological systems; self-organisation, evolution and ontogenesis. The volume includes contributions from leading scientists in these fields from the Soviet Union and other countries, therefore it presents the most important trends in modern studies of the thermodynamics of biological processes which deal with control and regulation in living systems.

The monograph will be of value to researchers in biophysics, physiology, developmental biology, as well as for physicists and mathematicians interested in modern thermodynamics, information and control theory, thermodynamics of biological processes and mathematical biology.

Preface

This book is the last in the series of monographs on the thermodynamics of biological processes which are the result of active cooperation between scientists of the USSR and many other countries.

The first monograph of the series entitled 'Thermodynamics of Biological Processes' appeared in Russian in 1976 (Nauka publishers, Moscow) and in English in 1978 (de Gruyter, Berlin). It dealt mainly with thermodynamic aspects of developmental biology, dissipative structures, classification and evolution of organisms.

The second monograph 'Thermodynamics and Kinetics of Biological Processes' appeared in Russian in 1980 (Nauka publishers, Moscow) and in English in 1982 (de Gruyter, Berlin). It was devoted to the kinetics of transition processes, the reaction of living systems to external influences, and to the adaptation of the organism to changing environmental conditions. Attention was paid mainly to purely thermodynamic problems concerned with the present state of phenomenological and statistical thermodynamics of non-equilibrium processes.

This third monograph 'Thermodynamics and Regulation of Biological Processes' deals with the processes most specific to the living systems, such as regulation and control, self-organisation, development and evolution. Perhaps the discussion of these problems in terms of thermodynamics is not fully justified as yet, since the thermodynamics of organised systems, i. e. those systems in which regulation and control are realised, has not been adequately developed. However, recently, along with thermodynamics of non-equilibrium processes, there has appeared such promising trends as synergetics (Haken, 1978) and thermodynamics of information processes (Poplavsky, 1981), which makes us believe that in the very near future the thermodynamics of organised systems will achieve a satisfactory basis. The formation of such trends as the thermodynamics of information processes seems most promising. Its main task is the thermodynamic analysis of such phenomena as: orderliness, organisation and control; energetic cost of information storage, transmission and usage; precision and uniformity in control systems. It is evident that such problems are of primary importance for biology, since a specific concern of biological objects is, primarily, the orderliness and organisation of all the processes proceeding in the living system. It is impossible to construct a thermodynamics of biological processes without taking into account these specific features of biological systems.

Notwithstanding the above, it is too early to declare the beginning of a new era in the thermodynamics of biological processes. The fact is, that, these thermodynamic trends have not merged organically in order to form a satisfactory theoretical foundation for this field of science. For example, the thermodynamics of information processes is not sufficiently associated with the thermodynamics of non-equilibrium processes. This is partly due to historical reasons, since the thermodynamics of information processes emerged on the basis of the analysis of Maxwell demons (Maxwell, 1872), i. e. long before the appearance of a thermodynamics of irreversible processes (Onsager, 1931 a, b). Later these two trends developed independently, and a thermodynamics of non-equilibrium processes elaborated its own language and yielded some relationships which differed from those of classical thermodynamics, whilst the thermodynamics of information processes preserved the language and methodology of classical thermodynamics. As a result both trends have different languages, although they both deal with the same open non-equilibrium systems. Almost the same can be said about the interrelations between these trends and the science of selforganisation - synergetics. This separation is fully reflected in the present volume and will be clear to the readers on comparison of different chapters. It is hoped that the book will promote closer connections and, eventual, amalgamation of these trends in thermodynamics.

Like the first two monographs of the series the present one has been published because of the active support of the Corresponding Member of the USSR Academy of Sciences T. M. Turpaev, Director of the Institute of Developmental Biology, USSR Academy of Sciences and with valuable assistance of I. A. Ilyuchenko, G. S. Lozovskaya, I. S. Nikolskaya, S. M. Malyghin, L. I. Radzinskaya and of E. V. Presnov especially.

We thank all these and, of course, the contributors for supporting the idea of publication of this monograph, for their cooperation and effort in preparing the monograph.

The English translation was again thoroughly checked by A. E. Beezer, London, to whom we are indebted very much.

Berlin, October 1984

I. Lamprecht
A. I. Zotin

Contents

V. Extremal Principles

VI. Synergetics - The Problems of Selforganisation

VII. The Problems of Evolution

General Problems

Information and Cybernetic Aspects of Biological Thermodynamics

D. Leuschner

Monographs such as those by Zotin (1976, 1980) and Lamprecht, Zotin (1978, 1982), have shown the importance of thermodynamics in biology. This is true also, for information theory (see e. g. Eigen, Winkler, 1973/74; Eigen, Schuster, 1977, 1978) and for cybernetics by which is meant control theory or regulation (Drischel, 1973; Hassenstein, 1967; Nicolis, Prigogine, 1977). There is neither a complete theory nor a concrete distinction between biology and thermodynamics, nor between information theory and cybernetics. The aim of this paper is not to describe an essential, correct theory of biothermodynamics, bioinformation or biocybernetics, nor shall we attempt to create such a theory. We shall only illustrate a few of their interrelations between biology (Bio), thermodynamics (Th), information theory (Inf) and cybernetics (Cyb). In this connection, the central subject of this work is an outline of two new concepts: Thermodynamic phoronomics (ThP) and thermodynamic taxonomy (ThT). The first is the prologue to a theory of creation and production of information, and the second to the concentration of information, both, of course, refer to biology. Furthermore we give an example: photosynthesis and its interrelations to thermodynamics, information theory and cybernetics at different levels of organisation in the green plant and also include ontogeny, phylogeny and genetics.

Thermodynamics and Regulation of Biological and Processes

General and specific interrelations

The structure of the interrelations between biology, thermodynamics, information theory and cybernetics is respresented in Fig. 1. The connectors are numbered ("   ") and a short explanation for each number appears in the text. All levels of organisation in biology (molecular, cellular, organismic, ecologic) can be considered.

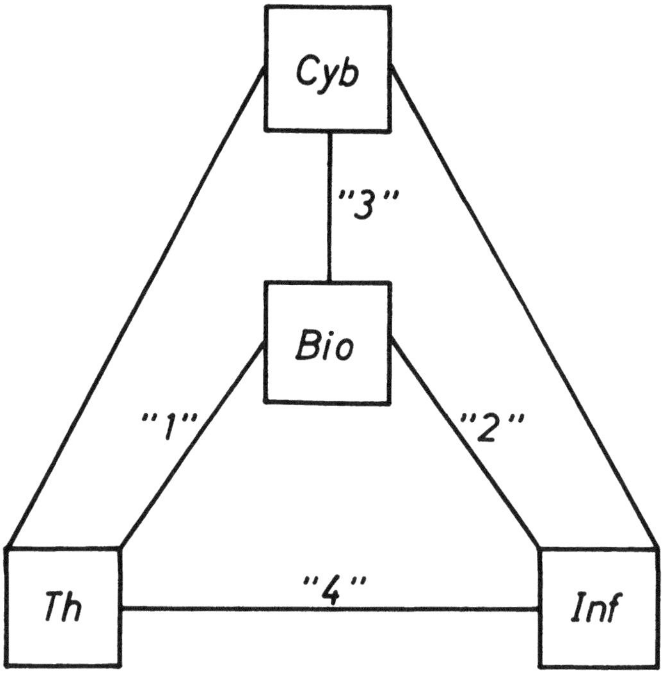

Fig. 1 Scheme of relationships between information theory (Inf), cybernetics (Cyb), thermodynamics (Th) and biology (Bio).

Thermodynamics and biology ("1")

The literature on this topic is large (see e. g. Zotin, 1974, 1976, 1982; Lamprecht, Zotin, 1978, 1980). Examples for different levels of organization have been discussed by Leuschner (1981 a). Consideration of examples was shown to be possible by Schrödinger (1951) who removed the apparent contradiction between the Second Law of Thermodynamics and phenomena associated with living matter. Biological phenomena do not exist in contradiction of this law; they exist on the basis of the Second Law which is also valid for open, living systems.

Modern biology includes study of the thermodynamics of equilibrium and of non-equilibrium (biological self-organisation see Ebeling, 1976; Ebeling, Feistel, 1981) states. The First Law can be **written** as

$$dE = đA + đQ \tag{1}$$

The change in internal energy E of the system (E is a state variable, and therefore dE a total differential) is equal to the sum of the work, $đA$, and the heat $đQ$ ($đ$ represents a non-exact differential, because A and Q are not state variables). We know no phenomenon in the field of biology for which the conservation law of energy is not valid. The Second Law is

$$dS = đ_e S + đ_i S \qquad đ_i S \geq 0 \tag{2}$$

where dS = the change of entropy in the system, $đ_e S$ = entropy flow from, or to, the system, $đ_i S$ = entropy production inside the system as a result of irreversible processes, and therefore $đ_i S$ is always positive for spontaneous processes.

The mathematical expression of the condition for maintenance of an open, isothermal system, i. e. of a living system, is

$$(-đ_e S) > đ_i S \geq 0$$

and from this we obtain

$$đ_e F > dE + T đ_i S \, ,$$

since

$$đ_e F = đ_e E - T đ_e S$$

and

$$dE = đ_e E + đ_i E \, , \quad đ_i E = 0$$

where F = free energy (subscripts e and i are analogous to those in (2) and T = absolute temperature). An organism is fed with free energy, $đ_e F$, from the surroundings and the value of this free energy underlies (2).

Information theory and biology ("2")

The term information as used in information theory is very general. In (3) we find only (see below) probabilities and reference to any phenomena (symbols, states or quantities). Therefore success in the application of information theory in Eigen's (1971) theory of self-organization and evolution was possible. In this respect we note: the information content of a DNA molecule and information creation (see (4) and Eigen, Winkler, 1973/74); and von Neumann's theory of games, see e. g. Wentzel, 1964).

It will be noticed that information possesses importance not only in the sphere of molecules, but also, for example, at the level of cells and of biocenoses. Two examples will illustrate this point. One is the greater volume of a eucaryotic cell in comparison with a procaryotic cell as a consequence of the greater information content of DNA in the former cell. The other is the information content in the manifold of a state of insects. Both examples demonstrate the generality of information in biology, but they also pose many questions that remain unsolved.

For example: The volume of a dinosaur is not equivalent to its analogous information content, and the information content of a state of insects is perhaps not best described through equations such as (3).

In spite of this we consider information to be the central theme of this article. Without information there can be no thermodynamics, no cybernetics and no biology (for the association between thermodynamics and information theory see e. g. Ebeling, 1976; Lamprecht, Zotin, 1978 and below; for the association between cybernetics and information theory see Laskowski, Pohlit, 1974 and below; last, but not least, DNA is information and, without DNA there is no life, and no biology).

Information theory investigates the transmission of information along communication channels and therefore we must discuss, in a mathematical term, the measure of information content, and its changes along communication channels.

Information content (or more briefly, information; sometimes named Shannon's entropy; see Meyer-Eppler, 1959; Schwartz, 1963) is defined by

$$I = -\sum_{i=1}^{n} p_i \, \mathrm{ld}\, p_i \, , \quad \sum_{i=1}^{n} p_i = 1 \, , \quad p_i > 0 \tag{3}$$

where p_i = probabilities for i = 1, ..., n arbitrary items (symbols, values of measure, objects, states and so on) and ld is the logarithim dualis (logarithim to base 2).

We assume two different distributions of probabilities (p_1, p_2, ..., p_n) and (q_1, q_2, ..., q_n) for the same set of objects. The difference in the two values of I_p and I_q for the p- and q- distribution, respectively, can be taken as a measure of the increase of information. This difference can be positive or negative and therefore a better measure of the information increase is (after Renyi, 1966))

$$I(q_i | p_i) = \sum_{i=1}^{n} q_i \, \mathrm{ld}\, (q_i / p_i) \, . \tag{4}$$

Trans-information is the information possessed by both the transmitter and the receiver, if there is noise in the communication channel. When the p_i are the probabilities of the transmitter, p_j those of the receiver and p_{ij} are the probabilities of a coupled event i and j in the transmitter and in the receiver the trans-information is

$$T = I(p_i) + I(p_j) - I(p_{ij}) \tag{5}$$

with

$$I(p_i) = -\sum_{i=1}^{n} p_i \, \mathrm{ld}\, p_i \, ,$$
$$I(p_j) = -\sum_{j} p_j \, \mathrm{ld}\, p_j \, ,$$
$$I(p_{ij}) = -\sum_{i} \sum_{j} p_{ij} \, \mathrm{ld}\, p_{ij} \, .$$

The ideal case for the transmission of information (ideal coupling) without noise is that where $T = I(p_i) = I(p_j) = I(p_{ij})$.

Cybernetics and biology ("3")

Control processes are very important in plants and animals at all levels of organization. Without control such dynamic systems could not exist under the continuous pressure from their surroundings. Because of the limited space available here we can only note some terms, describe some biological examples and give some important equations.

Firstly we have to distinguish a control chain (pilot circuit) from a control circuit (control loop). An example of the first is the control of photosynthesis in green plants by light, for the second the regulation of human body temperature. Secondly, there is regulation with constant command variation (e. g. the body temperature), according to a program (the ontogenesis of an organism), according to a set of unknown commands varying in time (the rhythm and deepness of respiration under physical stress) and, according to an optimum (it is diffcult to give biological examples because we do not know, for many cases, what is the optimum and what are the conditions of the optimum state; moreover many biological systems do not work at the optimum state). In spite of our slight knowledge about optimum regulation we have to remember that cybernetics alone has indicated what may be regarded as optimization in the evolution of macromolecules (Eigen, 1971; Eigen, Winkler, 1973/74).

The generality of the following equations is not as great as those of thermodynamics nor those of information theory. Firstly, we require a transfer function, f, between x_E (input) and x_A (output):

$$x_A = f(x_E) \tag{6}$$

which, in the case of linearity, can be written

$$x_A = K x_E \tag{7}$$

with K as the transfer coefficient. The quantities x_E and x_A can be absolute or difference values. A special case of such a transfer characteristic of a continuous, linear, constant value controller is the relation between the control value x and the controlled value u (difference between the command vaiable and the measured variable). This is (see Junge, 1975), for a proportional controller,

$$x = K_p u , \tag{8}$$

and for an integral controller,

$$\frac{dx}{dt} = K_I u , \tag{9}$$

and for a derivative controller

$$x = K_D \frac{du}{dt} . \tag{10}$$

Moreover for a proportional, two point controller we have an unavoidable control deviation, ΔY_u, of the controlled value Y (this is a measure of the fluctuation around the command value)

$$\Delta Y_u = \frac{dY}{dt} \cdot \tau \tag{11}$$

where dY/dt = velocity of change in the controlled value and τ = dead time (time from the registration of the "old" measured value to the establishment of the "new" measured value). A second relation in cybernetics concerns the co-work of several control loops (e. g. the homeostat considered by Ashby, see Beier, 1968). In the case of linerarity the co-operation of these controllers (each controller has an influence on each control element) is described by a system of ordinary differential equations

$$\dot{x}_i = \sum_{j=1}^{n} a_{ij} x_j \qquad i = 1, \ldots, n \tag{12}$$

(x_j = deviation of the controlled value of the control loop j from the command value, $\dot{x}_i$ = velocity of change of such a deviation in the control circuit i).

Consideration of the application of cybernetics in biology has an inverse aspect in bionics: biological experiences in the field of regulation can yield prototypes for application to physics and technology.

Cybernetics, information theory, thermodynamics and biology

Through statistical physics information theory can be related to thermodynamics ("4"); it is possible to calculate partition functions in information theory. The association between partition functions has been given by Brostow (1972) and it is a well known fact that the partition functions of statistical physics yield the thermodynamic parameters (see e. g. Kittel, 1973; Leuschner, 1979). Association of information theory with thermodynamics is also possible through entropy (see Ziesche, 1968) and this association is important in biology.

An understanding of living systems requires understanding of their motivation, their existence and of their regulation. The one yields thermodynamics and the other cybernetics. A synthesis of both, appropriate to biology, is given in the following part.

Without information there is no regulation. Therefore we find information theory in textbooks as part of cybernetics and *vice versa*. In information theory we are interested in the transfer of information from a transmitter to a receiver. Whereas, in cybernetics, we study the cooperation of several information channels and the utilization of the transferred information in regulation. For a biological example see Eigen's hypercycle (Eigen, Schuster, 1977, 1978).

Eigen's theory of the prebiotic phase of selforganization of macromolecules and of their evolution is the best illustration of the application of Inf, Cyb and Th to biology. It can be employed as a model from a physical point of view for the evolution of biological species (see Mayr, 1967). Our knowledge on this topic is very slight and for the future, the results of such a theory have to be clear in any synthesis as is discussed in the last part.

Thermodynamic phoronomics

This concept, established through consideration of embryogenesis and Prigogine's principle of minimum entropy production (Leuschner, 1981 b), is a contradiction of physical principles (dynamic and phoronomic or kinematic are different). Thermodynamics has however changed remarkably from the time of Carnot's paper on the steam engine (1824) to Prigogine's paper on the thermodynamics of irreversible processes in open systems (1947). During this period at best we had thermostatics. It is an open question as to whether we have 'real' thermodynamics today. There is no term in thermodynamics like, or similar to, the term 'force' in mechanics (see also Leuschner, 1975). The forces in the thermodynamics of irreversible processes are not forces in this sense. Therefore we refer to thermodynamic phoronomics, or thermodynamic kinematics. This requires a synthesis of thermodynamics, cybernetics and 'kinetics' with relevance for biology.

Kinematics (this term is convenient in connection with the technology (see bionics) or phoronomics (this term is convenient in theoretical considerations) deals with the interrelations between, on the one hand, the trajectory and its curvature, and, on the other hand, velocity and acceleration (Lenk et al., 1972). In the case of ThP the trajectory is a trajectory in any space of parameters or characteristics (see e. g. Lerner, 1970; Leuschner, 1974 1974 a, b). Motion along this trajectory corresponds to a change. In biology this change can be physiological, ontogenetic or phylogenetic. The curvature of the trajectory is a measure of the 'strength' of such a change and we assume that there exists an interrelation between the kinematic quantities (e. g. velocity, acceleration etc.) and the geometric quantities (e. g. curvature). We therefore only investigate kinematic quantities, especially those quantities which refer to entropy, S.

To this end we distinguish between processes and states. From the point of view of THP, in a system, a process takes place, when

$$d^2S/dt^2 \neq 0 . \qquad (13)$$

On the other hand we speak about a state when

$$d^2S/dt^2 = 0 . \qquad (14)$$

This definition of a state includes alongside

$$dS/dt = c \qquad (c = \text{constant}) \qquad (15)$$

the case of the steady state with

$$c = 0 \qquad (15\text{ a})$$

and the case of equilibrium (S itself is a constant, namely the maximum value when referred to other states). The steady state, as a specific, non-equilibrium state is also a state by this definition, and situations of non-equilibrium with variable $\frac{dS}{dt}$ are processes in a real sense.

Moreover we distinguish between optimum states and optimocline processes (those directed toward an optimum state). We give an example from physics: The 'construction' of a parallelogram of forces corresponds to an optimocline process and the end point of the resultant is an optimum state in comparison to all other possible points in the plane, Equilibrium and steady state are optimum states in ThP and a optimocline process is a process directed toward a new optimum state. Non-linear processes very far from equilibrium can be regarded as such processes. The rules and criteria for further characterization of states and processes in thermodynamic phoronomics can be given. First, we consider the decomposition of entropy production as given by Prigogine (see (2))

$$\frac{dS}{dt} = \frac{d_iS}{dt} + \frac{d_eS}{dt} \qquad (16)$$

($\frac{d_iS}{dt}$ = entropy production in the system, $\frac{d_eS}{dt}$ = entropy flux from, or into the system). Corresponding to this decomposition we write the following time derivatives of entropy (first, second, third moment of entropy)

$$C \equiv \dot{S} = \frac{dS}{dt} \qquad P \equiv C_P \equiv \frac{d_iS}{dt} \qquad C_F \equiv \frac{d_eS}{dt} \qquad (17)$$

$$\mathcal{B} \equiv \ddot{S} = \frac{d^2S}{dt^2} \qquad B_P \equiv \frac{dP}{dt} = \frac{d_i^2S}{dt^2} \quad B_F \equiv \frac{dC_F}{dt} = \frac{d_e^2S}{dt^2} \tag{18}$$

$$R \equiv \dddot{S} = \frac{d^3S}{dt^3} \qquad R_P \equiv \frac{d^2P}{dt^2} = \frac{d_i^3S}{dt^3} \quad R_F \equiv \frac{d^2C}{dt^2} = \frac{d_e^3S}{dt^3}. \tag{19}$$

At present there is no necessity to consider higher time derivatives than entropy jerk (third time derivative); neither from a theoretical, nor, from a practical point of view. What do we know about these quantities?

The Second Law of Thermodynamics (for a simple biological proof see Leuschner, 1981 c) states

$$P \geqslant 0. \tag{20}$$

Prigogine's minimum principle in the linear region of the thermodynamics of irreversible processes (Leuschner, 1979 l.c.) states

$$B_P \leqq 0. \tag{21}$$

Prigogine and Glansdorff's principle in the non-linear region of the thermodynamics of irreversible processes ($d_X P$ = variation of the entropy production associated with variation in the thermodynamic forces, X, and not with that of thermodynamic fluxes J; Glansdorff, Prigogine, 1971) with constant boundary conditions give

$$d_x P/dt \leqq 0. \tag{22}$$

The last three equations are either evolution criteria (see Ebeling, 1976) or extremal principles. It is possible to complement these equations by criteria of stability which are statements about second order variation in entropy (see Leuschner, 1979).

Of the other quantities C, B, R, R_p, C_F, B_F, R_F we only know that they can be greater, or less than zero or equal to zero. For example the statement

$$C > 0 \tag{23}$$

i. e.

$$\left|\frac{d_e S}{dt}\right| < \frac{d_i S}{dt}\,, \qquad \frac{d_e S}{dt} < 0 \tag{24}$$

always means the possibility of the death of an organism. Further we are able to calculate the value of P

$$P = \sum_k J_k X_k \quad \text{with} \quad J_k = \sum_l L_{kl}(X_m) X_l \tag{25}$$

(X_k = thermodynamic force, J_k = thermodynamic flux) and to decompose the quantity $\frac{dP}{dt}$ (see (22))

$$\frac{dP}{dt} = \frac{d_X P}{dt} + \frac{d_J P}{dt} \tag{26}$$

with

$$d_X P = \sum_k J_k \, dX_k \tag{27}$$

$$d_J P = \sum_k X_k \, dJ_k \tag{28}$$

Now we come back to the extremal principles mentioned above (Gyarmati's principle and Ziegler's principle are also of this type, Gyarmati, 1965; Zotin, 1976; Ziegler, 1977; Lamprecht, Zotin, 1978). Such principles have to be complemented by others in defining not only direction but also change in metabolism, in ontogenesis and in phylogenesis. For phylogenesis which is related with genetics, we briefly indicate the general idea. A certain value of the second entropy moment, $\frac{d^2S}{dt^2}$, in relation to other values determines 'selection, i. e. the 'survival of the fittest'. The third entropy moment $\frac{d^3S}{dt^3}$ determines a 'mutation' when a threshold is exceeded. The threshold can be zero, i. e. a thermodynamic or biological new quality could arise when $\frac{d^3S}{dt^3} \neq 0$; $\frac{d_i^3S}{dt^3} \neq 0$ reflects internal fluctuations, and $\frac{d_e^3S}{dt^3}$ external influences. These conditions are not necessary, but sufficient. This general 'selection-mutation law' is one of the centres of ThP, because it provides for the possibility of a creation of new optimocline processes and new optimum states.

In conclusion the existence of dualities in thermodynamic extremal principles (Leuschner et al., in preparation) should be noted. Here we have a possibility for thermodynamic motivation of the term 'antiorganism' (see Soka, Sneath, 1963).

At present there is no understanding of the relationship between equations (20) to (22) and for example, Liebig's minimum law (see Stugren, 1974), the bottle neck principle (see Romanovsky et al., 1974) and Zotin's principle of minimum external entropy production (Zotin, 1974; Zotin, Zotina, 1977; Zotin et al., 1978)

$$P_d \rightarrow min \qquad (29)$$

where total entropy production P is decomposed in

$$P = P_u + P_d \qquad (30)$$

(P_d = the proportion of the entropy production which flows into the surroundings and P_u = the proportion of the entropy production which remaines within the system). We know that thermodynamic optimisation (represented by T in the following) is characterized by a certain value of the quantities in equations (17), (18), (19) and also in (22), (29) by reference to a constant value (for instance zero).

Thermodynamic taxonomy

At the beginning of the last part we established the relationship between entropic velocities and accelerations on one hand side and the properties of the trajectories in a parameter space (for instance a taxonomic character space) on the other hand. However, the thermodynamic extremal principles of ThP are more important for ThT than this relationship. Therefore, firstly, we represent the foundation of ThT via information theory and secondly, we take into account the importance of the thermodynamic optimisation for ThT. The second point corresponds to consideration of the thermodynamic extremum principles of thermodynamic phoronomics in its relevance to thermodynamic taxonomy, i. e. we also have, in our opinion, a relationship with cybernetics.

The basis of thermodynamic taxonomy can be found in Leuschner (1981 d). In that paper we showed that thermodynamic knowledge was important for the biological manifold and how to proceed in the application of thermodynamics to the systematisation of this manifold (the taxonomy in the real sense of this term; see Sneath, Sokal, 1973). Three statements are necessary

- Thermodynamics and taxonomy have a common root in information theory (for the last, see Völz, 1982)
- Thermodynamics and taxonomy have many common properties
- Taxonomic similarity corresponds to, the related, thermodynamic optimisation.

We have already cited Browstow's concept of the deduction of the thermodynamics from information theory and we have also noted the generality of information theory; especially in its usefulness in biology. The last point has been demonstrated for taxonomy by Leuschner (1974). Therefore, not only is a common root shown, but also the synthesis of thermodynamics and taxonomy is made through thermodynamic taxonomy.

Thermodynamics and taxonomy have the following common characteristics. They are sciences

- of macroscopic or phenomenological nature
- of an axiomatic, but empirical type
- of universal type
- with the effect of classification
- with the possibility of microscopic foundation
- with utilitarian aims

- dealing with complex systems
- dealing with the creation of a natural system.

Items referring to these points; to appropriate characteristics; to handling and processing in ThT; and to examples (e. g. Zotin's orderliness criterion) can be found in Leuschner (1981 d). We cannot and will not discuss the large number of measures of similarity (see Sokal, Sneath, 1963; Sneath, Sokal, 1973) however, we represent any such measure by S. Further for two taxonomic units (e. g. two organisms) with thermodynamic optimal T_1 and T_2, respectively the taxonomic similarity S_{12} of both these units is given by

$$S_{12} = A(T_1 \cap T_2) \tag{31}$$

where A is an operator, a function or a constant and $\cap$ is the set-theoretical disjunction.

If the relation

$$A(T_1 \cap T_2) \approx A(T_1) \approx A(T_2) \tag{32}$$

is valid, then both units are very similar.

Physics is concerned with thermodynamics and electromagnetism. The first includes (statistical) mechanics. Biology is concerned with physics, information theory and cybernetics. Is this all? We do not know. We could answer, if we knew what ordering is, especially in biology. But ordering is a very difficult problem. This is true for non-living matter (e. g. turbulence, Leuschner, 1978, 1981), is even more so for life. Zotin and his group (Konoplev, Zotin, 1975; Zotin, 1976, 1981; Zotin, Krivolucky, 1982) have made a start in attempts to define biological ordering and this is continued in this book. Thus our knowledge of biology will become more comprehensive through application of thermodynamics, information theory and cybernetics theory. But there remains a question. What is the relationship between thermodynamics of the life, in general, and especially of thought (Kobozev, 1971)? We have to answer this question on the basis of a definite criterion. A criterion that enables one to distinguish between real and chance coincidence between a biological phenomenon and our thought. Is the criterion of practice in this question necessary and sufficient? This last should not be theoretical, but practical. Energy? Raw materials? Food stuff? Environmental protection? Interplanetary travel? None of these is possible without biology; biology helps to solve problems. Partly, by means of biological laws which have their basis in thermodynamics (physics), information theory and cybernetics.

Some Relations of Non-Equilibrium Thermodynamics of Open Systems

R. L. Stratonovich

As may be seen from reviews (e. g. Nicolis, Prigogine, 1977; Lamprecht, Zotin, 1978), the application of the thermodynamics of non-equilibrium processes to strongly non-equilibrium, open systems is now of great interest. Such states are often encountered in physics, chemistry, and biology and no single view exists of the content of a thermodynamic of non-equilibrium processes and its application to such states.

The purpose of non-equilibrium thermodynamics is the derivation of universal relations that describe the characteristics of macroscopic dissipative and fluctuative processes in complicated systems. The Onsager relations and the fluctuation-dissipation theorem are examples of such relations. The results of a non-equilibrium thermodynamic analysis of closed systems are a consequence of principles such as detailed equilibrium and microscopic time reversibility. The principle of detailed equilibrium we understand, in particular, as application of equilibrium distributions: Gibbs or microcanonical distributions.

Non-equilibrium thermodynamic analysis of open systems is poorer, i. e. there are fewer results than arise from thermodynamic analysis of closed systems. The point is that in strongly non-equilibrium states one cannot effectively use the principle of the reversiblity of microscopic processes. In addition, far from equilibrium one cannot use canonical distributions, or, arising from them distributions of the form

$$w(B) = \text{const}\, e^{S(B)/k} ; \quad w(B) = \text{const}\, e^{-F(B)/kT} \qquad (1)$$

for the internal parameters $B_\alpha = B_\alpha(q,p)$ (q, p are dynamic variables of the system).

Thermodynamics and Regulation of Biological and Processes

Some of the universal results of non-equilibrium thermodynamics of open systems are considered here. They include the H-theorem and fluctuation-dissipation relations. The subject of the H-theorem is the increase (non-decrease) in quasi-entropy which is a function of the internal parameters. This differentiates our result from that which appeared in many papers previously, namely, a decrease (non-increase) in the functional

$$\int w(B) \ln [w(B)/w_{cm}(B)]\, dB$$

i. e. a functional of non-stationary distribution density ($w(B)/w_{cm}(B)$ is the stationary distribution). The H-theorem under consideration here is closer to the conventional Second Law of Thermodynamics since this conventional entropy is not at all equal to the indicated functional.

The H-theorem

Let us consider the non-equilibrium stationary state of an open system characterized by the non-equilibrium stationary distribution, $w_{cm}(B)$, which cannot be represented by formula (1). By writing

$$w_{cm}(B) = \text{const} \exp[-\Psi(B)/\varkappa], \tag{2}$$

similarly to (1), one can, however, introduce a quasi-free energy $\Psi(B)$ or a quasi-entropy $S(B) = -\Psi(B)$. The parameter $\varkappa$ in (2) characterizes the intensity of fluctuations in the system. Equation (2) can be considered a non-equilibrium generalisation of equation (1).

The potential conjugated with $\Psi(B)$:

$$\Phi(x) = -\varkappa \ln \int \exp[-\varkappa^{-1}(\Psi(B) - Bx)]\, dB \tag{3}$$

also has the meaning of quasi-free energy or quasi-entropy, but, as distinguished from $\Psi(B)$, it corresponds to fixed additional external forces x, whereas $\Psi(B)$ corresponds to fixed internal parameters $B = \{B_\alpha\}$. It is easy to verify that the derivative, $-\partial\Phi(x)/\partial x_\alpha$, of (3) is none other than the non-equilibrium mean,

$$-\frac{\partial\phi_x}{\partial x_\alpha} = \int B_\alpha \exp[\varkappa^{-1}(Bx - \Psi(B))]\,dB/$$
$$/\int \exp[\varkappa^{-1}(Bx - \Psi(B))]\,dB = \langle B_\alpha\rangle_x , \tag{4}$$

which appears when the additional external forces $x = \{x_\alpha\}$ are switched on. Higher derivatives opposite in sign are proportional to the correlators

$$\langle B_{\alpha_1}, \ldots, B_{\alpha_n}\rangle_x = -\varkappa^{n-1}\partial^n\phi(x)/\partial x_{\alpha_1}\ldots\partial x_{\alpha_n}; \tag{5}$$

which correspond to the biased distribution

$$w_x(B) = \text{const}\exp[-(\Psi(B) - Bx)/\varkappa]. \tag{6}$$

To obtain the mean expectation value and correlators for the initial stationary distribution (2), one should assume in (4) and (5) $x = 0$.

If now one introduces the function

$$\Psi(A) = \phi(x(A)) + Ax(A) \quad \left(-\frac{\partial\phi}{\partial x}(x(A)) = A\right), \tag{7}$$

which is a Legendre transformation of the potential (3), then, as can be easily verified, the dependence

$$\frac{\partial\Psi(A)}{\partial A_\alpha} = x_\alpha \tag{8}$$

will be the reverse of the dependence $-\partial\phi(x)/\partial x_\alpha = A_\alpha$, i. e. the dependence (4) at $A_\alpha = \langle B_\alpha\rangle$.

Now consider the time derivative of the function $\overline{\Psi}(A)$ caused by change in time of the mean internal parameters A(t). Substituting in (8) the equation $d\overline{\Psi}(A)/dt = (\partial\overline{\Psi}/\partial A)dA/dt$ ..., we obtain

$$d\overline{\Psi}(A)dt = x_\alpha dA_\alpha/dt \qquad x = \partial\overline{\Psi}(A)/\partial A. \tag{9}$$

Assume that time evolution of the non-equilibrium means $A_\alpha = \langle B_\alpha\rangle$ is described by the inertia-less, phenomenological equation

$$dA_\alpha/dt = \varphi_\alpha(A), \qquad \alpha = 1, \dots, r. \tag{10}$$

Random values $B_\alpha(t)$ of the internal parameters satisfy some stochastic equation - the Langevin equation

$$dB_\alpha/dt = f_\alpha(B) + \xi_\alpha(t), \qquad \alpha = 1, \dots, r.$$

Since the phenomenological equation (10) is intertia-less, the random process B(t) is a Markov process. The synchronous probability distribution density $w_t(B)$ satisfies the master equation

$$\frac{\partial w_t(B)}{\partial t} = \sum_{s=1}^{\infty} \frac{1}{s!} (-1)^s \frac{\partial^s}{\partial B_{\alpha_1} \dots \partial B_{\alpha_s}} [K_{\alpha_1 \dots \alpha_s}(B) w_t(B)] \tag{11}$$

The functions involved $K_{\alpha_1 \dots \alpha_s}(B)$ are known to be determined by the equations

$$K_{\alpha_1 \dots \alpha_s}(B) = \lim_{\Delta t \to 0} (\Delta t)^{-1} \langle \Delta B_{\alpha_1} \dots \Delta B_{\alpha_s} \rangle_B, \tag{12}$$

where ΔB_α is the increment in parameter B_α for time Δt; the lower index B implies that we take a conditional mean value corresponding to fixed values of the parameters B before the increase.
Assuming in (12) $s = 1$, we have

$$K_\alpha(B) = \lim_{\Delta t \to 0} \langle \Delta B_\alpha / \Delta t \rangle_B . \tag{13}$$

The meaning of the derivative in (10) should be specified. Suppose that the derivative in (10) corresponds to averaging the derivative in (13) with biased distribution (6), where x is connected with A by the equality $-\partial\phi/\partial x = A$, i. e by equation (4). Then from (10) and (13) we have

$$dA_\alpha/dt = \lambda_\alpha(x) \qquad x = \partial\tilde{\Psi}(A)/\partial A,$$

where

$$\lambda_\alpha(x) = \int K_\alpha(B) w_x(B) dB \equiv \langle K_\alpha(B) \rangle_x . \tag{14}$$

Substituting this equation into (10), we find

$$d\bar{\Psi}(A)/dt = x_\alpha \lambda_\alpha(x) \qquad x = \partial\bar{\Psi}(A)/\partial A . \tag{15}$$

To prove that this derivative is not positive we take into account that distribution (2) is a stationary solution of (11), i. e. at all B the following equation holds

$$\sum_{s=1}^{\infty} \frac{(-1)^s}{s!} \frac{\partial^s}{\partial B_{\alpha_1} \dots \partial B_{\alpha_s}} \{K_{\alpha_1 \dots \alpha_s}(B) \exp[-\Psi(B)/\varkappa]\} = 0 \tag{16}$$

Multiplying this equation by $\exp(Bx/\varkappa)$ and taking the exponent from the left to the right through the differentiation signs using the identity

$$e^{Bx/\varkappa} f(\partial/\partial B) = f(\partial/\partial B - x/\varkappa)\, e^{Bx/\varkappa} ,$$

where f is any function we obtain

$$\sum_{s=1}^{\infty} \frac{1}{s!} \left(\frac{x_{\alpha_1}}{\varkappa} - \frac{\partial}{\partial B_{\alpha_1}}\right) \dots \left(\frac{x_{\alpha_s}}{\varkappa} - \frac{\partial}{\partial B_{\alpha_s}}\right) \{K_{\alpha_1 \dots \alpha_s}(B) w_x(B)\} = 0 .$$

If now we integrate this equation over infinite limits, the derivatives with resprect to B_α drop out and we are led to

$$\sum_{s=1}^{\infty} \frac{1}{s!} \varkappa^{-s} \lambda_{\alpha_1 \dots \alpha_s}(x)\, x_{\alpha_1} \dots x_{\alpha_s} = 0 , \tag{17}$$

where by analogy with (14) the notation

$$\lambda_{\alpha_1 \dots \alpha_s}(x) = \langle K_{\alpha_1 \dots \alpha_s}(B) \rangle_x \tag{18}$$

is introduced. By virtue of (17) equation (15) can be reduced to the form

$$\begin{aligned} &d\bar{\Psi}(A)/dt = -\varkappa^{-1} \sum_{s=2}^{\infty} \frac{1}{s!} \varkappa^{-s} \lambda_{\alpha_1 \dots \alpha_s} x_{\alpha_1} \dots x_{\alpha_s} \\ &x = \partial\bar{\Psi}(A)/\partial A . \end{aligned} \tag{19}$$

Substituting (12) into (18), we find

$$\begin{aligned} \lambda_{\alpha_1 \dots \alpha_s}(x) &= \lim_{\Delta t \to 0} (\Delta t)^{-1} \int \langle \Delta B_{\alpha_1} \dots \Delta B_{\alpha_s} \rangle_B\, w_x(B)\, dB \\ &= \lim_{\Delta t \to 0} (\Delta t)^{-1} \langle \Delta B_{\alpha_1} \dots \Delta B_{\alpha_s} \rangle_x . \end{aligned}$$

Having substituted the latter equation into (19), we obtain

$$d\bar{\Psi}(A)/dt = -\varkappa^{-1} \lim (\Delta t)^{-1} \langle g(x_\alpha \Delta B_\alpha / \varkappa) \rangle , \tag{20}$$

where

$$g(z) = \sum_{s=2}^{\infty} \frac{1}{s!} z^s = e^z - 1 - z . \tag{21}$$

Function (21) is everywhere non-negative, and, moreover, it is positive at $z \neq 0$. The mean of a non-negative value is non-negative. The limit of a non-negative value is non-negative. Therefore, from (20) we have

$$d\bar{\Psi}(A)/dt \leq 0 . \tag{22}$$

This equation is the essence of the H-theorem. The derivative $d\bar{\Psi}(A)/dt$ can be proved to be negative in nontrivial cases.

According to the proof, the result, equation (22), is exact if the phenomenological evolution equation is understood in the sense outlined above. From this result some asymptotic consequences can be obtained. If fluctuations in the system are small, i. e. the fluctuation parameter $\varkappa$ can be considered small and, if the equation $\partial \Psi(A)/\partial A = X$ at different x has a single root, then, as can be proved, function (7) is close to function $\Psi(A)$, which appears in (2), so that instead of (22) one can deduce the inequality

$$d\Psi(A)/dt \leq 0 .$$

Further, smallness of the parameter $\varkappa$ is an additional argument in favour of interpereting the function $\varphi_\alpha(A)$ in (10) as $\lambda_\alpha(x(A))$. Indeed, for small fluctuations equation (10) holds for any interpretation of function $\varphi_\alpha(A)$, including the one indicated.

If we substitute the equilibrium distribution (1) for the stationary distribution (2), then from the H-theorem we obtain the Second Law of Thermodynamics. Another approach to the problem considered above is presented in the paper by Schlögl (1980).

Markov's fluctuation-dissipation relations in the absence of microscopic time reversibility

Equation (17) can be considered a generating equality for a system of relations obtained by differentiating (17) with respect to $x_{\alpha_1}, x_{\alpha_2} \ldots$ at a zero point. It is convenient to introduce the notation

$$l_{\alpha_1 \ldots \alpha_n \beta_1 \ldots \beta_m} = \partial^m \lambda_{\alpha_1 \ldots \alpha_n}(x) / \partial x_{\beta_1} \ldots \partial x_{\beta_m} \quad x = 0 \tag{23}$$

Differentiating (17) with respect to x and assuming x = 0 we find

$$l_\alpha = 0 . \tag{24}$$

Further, differentiating (17) with respect to x_α and x_β and assuming x = 0, with account taken of (23), we find

$$l_{\alpha,\beta} + l_{\beta,\alpha} + \varkappa^{-1} l_{\alpha\beta} = 0 . \tag{25}$$

Triple differentiation gives

$$l_{\alpha,\beta\gamma} + l_{\beta,\alpha\gamma} + l_{\gamma,\alpha\beta} + \varkappa^{-1}(l_{\alpha\beta,\gamma} + l_{\beta,\alpha\gamma} + l_{\gamma,\alpha\beta}) + \varkappa^{-2} l_{\alpha\beta\gamma} = 0 , \tag{26}$$

and quadruple differentiation at a zero point leads to the relation

$$\begin{aligned} & l_{\alpha,\beta\gamma\delta} + l_{\beta,\alpha\gamma\delta} + l_{\gamma,\alpha\beta\delta} + l_{\delta,\alpha\beta\gamma} + \varkappa^{-1} (l_{\alpha\beta,\gamma\delta} + \\ & + l_{\alpha\gamma,\beta\delta} + l_{\alpha\delta,\beta\gamma} + l_{\beta\gamma,\alpha\delta} + l_{\beta\delta,\alpha\gamma} + l_{\gamma\delta,\alpha\beta}) + \varkappa^{-2} \cdot \\ & \cdot (l_{\alpha\beta\gamma,\delta} + l_{\alpha\beta\delta,\gamma} + l_{\alpha\gamma\delta,\beta} + l_{\beta\gamma\delta,\alpha}) + \varkappa^{-3} l_{\alpha\beta\gamma\delta} = 0 . \end{aligned} \tag{27}$$

We shall not present relations with a larger number of indices.

Relations (24) to (27) are exact and from them one can obtain approximate, asymptotic relations valid at small $\varkappa$. To derive these relations, we write the function $K_{\alpha_1 \ldots \alpha_s}(B)$ as an expansion into a Taylor series viz

$$K_{\alpha_1 \ldots \alpha_s}(B) = \sum_{n=0}^{\infty} \frac{1}{n!} K_{\alpha_1 \ldots \alpha_s, \beta_1 \ldots \beta_n} B_{\beta_1} \ldots B_{\beta_n} .$$

Averaging in accordance with (18) gives

$$\lambda_{\alpha_1 \ldots \alpha_s}(x) = \sum_{n=0}^{\infty} \frac{1}{n!} K_{\alpha_1 \ldots \alpha_s, \beta_1 \ldots \beta_n} \langle B_{\beta_1} \ldots B_{\beta_n} \rangle_x .$$

Expressing the involved non-stationary momenta through the correlators (5), we find,

$$\lambda_\alpha(x) = K_\alpha - K_{\alpha,\beta}\frac{\partial\phi(x)}{\partial x_\beta} + \frac{1}{2}K_{\alpha,\beta\gamma}\left(\frac{\partial\phi}{\partial x_\beta}\frac{\partial\phi}{\partial x_\gamma} - \varkappa\frac{\partial^2\phi}{\partial x_\beta\partial x_\gamma}\right) + \ldots$$

$$\alpha_{\alpha\beta}(x) = K_{\alpha\beta} - K_{\alpha\beta,\gamma}\frac{\partial\phi(x)}{\partial x_\gamma} + \ldots$$

We get asymptotic formulae if, in these and similar equalities, we drop terms of the order of $\varkappa$, $\varkappa^2$, etc. Differentiating the obtained equalities in accordance with (23) and taking into account formula (5) taken at the point $x = 0$, we have

$$\begin{aligned} &l_{\alpha,\beta} = -K_{\alpha,\sigma}\frac{\partial^2\phi}{\partial x_\sigma\partial x_\beta} = \varkappa^{-1}K_{\alpha,\sigma}\mu_{\sigma,\beta}; \quad l_{\alpha\beta} = K_{\alpha\beta};\\ &l_{\alpha,\beta\gamma} = \varkappa^{-2}K_{\alpha,\sigma}\mu_{\sigma\beta\gamma} + \varkappa^{-2}K_{\alpha,\sigma\tau}\mu_{\sigma\beta}\mu_{\tau\gamma};\\ &l_{\alpha\beta,\gamma} = -K_{\alpha\beta,\tau}\frac{\partial^2\phi}{\partial x_\sigma\partial x_\gamma} = \varkappa^{-1}K_{\alpha\beta,\sigma}\mu_{\sigma\gamma}; \quad l_{\alpha\beta\gamma} = K_{\alpha\beta\gamma}. \end{aligned} \tag{29}$$

Here we introduce a special notation for stationary correlators

$$\mu_{\alpha_1\ldots\alpha_n} = \langle B_{\alpha_1}, \ldots, B_{\alpha_n}\rangle_{cm}. \tag{30}$$

We assume that $\langle B\rangle_{cm} = 0$. This condition determines the choice of the origin in the range of the values of B. Substituting (29) into (25), (26), we get

$$\begin{aligned} &K_{\alpha\beta} + K_{\alpha,\sigma}\mu_{\sigma\beta} + K_{\beta,\sigma}\mu_{\sigma\alpha} = 0;\\ &K_{\alpha\beta\gamma} + K_{\alpha\beta,\sigma}\mu_{\sigma\gamma} + K_{\alpha\gamma,\sigma}\mu_{\sigma\beta} + K_{\beta\gamma,\sigma}\mu_{\sigma\alpha} + K_{\alpha,\sigma\tau}\mu_{\sigma\beta}\mu_{\tau\gamma}\\ &+ K_{\beta,\sigma\tau}\mu_{\sigma\alpha}\mu_{\tau\gamma} + K_{\gamma,\sigma\tau}\mu_{\sigma\alpha}\mu_{\tau\beta} + K_{\alpha,\tau}\mu_{\sigma\beta\gamma} +\\ &+ K_{\beta,\sigma}\mu_{\sigma\alpha\gamma} + K_{\gamma,\sigma}\mu_{\sigma\alpha\beta} = 0. \end{aligned}$$

These are just the desired asymptotic relations. Note that in the chosen approximation the derivatives $K_{\alpha,\beta_1\ldots\beta_n}$ of the functions $K_\alpha(B)$ at a zero point coincide with the derivatives of the functions $\varphi_\alpha(B)$ which enter the phenomenological equation (10), i. e.

$$K_{\alpha,\beta_1\ldots\beta_n} = \partial^n\varphi(A)/\partial A_{\beta_1}\ldots\partial A_{\beta_n} \qquad A = 0. \tag{32}$$

Relation (31) and other relations with a different number of indices connect three types of quantities:

1. The derivatives (32) characterising a relaxation-dissipation process
2. The correlators $\mu_{\alpha_1\ldots\alpha_n}$ characterising only stationary fluctuations
3. The derivatives of the functions $K_{\alpha\beta}(B)$, $K_{\alpha\beta\gamma}(B)$, characterizing a transition fluctuation process. That is why the indicated relations may

be called fluctuation-dissipation relations. They can be applied in two ways.

I. If the operator of the master equation (11) is known, they can be used to find synchronous stationary correlators.

II. If the stationary distribution (2) is known (for example, from experiment), they can be used to investigate limitations imposed on the operator of a master equation.

Relations valid near non-equilibrium phase transitions

Designate by $\mathcal{B}_i^{\circ}$ the isolated points of the minimum of the function $\Psi(\mathcal{B})$, i. e. the points of the local minimum of the distribution (2). Additionally, introduce the following notation for the derivatives at these points:

$$\Psi_{\alpha_1 \dots \alpha_n} = \partial^n \Psi(\mathcal{B}) / \partial \mathcal{B}_{\alpha_1} \dots \partial \mathcal{B}_{\alpha_n} \qquad \mathcal{B} = \mathcal{B}_i^{\circ} \,. \tag{33}$$

It is obvious that the equalities $\Psi_\alpha = 0$ must hold. Suppose that the operator of a master equation and the function Ψ depend on an external parameter, or parameters, Θ. If for some of the points, $\mathcal{B}_i^{\circ}$, there occurs a limiting transition

$$\det \|\Psi_{\alpha\beta}\| \to 0$$

as Θ tends to some value Θ_c, we say that in this system there occurs a phase transition or a catastrophe. The $\mathcal{B}_i$ corresponding to this value is called the critical value.

Near a phase transition (31) no longer determines the non-infinite correlators (30) at known $K_{\alpha_1 \dots \alpha_s, \beta_1 \dots \beta_n}$. This shows that the approximation corresponding to dropping terms of the order of $\varkappa$ and higher in (28) becomes invalid near a phase transition. The exact relations (24) to (27) remain, of course, valid near phase transitions. The invalidity of relation (31) near a phase transition is due to the fact that the correlators (30) depend non-analytically on $\varkappa$ at the point $\varkappa = 0$.

Another method should be applied to obtain simplified asymptotic relations, i. e. conditioned by the smallness of $\varkappa$ relations valid near a phase transition. Now we shall use not (17), but the asymptotic equation

$$\sum_{s=1}^{\infty} \frac{\varkappa^{-s}}{s!} K_{\alpha_1\ldots\alpha_s}(B)\, \frac{\partial\Psi(B)}{\partial B_{\alpha_1}}\cdot\ldots\cdot\frac{\partial\Psi(B)}{\partial B_{\alpha_s}} = \sigma$$

derived with the help of (16). From this equation, by differentiating with respect to B at the point $B = B_i^\circ$, one can derive different relations. So, after double differentiating with respect to B_α and B_β we have

$$\Psi_{\alpha\sigma}\overset{\circ}{K}_{\sigma\beta} + \Psi_{\beta\sigma}\overset{\circ}{K}_{\sigma\alpha} = -\Psi_{\alpha\sigma}\overset{\circ}{K}_{\sigma\tau}\Psi_{\tau\beta}\,. \tag{34}$$

Triple differentiation gives

$$\begin{aligned} 3\{\Psi_{\alpha\sigma}\overset{\circ}{K}_{\sigma,\tau}\Psi^{-1}_{\tau\rho}\Psi_{\rho\beta\gamma}\}_{sym} &= 3\{\Psi_{\alpha\sigma}\overset{\circ}{K}_{\sigma,\beta\gamma}\}_{sym} + \\ &+ 3\{\Psi_{\alpha\sigma}\Psi_{\beta\tau}\overset{\circ}{K}_{\sigma\tau,\gamma}\}_{sym} + \Psi_{\alpha\sigma}\Psi_{\beta\tau}\Psi_{\gamma\rho}\overset{\circ}{K}_{\sigma\tau\rho}\,. \end{aligned} \tag{35}$$

We use here the following notation

$$\overset{\circ}{K}_{\alpha_1\ldots\alpha_s,\beta_1\ldots\beta_n} = \varkappa^{1-s}\,\partial^n K_{\alpha_1\ldots\alpha_s}(B)/\partial B_{\beta_n} \qquad B = B_i^\circ\,.$$

The brackets $\{\ \}_{sym}$ indicate the symmetry operations on the derived expression in the indices over which summation has not been carried out (in this case indices α, β, γ).

The four-index relation has the form

$$\begin{aligned} &4\{\Psi_{\alpha\sigma}\overset{\circ}{K}_{\sigma,\tau}\Psi^{-1}_{\tau\rho}\Psi_{\rho\beta\gamma\delta}\}_{sym} = 4\{\Psi_{\alpha\sigma}\overset{\circ}{K}_{\sigma,\beta\gamma\delta}\}_{sym} + \\ &6\{\Psi_{\alpha\sigma}\Psi_{\beta\tau}(\overset{\circ}{K}_{\sigma\tau,\gamma\delta} + \overset{\circ}{K}_{\sigma\tau,\mu}\Psi^{-1}_{\mu\nu}\Psi_{\nu\gamma\delta})\}_{sym} + \\ &4\{\Psi_{\alpha\sigma}\Psi_{\beta\tau}\overset{\circ}{K}_{\sigma\tau\rho}\Psi_{\rho\gamma\delta}\}_{sym} + 4\{\Psi_{\alpha\sigma}\Psi_{\beta\tau}\Psi_{\gamma\rho}\overset{\circ}{K}_{\sigma\tau\rho,\delta}\}_{sym} + \\ &\Psi_{\alpha\sigma}\Psi_{\beta\tau}\Psi_{\gamma\rho}\Psi_{\delta\mu}\overset{\circ}{K}_{\sigma\tau\rho\mu}\,. \end{aligned}$$

Equations (34) to (36) are simply the asymptotic relations valid near a phase transition. By means of these relations one can also find the derivatives (33) in the critical point. These equations may be solved step by step: first, using (34), find $\Psi_{\alpha\beta}$, then, using (35), find $\Psi_{\alpha\beta\gamma}$, and then solve (36) for finding $\Psi_{\alpha\beta\gamma\delta}$. The right-hand sides of (35) and (36) should be known.

It should be noted that relations (35) and (34) are equivalent to (31) if, and only if, the function $\Phi(x)$ in (5) is associated with $\Psi(B)$ by the Legendre transformation. In reality, however, (5) involves the function (3) which differs essentially from the Legendre transformation near a phase transition.

Peculiarities of fluctuations of internal parameters in the critical region - the single-component case

In a single-component case the solution of (34) to (36) is trivial. So, from (36) we obtain directly

$$\Psi_2 \equiv \Psi_{11} = -2K^{\circ}_{1,1} / K^{\circ}_{11} . \tag{37}$$

Taking into account this result and using (35), we find

$$\Psi_3 \equiv \Psi_{111} = -(2/K^{\circ}_{11})(K^{\circ}_{1,11} + \Psi_2 K^{\circ}_{11,1} + \Psi_2^2 K^{\circ}_{111}/3). \tag{38}$$

According to the above definition of a phase transition, the second derivative $\Psi_{11} \equiv \Psi_2$ tends to zero when a phase transition is approached. With a limit from (38) we obtain, at the critical point,

$$\Psi_3^c = -2K^{\circ}_{1,11} / K^{\circ}_{11} . \tag{39}$$

At the critical point we have, from (36),

$$\Psi_4^c = -2K^{\circ}_{1,111}/K^{\circ}_{11} - 3K^{\circ}_{11,1}\Psi_3^c/K^{\circ}_{11} . \tag{40}$$

After the values (37), (39), (40) are found, one can write an approximate probability distribution

$$\begin{aligned} w(B) = \text{const} \exp\{ -\tfrac{1}{2\varkappa} [\Psi_2 (B-B^{\circ})^2 + \\ + \tfrac{1}{3}\Psi_3^c (B-B^{\circ})^3 + \tfrac{1}{12}\Psi_4^c (B-B^{\circ})^4] \} \end{aligned} \tag{41}$$

which is an approximation of distribution (2) and is valid near a phase transition. Approximation of a free energy or an entropy by a polynomial of the type used in the exponent is usual in the phenomenological theory of thermodynamic phase transitions proposed by Landau.

Solutions in particular cases are possible. Consider them in turn.
a) The case where $\Psi_3^c = 0$, $\Psi_4^c > 0$. According to (39), (40), these conditions are associated with equations $\overset{\circ}{K}_{1,11}=0, \overset{\circ}{K}_{1,111}<0$ and in this case we deal with a second-order phase transition. Introducing the notation

$$y = (12\,\varkappa)^{-1/4}(\Psi_4^c)^{1/4}(B - B^\circ), \tag{42}$$

we reduce distribution (41) to the form

$$w(y) = \text{const}\,\exp(-\alpha y^2 - y^4/2), \tag{43}$$

where

$$\alpha = (3/\varkappa)^{1/2}(\Psi_4^c)^{-1/2}\Psi_2 .$$

The critical region is defined by the condition $\alpha \sim 1$. Using distribution (43) and equation (42), it is easy to find the correlator

$$\langle B, B\rangle_{cm} \equiv \mu_{11} = (3\varkappa/\Psi_4^c)^{1/2} D_{-3/2}(\alpha)/D_{-1/2}(\alpha)$$

It is readily seen that in the critical region the double correlator has the order $\varkappa^{1/2}$, whereas far from phase transition it has the normal order $\varkappa$. Consequently, at small $\varkappa$ fluctuations in the parameter B are anomalously large as compared with the usual ones.

In this case it is easy to calculate other correlators too. We shall restrict ourselves to discussing the value of the asymmetry and excess coefficients determined by the formulas

$$\gamma_a = \langle B, B, B\rangle/\langle B, B\rangle^{3/2};\quad \gamma_э = \langle B, B, B, B\rangle/\langle B, B\rangle^2$$

and, characterizing the degree of non-Gaussian character of the random quantity B. Far from a phase transition the correlators $\langle B, B, B\rangle$, $\langle B, B, B, B\rangle$ have the order $\varkappa^2$ and $\varkappa^3$, respectively. Therefore, the indicated coefficients have the following order of magnitude outside the critical region

$$\gamma_a \sim \varkappa^{1/2} \qquad \gamma_э \sim \varkappa .$$

Accordingly, at small $\varkappa$ fluctuations are close to Gaussian.

Inside the critical region it is easy to obtain

$$\gamma_a = 0, \quad \gamma_3 = 3 D_{-5/2}(\alpha) D_{-1/2}(\alpha) / D^2_{-3/2}(\alpha) - 3 .$$

This testifies to a strong non-Gaussian character of the internal parameter fluctuations near the phase transition. The quadruple correlator is estimated here as: $\langle B, B, B, B \rangle \sim \varkappa$, whereas the triple correlator is equal to zero.

b) The case where $\Psi_3^c = 0$, $\Psi_4^c = 0$, i.e. $K^0_{1,11} = 0$, $K^0_{1,111} > 0$.
We deal here with a first-order phase transition. The function

$$u(B) \equiv \Psi(B) - \Psi(B_0) = \frac{1}{2}\Psi_2 (B - B^0)^2 + \frac{1}{24}\Psi_4^c (B - B^0)^4$$

at $\Psi_2 > 0$ has a local minimum at the point $B = B^0$ and a local maximum at the points $B = B_0 \pm (-6\Psi_2/\Psi_4^c)^{1/2}$. The difference between the maximum and the minimum is given by

$$u_{max} = -\frac{3}{2}\Psi_2^2 / \Psi_4^c > 0 .$$

As Θ approaches the critical point Θ_c, this difference becomes smaller and smaller and, therefore, the point B = B° becomes more and more unstable. At a certain moment the parameter B changes hugely and thus the transition is called first-order transition. The degree of instability of the point B° is characterized by the relative difference

$$\lambda = u_{max}/\varkappa = -3\Psi_2^2/(2\varkappa\Psi_4^c) . \tag{44}$$

We consider the state B° to be sufficiently stable if λ exceeds some constant λ_{rp} , say, λ_{rp} = 3.

At the stability boundary we have $\lambda = \lambda_{rp}$ or, by virtue of (44)

$$\Psi_2 = 3^{-1/2}(-2\varkappa\lambda_{rp}\Psi_4^c)^{1/2} . \tag{45}$$

The correlators of the internal parameter B near the stability boundary can be obtained by means of the distribution

$$w(B) = \begin{cases} \text{const} \exp -\frac{1}{2\varkappa}\left[\Psi_2(B-B^\circ)^2 + \frac{1}{12}\Psi_4^c(B-B^\circ)^4\right] & |B-B^\circ| < \left(\frac{6\Psi_2}{-\Psi_4^c}\right)^{1/2} \\ 0 & |B-B^\circ| > \left(\frac{6\Psi_2}{-\Psi_4^c}\right)^{1/2} \end{cases}$$

and equation (45). A simple analysis gives the following correlators

$$\langle B,B\rangle = (3\varkappa)^{1/2}(-2\Lambda\Psi_4^c)^{-1/2}\left[1+3/(4\Lambda)+O(\Lambda^{-2})\right]; \quad \langle B,B,B\rangle = 0$$

$$\langle B,B,B,B\rangle = 3\varkappa(-2\Lambda\Psi_4^c)^{-1}\left[3/(2\Lambda)+O(\Lambda^{-2})\right].$$

From this it is seen that the order of smallness of these correlators in the parameter $\varkappa$ is the same as in the previous case. The fluctuations are also relatively large and strongly non-Gaussian.

c) The case where $\Psi_3^c \neq 0$, i. e. $\overset{\circ}{K}_{1,11} \neq 0$. In this case we deal with a first-order phase transition, but one that differs from the previous case.

At small $\Psi_2 > 0$ one can consider B-B° small. The term $\Psi_4^c(B-B^\circ)^4$ has, here, a small effect. Discarding this term, we have the function

$$u(B) = \frac{1}{2}\Psi_2(B-B^\circ)^2 + \frac{1}{6}\Psi_3^c(B-B^\circ)^3$$

which has a local minimum at the point B = B° and a local maximum at the point B = B° - $2\Psi_2/\Psi_3^c$. The excess of the maximum over the minimum is now equal to $u_{max} = (2/3)\Psi_2^3(\Psi_3^c)^{-2}$. The degree of instability of the point B° is characterized by the relative excess

$$\Lambda = 2(3\varkappa)^{-1}\Psi_2^3(\Psi_3^c)^{-2}.$$

At fixed Λ the quantity Ψ_2 is a function of $\varkappa$

$$\Psi_2 = (3\varkappa\Lambda/2)^{1/3}(\Psi_3^c)^{2/3}.$$

Using this equality, it is easy to find correlators in the critical region:

$$\langle B,B\rangle = \varkappa^{2/3}\left(\frac{2}{3\Lambda}\right)^{1/3}(\Psi_3^c)^{-2/3}\left[1+\frac{2}{3\Lambda}+O(\Lambda^{-2})\right];$$

$$\langle B,B,B\rangle = \frac{2\varkappa}{3\Lambda}(\Psi_3^c)^{-1}\left[1+O(\Lambda^{-1})\right];$$

$$\langle B,B,B,B\rangle = \varkappa^{4/3}\left(\frac{2}{3\Lambda}\right)^{2/3}(\Psi_3^c)^{-4/3}\left[\frac{2}{\Lambda}+O(\Lambda^{-2})\right].$$

We see that the exponents of the parameter $\varkappa$ differ from the two previous cases. Due to this, the fluctuations in the critical region are somewhat smaller, but all the same they are anomalously large and strongly non-Gaussian. This non-Gaussian character disappears with movement away from the critical point. There is no doubt that phenomena similar to those described above also take place in the case of multi-component phase transitions, although the number of different types of phase transitions will increase in these cases.

The Statistical Background of the Glansdorff-Prigogine Criterion

F. Schlögl

Introduction

The development of thermodynamics in the last few decades has been characterized by the inclusion of non-linear processes. Previously thermodynamics had been chiefly concerned with thermal equilibrium states, and with non-equilibrium processes occurring 'near' an equilibrium, in the sense that the dynamical equations can be linearized with respect to the quantities ('forces' and 'fluxes') which describe the deviation from this equilibrium. By restriction to this type of non-equilibrium process, so called 'linear thermodynamics' is defined. The field of thermodynamics which is concerned with non-equilibrium processes describable by macroscopic 'thermodynamics' quantities, yet not occurring 'near' an equilibrium in the above sense, is called 'non-linear' thermodynamics. Most chemical reactions belong to this field and thus so do most of the thermodynamic processes important for biology. In non-equilibrium thermodynamics 'open' systems are considered which exchange energy or matter or both with their environment. Of particular interest are 'steady' or 'stationary' states, in which the system does not change its macroscopic features, whereas an exchange of energy or matter occurs steadily. A typical example of such a type is a burning flame which does not change its shape but nevertheless has a permanent input and output of matter and energy. Such steady states often show a distinct individual structure in space which restores itself if not disturbed too much. We speak of self-organization of 'dissipative' structures (Glansdorff, Prigogine, 1971; Nicolis, Prigogine, 1977). Standard examples of such structures are the hydrodynamic patterns of the Bénard flow and the Taylor flow which were known already in the time before non-linear thermodynamics became a field of its own. These patterns are the result of a competition between temperature gradient as driving force for heat flow and gravitation in the

Thermodynamics and Regulation of Biological and Processes

Bénard system. They are the result of a competition between friction and centrifugal force in the Taylor system. Friction as well as mixing has to be considered as a typical thermodynamic phenomenon, even if not connected with change of temperature, because it is an irreversible process. Some systems also show dissipative structures in time instead of space. They return periodically to the same states once they have reached the so called 'limit cycle'. In the following, however, we shall not be concerned with them, but with the steady states.

The development in time of macroscopically described thermal states of the considered - in general open - system will be given by the solutions of differential equations in time for the respective macroscopic thermal variables. These may be called the macroscopic dynamic equations. The steady solutions are those for which the thermal variables of the open system itself do not change in time. However, only the stable, steady solutions which restore themselves after being perturbed within certain bounds are practically realisable. A characteristic feature of non-linear thermodynamics is the occurrence of unstable steady solutions. And this is closely related to the formation of structures. A steady state belongs to certain constraints of the environment. It can happen that an unstable solution becomes stable or vice versa by a continuous change of a parameter characterizing the constraints. A plot of the steady states of the system in a diagram of the macrovariables is an analogue to a state diagram of equilibrium states of a system in equilibrium thermodynamics. The change from stability to instability can be compared, for instance, with the change of the unmagnetized state of a ferromagnet passing with decreasing temperature through the critical temperature. Above the critical point this state is stable but becomes unstable below. The change is, moreover, connected with a bifurcation into this unstable state and into stable states with magnetization of different orientation. Similar bifurcations occur in the diagram of steady states in non-linear thermodynamics, and the analogy with the situation in equilibrium systems in yet further respects has led to the concept of so called 'non-equilibrium phase-transitions'. These are phenomena far away from equilibrium. As will not be explained in detail, the formation of patterns is, in a more complex way, connected with bifurcation into competing states. Here it is sufficient to stress that criteria for the stability of a steady solution of the dynamical equations are of real physical interest.

Of course there are various possibilities to check the stability of a steady state. One is the study of the solutions of the dynamics in the neighbourhood of this state. More advantageous, however, are criteria which can be applied without solving the dynamic equations and which require the knowledge of these equations only. Such criteria are given by the Lyapunov theory. For a given system various different Lyapunov functions can be found, dependent on individual features of the system. Glansdorff and Prigogine (1970), however, have given a criterion, which is based on a general thermodynamic quantity taken as Lyapunov function (La Salle, Lifshitz, 1961), and, which thus is valid in the same form for all systems under consideration. So it can be called a thermodynamic theorem. The value of such a theorem is not only to give a direct answer to the original question concerning stability, but to allow general conclusions with general results which often appear relatively independent of this question. Let this be compared, for instance, with the positiveness of specific heat as a result of thermostatic stability. The criterion of Glansdorff and Prigogine has been given on the basis of macroscopic arguments. As it is a really new theorem, it was found in a rather intuitive and inductive way and not by deduction from known principles. Thus it is the result of a long development in which the exact conditions for validity had to be found by degrees. This development starts with the formulation of the principle of minimum entropy production by Glansdorff and Prigogine (Prigogine, 1947; Prigogine, 1949) which - as we know today - distinguishes steady states in the framework of linear thermodynamics if the system possesses Onsager symmetry. Later is was suspected that steady states would be distinguished in non-linear thermodynamics in a similar way by a vanishing, non-integrable differential form which would have led to a general evolution criterion (Glansdorff, Prigogine, 1964). The final step was performed by Glansdorff and Prigogine (1970) who made clear that the criterion distinguishes stable steady states.

In the following we shall show how this stability criterion, the 'Glansdorff-Prigogine criterion' which was given in the frame of macroscopic thermodynamics can be understood on the basis of statistical mechanics (Schlögl, 1967, 1971). This derivation does not depend on individual features of the system and shows in its generality that the criterion in macroscopic thermodynamics is indeed an independent theorem and not a consequence of already known thermodynamic principles. It is the only theorem in non-linear thermodynamics which had not been deducted earlier and is so general that it can be called a thermodynamic theorem. In comparison, all other recent results in non-linear thermodynamics are valid for more special classes of systems.

The self reproduction of an equilibrium structure, for instance a crystal, is based on minimum energy in this state, or more precisely, on minimum free energy. Therefore such a structure is stable. The self reproduction of a permanently fed dissipative structure, like that of a living organism, however, is of totally different character. It is the stability of the steady state far away from equilibrium that is characterized by the Glansdorff-Prigogine criterion.

The foundation of this theorem, on the basis of statistical mechanics, does not have an inductive character in the same sense as in macroscopic physics. Therefore in the following this will be discussed first, and the macroscopic formulation will then be given as the result.

Stability of stochastic dynamics

Whereas in macroscopic thermodynamics only 'macrostates' occur, i. e. thermal states of a physical system which are described by relatively few macroscopic thermal variables such as temperature, pressure, mass-densities of chemical components, magnetization and others, in statistical mechanics we use 'microstates' which are described by all degrees of freedom of all particles and fields which form the system. A central question is: What probability distribution over these microstates corresponds to a given macrostate? For equilibrium states the answer is standard. In non-equilibrium states this is a question which we shall deal with later on.

First we shall discuss a stability criterion for time dependent probability distributions in general. Let us consider an experiment which allows exactly n different results. These results form a so called 'sample set' of events. This means, they exclude one another and it is certain that one of these events will occur. Let the events of the sample set be numbered by the subscript i . We call the whole set of probabilities, p_i , of these events the probability distribution p . Later on these events will be replaced by the microstates of a physical system and p will give the statistical description of a macrostate. We shall be concerned with the situation that p is dependent on time, t , and that differential equations of first order in t determine $p(t)$. We shall call these differential equations a 'stochastic dynamics'. We can represent all possible distributions p as points in a space of the n positive rectangular coordinates p_i which lie in the subspace (hyper-plane) given by the normalization condition for all p :

$$\sum_{i=1}^{n} p_i = 1 . \qquad (1)$$

We call this subspace the 'probability space'. A special solution of a given stochastic dynamics is called 'steady' or 'stationary' if it does not change with time. We are interested in the stability of such a steady solution p^o and study therefore the behaviour of solutions $p(t)$ which start in the neighbourhood of p^o . We call p^o stable, or more precisely 'Lyapunov stable', if there exists an arbitrarily small neighbourhood of p^o which is never left by any $p(t)$ if once in the interior.

Let us consider a function $L(p,p^o)$ of the p which has the features that it vanishes for p equal to p^o , that it is positive for all other p , and that, in the probability space all points at which L has a constant value C form a closed surface in such a way that all points with smaller values of C lie in the interior of this surface whatever the value of C . We call such a function L of p monotonically increasing near p^o .

If now for a given stochastic dynamics there exists such a monotonic function L which never increases with time, that means that the time derivative $\dot{L}$ of L is never positive,

$$\dot{L} \leq 0 \tag{2}$$

in the interior neighbourhood of p° in the probability space

$$L \leq C, \tag{3}$$

then a $p(t)$, once in its interior, will stay there. Moreover this will then hold for all smaller neighbourhoods

$$L \leq C' \tag{4}$$

with $C' < C$. That means that then p° is Lyapunov stable. L is called a 'Lyapunov function'.

A special class of such functions (Csiszard, 1967) is given by

$$J(p, p^{\circ}) = \sum_i p_i^{\circ} f(p_i / p_i^{\circ}) \tag{5}$$

where, for positive u, the function $f(u)$ is convex; which for our purpose can be defined by the positive nature of the second derivative:

$$f''(u) \geq 0. \tag{6}$$

The variation, δJ, of J resulting from small changes, δp_i, of the probabilities p_i can be ordered by the powers of δp_i:

$$\delta J = \delta^{(1)} J + \delta^{(2)} J + \ldots \tag{7}$$

$$\delta^{(1)} J = \sum_i f'(p_i / p_i^{\circ}) \, \delta p_i \tag{8}$$

$$\delta^{(2)} J = \frac{1}{2} \sum_i (1/p_i^{\circ}) f''(p_i / p_i^{\circ}) (\delta p_i)^2 \tag{9}$$

The variation of second order, $\delta^{(2)} J$, is always positive and thus J is a convex function of p . The variation of first order $\delta^{(1)} J$ vanishes if p becomes equal to p° because

$$\sum \delta p_i = 0. \tag{10}$$

Thus J as a function of p is minimum at p° and increases monotonically near p° .

In thermodynamics we are interested in parameters of statistical theory which have simple behaviour in the thermodynamic limit, i. e. for unlimited increase of volume of the system at fixed densities. Therefore we seek for a J which is additive for independent, partial systems. Such a J will then be an extensive quantity in macroscopic thermodynamics in which macroscopic volume elements are practically uncorrelated. If two subsystems, denoted by subscripts I and II, are stochastically independent, then the probability of a state of the joint system is the product of the probabilities of the two partial systems

$$p_i = p_{iI} \, p_{iII}. \tag{11}$$

The requirement that in this case J becomes additive

$$J = J_I + J_{II} \tag{12}$$

leads

$$f(u) = u \log u \tag{13}$$

The corresponding J is called the 'Kullback measure' (Kullback, 1951):

$$K(p, p^{\circ}) = \sum_i p_i \log (p_i / p_i^{\circ}), \tag{14}$$

which is positive, vanishing only if p is identical to p° .

Using K as Lyapunov functions we obtain the stability criterion in the following form (Schlögl, 1971): If in the interior of a neighbourhood of p_o

$$K(p, p^{\circ}) \leq C \tag{15}$$

and

$$\dot{K}(p, p^\circ) \leq 0 \tag{16}$$

holds always and everywhere, then p will never leave this neighbourhood. This is a sufficient but not necessary condition for Lyapunov stability of the steady distribution p° .

We shall show that this criterion yields the thermodynamic macroscopic criterion of Glansdorff and Prigogine. First, however, we shall discuss the role of the Kullback measure in information theory.

Relations to information measures

In the 1950's when information theory was developed, a close association between information and entropy was revealed. Information theory uses information measures which can be based on the length of a communication. This length is expressed by the number of bits (yes-no decisions) which are necessary to convey the communication. If an observer already knows the probability distribution, p , over a sample set, then the length of the communication that the event does occur is (up to a numerical factor defining a unit) given by

$$b_i = - \log p_i . \tag{17}$$

Its negative measures the knowledge which the observer already had before the observation concerning the question of whether i will occur.

The average of $-b_i$ is called the 'Shannon-information' (Shannon, Weaver, 1949)

$$I(p) = \sum_i p_i \log p_i \tag{18}$$

and is a measure of the knowledge contained in the distribution p .

If, originally, the distribution p° was known, and afterwards was changed to p by additional information, then this information can be measured by the mean value of the difference between b_i° and b_i in the correct distribution p. This is the 'Kullback measure' (14).

It is also called 'information gain' or 'excess information' of p with respect to p°.

The additivity for independent subsystems, which, in thermodynamics, is characteristic of extensive quantities, is considered also in information theory as a characteristic feature which has to be known for an information measure. This requirement, however, allows also an alternative to the Kullback measure, that is, the Renyi-information (Renyi, 1966).

$$R(p, p^\circ) = (\alpha - 1)^{-1} \log J(p, p^\circ) \tag{19}$$

with $\alpha > 1$ and

$$f(u) = u^\alpha . \tag{20}$$

This is monotonic in p near p° as well, and can also be used as a Lyapunov function. For α equal to one it becomes equal to the Kullback measure K. Also, for small deviations

$$\delta p_i = p_i - p_i^\circ \tag{21}$$

of quadratic order, it becomes equal to K up to the numerical factor $\alpha/2$.

We can write

$$K(p, p^\circ) = \Delta I - \sum_i (\partial I(p^\circ)/\partial p_i^\circ) \Delta p_i \tag{22}$$

where

$$\Delta I = I(p) - I(p_i^\circ) \tag{23}$$

and

$$\Delta p_i = p_i - p_i^\circ . \tag{24}$$

This means K is that part of the power expansion of ΔI with respect to the variations Δp_i which remains if the linear contribution is ignored. It is the non-linear variation, $\Delta_{NL} I$, of I in p-space.

In macroscopic thermodynamics the states can be described uniquely by a set of variables which, in statistical theory, are mean values

$$\langle m^\nu \rangle = \sum_i m_i^\nu p_i \tag{25}$$

of observables m_i^ν , defined in the microstates. We shall call such quantities 'direct variables'. In the canonical distribution, for instance, energy is a direct variable, but temperature is not. Now, if I is given as a function of the variables $\langle m^\nu \rangle$, the non-linear variation in p-space is equal to that in the space of the macrovariables $\langle m^\nu \rangle$ because finite variations in both spaces are also connected linearly:

$$\Delta \langle m^\nu \rangle = \sum m_i^\nu \Delta p_i \, . \tag{26}$$

So we can write also in terms of the macroscopic variables $\langle m^\nu \rangle$:

$$K = \Delta_{NL} I \, . \tag{27}$$

It should be stressed that this is true only if I is considered as a function of direct variables (Schlögl, 1971). This restriction, however, is no longer important if the variations of p become infinitesimal, δp . Thus for any kind of macroscopic variables

$$K = \delta^{(2)} I \tag{28}$$

holds.

This means that K is then equal to the term of δI which is of second order in the variation of the macrovariables used for the description (the second order variation of I).

$$\delta^{(2)} \dot{I} \leq 0 \tag{29}$$

then is sufficient for stability of the steady state if it holds for any infinitesimal deviation from this state.

Thermodynamic probability distributions

A fundamental question in statistical physics is, which probability distribution over the microstates of a physical system is adequate to characterize a given macroscopic state described by comparatively few thermodynamic variables? Comparatively few is meant in comparison to the enormously high number of variables which describe a microstate. In classical mechanics these microvariables are the canonical coordinates and their conjugate momenta. They have continuous ranges of values and form the high-dimensional Gibbs phase space. Let the whole set of these variables, describing a microstate by one point in this space, be called ξ , which takes over the role of the subscript i , in the preceding considerations but assumes, in contrast to i , continuous values. Therefore a probability distribution in Gibbs phase space is described by a probability density $w(\xi)$. To p_i there corresponds the probability $w d\xi$ of finding the system in a microstate which lies in a volume element with the volume $d\xi$ at the point ξ of the phase space. The summation over i has to be replaced by integration over ξ .

First we consider thermal equilibrium states. In macroscopic physics such a state is completely described by a set of direct variables which in statistical theory are mean values

$$\langle M^\nu \rangle = \int d\xi \, w(\xi) M^\nu(\xi) \tag{30}$$

of phase space functions $M^\nu(\xi)$. The corresponding phase space distribution is given by

$$w(\xi) = exp\,[\Phi - \sum_\nu \lambda_\nu M^\nu(\xi)] \, . \tag{31}$$

The parameters λ_ν , are implicitly determined by the mean values $\langle M^\nu \rangle$, and the parameter Φ by the normalization of the distribution. The most familiar case is that in which only one value $\langle M^\nu \rangle$ for the mean value of energy is given. This leads to the canonical distribution where the only λ_ν occurring is, except for Boltzmann's constant, the reciprocal temperature. If, in addition for instance, the mean value of a particle number is given, we obtain the grandcanonical disbribution, where a second λ_ν is the negative of the chemical potential divided by temperature and Boltzmann's constant. The general distribution (4.2) will be called a 'generalized canonical distribution'.

The macroscopic entropy of such an equilibrium state is, up to a numerical factor, equal to the negative of the Shannon information, I , of the distribution w. This general connection of entropy with w was already known by Boltzmann, long before the introduction of information measures. The canonical distribution, then, was defined by the requirement that it corresponds to maximum entropy under the restriction that the mean value of energy be equal to the macroscopic thermal variable 'internal energy'. More generally, the generalized canonical distribution corresponds to maximum entropy with given values $\langle M^{\nu} \rangle$. Jaynes (1967) interpreted this prescription by minimizing information, I , to avoid any unjustified prejudice, and called the prescription the method of 'unbiased guess'.

The question of which distributions w correspond to non-equilibrium states brings new problems. The macroscopic description of thermodynamic non-equilibrium states can be given by local time dependent direct variables. They are mean values of phase space functions $m^{\nu}(x; \xi)$ which are local densities of quantities $M^{\nu}(\xi)$ in three-dimensional space x. They become time-dependent through the time-dependence of w .

If we construct w for given mean values of m^{ν} in the same way as we did for the equilibrium case, we obtain the so called local equilibrium disbribution:

$$w_{\ell}(\xi, t) = \exp[\phi(t) - \lambda(t).m(\xi)] \qquad (32)$$

with the abbreviation for the summation over the macroscopic local quantities:

$$\lambda.m = \sum_{\nu} \int d^3x\, \lambda_{\nu}(\underline{x})\, m^{\nu}(\underline{x}). \qquad (33)$$

This local equilibrium distribution supresses certain correlations in space. Therefore such a distribution is not suitable for an adequate description of transport phenomena. The reason is that the method of unbiased guess, if based on the instantaneous mean values, ignores the knowledge of earlier values which is contained in the dynamic features of the system.

An adequate thermodynamic description by local variables holds only if the system shows a distinct separation into a microscopic and a macroscopic time scale. That means that the changes of the macroscopic variables are very slow in comparison with the microscopic processes between particles which give rise to changes of the microstates. For these systems a construction of distributions w which corrects the above shortcomings of w_ℓ was first given by Mori (1959). We call them Mori-distributions. They have been given by Mc Lennan (1963) and Zubarev (1965) too by different arguments but with the same result. For our purposes it is not necessary to know the result, but only the manner of construction. In this method $w(t)$ is obtained by application of microscopic dynamics to $w_\ell(t-\tau)$ over a time interval τ, which is large on the microscopic, and small on the macroscopic, time scale. In this time interval, the mean values of the m remain unchanged but the distribution changes fundamentally because all correlations which were suppressed in w_ℓ are fully realised in w by the reversible microscopic dynamics during the time τ. These dynamics, however, do not change the measure I, because this measure has the specific feature of being invariant with respect to reversible motion. The changes in the system due to the microscopic dynamics during time τ are essentially changes in an isolated system, because the influences of the surroundings do not become effective in this short time except within the immediate surface region. So, practically, the change from w_ℓ to w during the time τ is reversible. Thus the measure $I(w)$ at time t is essentially equal to $I(w_\ell)$ at time $t-\tau$. And $I(w_\ell)$ as a function of macroscopic variables changes essentially only on the macroscopic scale. That means, that the measure $I(w)$ is practically equal to $I(w_\ell)$ at any time (Schlögl, 1975).

It is possible to define the entropy of a non-equilibrium distribution as being proportional to the negative of information with the same numerical factor as in the equilibrium case. The numerical factor is fixed by the unit of energy and temperature. Total entropy has the form

$$S = \int d^3x \; s(\underline{x}) \tag{34}$$

where entropy density $s(\underline{x})$ is the same function of the local variables as in the equilibrium case. The entropy is correctly given by the local equilibrium distribution. It should be stressed that this is not true for other thermal quantities such as, for instance, transport coefficients.

The macroscopic criterion

Now we can give the macroscopic form of the stability criterion for systems with local variables. Any neighbourhood of a steady state in which the non-linear part of the variation of entropy S as a function of direct variables $\langle m^\nu \rangle$ fulfills

$$\Delta_{NL} S = \int d^3x \; \Delta_{NL} s(\underline{x}) \geq \text{const} \tag{35}$$

$$\Delta_{NL} \dot{S} = \int d^3x \; \Delta_{NL} \dot{s}(\underline{x}) \geq 0 \tag{36}$$

will never be left because of thermodynamic changes.

The Glansdorff-Prigogine criterion is concerned with infinitesimal variations and has the following form: A steady state is stable if the second order variation of entropy in this state always increases

$$\delta^{(2)} \dot{S} \geq 0 \tag{37}$$

That is

$$\int d^3x \sum_\nu \frac{\partial}{\partial t} \langle m^\nu \rangle \, \delta\lambda_\nu \geq 0 . \tag{38}$$

In linear thermodynamics Onsager's symmetry law is formulated by the introduction of 'forces'

$$\mathfrak{X}_\nu = \lambda_\nu - \lambda_\nu^e \tag{39}$$

describing the deviation from the neighbouring equilibrium state; and of 'fluxes'

$$J^\nu = \frac{\partial}{\partial t} \langle m^\nu \rangle \tag{40}$$

describing the dynamic response to these forces. It is not necessary to fix the equilibrium state uniquely to identify the variation $\delta\lambda_\nu$ of the steady state with $\delta \mathfrak{X}_\nu$. Then (38) can be written in the standard form

$$J . \delta \mathfrak{X} \geq 0 . \tag{41}$$

(In the original work by Glansdorff and Prigogine this inequality is expressed by the deviation of the steady state values $\mathfrak{X}^0$ from the momentary state $\mathfrak{X}$, which gives the opposite sign in the inequality.)

Often it is convenient to introduce local flow density vectors $\underline{j}^\nu$ and source densities ω^ν of the quantities with densities $\langle m^\nu \rangle$ by the equation

$$\frac{\partial}{\partial t} \langle m^\nu \rangle + \underline{\nabla}\, \underline{j}^\nu = \omega^\nu . \tag{42}$$

If such a quantity is conserved, like energy, or the number of particles of a component not taking part in chemical reactions, then its source density ω^ν is zero. Generally. however, this is not the case and ω^ν is the local production of this quantity per unit time and volume. In general the flow densities $\underline{j}^\nu$ and source densities ω^ν are independent quantities of different character. This can be demonstrated through the example of a many component mixture, in the interior of which, diffusion and chemical reactions take place. Flow densities $\underline{j}^\nu$ of particle numbers of different components will in general be independent. The production rates ω^ν , however, are dependent on one another through chemical reactions. They are linear functions of reaction rates ξ^s the number of which is equal to the number of independent reaction equations and smaller than the number of components:

$$\omega^\nu = \sum_s \alpha_s^\nu \xi^s . \tag{43}$$

α_{ρ}^{ν} is the stochiometric coefficient of the ν -th component in the ρ-th reaction. Equation

$$\sum_{\nu} \omega^{\nu} X_{\nu} = \sum_{\rho} \xi^{\rho} A_{\rho} \tag{44}$$

introduces the affinities A_{ρ} . A generalisation of this definition is possible for any association between rates ω^{ν} and independent rates ξ^{ρ}

As we have to compare the steady state with non-equilibrium states in which the state of the surroundings is always the same, we admit only variations δX^{ν} which vanish at the surface of the system. Therefore a surface contribution vanishes when the integral in (38) is transformed by Gauss' theorem and we obtain

$$\int dx^3 \left(\sum_{\rho} \xi^{\rho} \delta A_{\rho} - \sum_{\nu} \underline{j}^{\nu} \delta \underline{\nabla} X_{\nu}\right) \geq 0 . \tag{45}$$

In traditional linear thermodynamics, the quantities A_{ρ} and $-\underline{\nabla} X_{\nu}$ are present as 'forces' and the quantities ξ^{ρ} and $\underline{j}^{\nu}$ as the corresponding 'fluxes' (de Groot, 1952; Meixner, Reitz, 1959) and thus we can also interpret this inequality in terms of the general form of (41). The essential point is that the fluxes J have to be independent. Only the corresponding X must be varied independently.

The principle of minimum entropy production

Using the shorthand of (41) we can write, generally, for entropy production

$$P = J . X . \tag{46}$$

Its variation is given by

$$\delta P = J . \delta X + \delta J . X . \tag{47}$$

Glansdorff and Prigogine already pointed out that the expression which occurs in their criterion viz.

$$\delta_X P = J \cdot \delta X \tag{48}$$

relates to the contribution to δP which derives from the variation of forces only, i. e. the fluxes J are fixed. Let us call this expression the Glansdorff-Prigogine variation.

In linear thermodynamics the forces X and fluxes J are connected by linear equations (de Groot, 1952; Meixner, Reitz, 1959), which are often called transport equations; even if no real transport in space occurs. Let these equations formally be written as

$$J^{\nu} = \sum_{\mu} L^{\nu\mu} X_{\mu} \tag{49}$$

with the so called Onsager matrix L . The expression is formal as the super- and subscripts like ν are abbreviations for the pairs (ν, x) . The summation correspondingly contains integration in space. The matrix L is independent of all X and J . Only its symmetric element, L_+ , enters into

$$P = \sum_{\mu} L_+^{\mu\nu} X_{\mu} X_{\nu} = (L_+ X) \cdot X \tag{50}$$

$$\delta P = 2 (L_+ X) \cdot \delta X \,. \tag{51}$$

We have to distinguish between systems with symmetric L , the so called 'Onsager symmetric' systems, and systems which show a so called 'Casimir symmetry'. In the latter L is not a symmetric matrix. Only in linear thermodynamics, i. e. in the neighbourhood of an equilibrium, and, only for systems with Onsager symmetry do we obtain, as a consequence of the Glandorff-Prigogine criterion, that

$$\delta_X P = \frac{1}{2} \delta P \geq 0 \tag{52}$$

for variations in a steady state always holds. In these cases a steady state is distinguished by the extremum of a thermodynamic quantity - the minimum in entropy production (Prigogine, 1947, 1949). This is similar to the distinction of an equilibrium state by minimum energy or related thermodynamic potentials like free energy.

Concluding remarks

Variational principles by which certain states, like equilibrium states, or steady, non-equilibrium states, are distinguished by extrema of state functions bring not only an essential simplification to the description of dynamic processes, and thus a deeper insight into nature, but are also important methodiological means for distinguishing these states as mathematical solutions of a variational problem. The principle of minimum entropy production is of such kind. Its impressive simplicity, however, has often led to wrong conclusions - in particular in more general discussions - when the restrictive conditions of validity were not correctly observed. It is valid in linear thermodynamics for systems with Onsager symmetry.

This general criterion which is valid also for systems far from equilibrium permits a distinction between stable steady states by the sign of the Glansdorff-Prigogine variation. This variation, however, is in general not integrable. So we can not give a general function of thermodynamic variables which becomes extremum in steady states.

Also, the connection between the Glansdorff-Prigogine variation and excess information can bring similar confusions like the general connection of entropy with information, when interpreted in an incorrect way. It should be stressed that these thermodynamic quantities are information measures concerning only the question 'In which Gibbs phase space state is the system?'. They should not be confused with information about other questions, for instance with information in a genetic code. The information measures are not the contents of a message but measures of probability distributions. They yield no more than a probability distribution can give, even less. Above all, they say nothing about the individual features of the events over which the probabilities are considered. It is the well-defined measure for surplus knowledge in the varied macrostates in comparison to the steady state the time derivative of which becomes the Glansdorff-Prigogine variation if expressed by macroscopic variables.

An equilibrium state is a special stable steady state in which the Glansdorff-Prigogine variation is the time derivative of a thermodynamic potential. Which one? This depends upon the set of the m^ν occurring in the inequality (38). The criterion then gives a form of the Second Law dependent on the boundary conditions of an open system. The special case of an isolated system gives the increase of entropy. We see that the Second Law is contained in the criterion. The opposite is not true as the derivation of the criterion in statistical theory demonstrates. This shows that the Glansdorff-Prigogine criterion is indeed an independent fundamental theorem of thermodynamics.

The deduction of the criterion from the statistical basis is represented in this article in the frame of classical physics. It should, however, be stressed that a quantum theoretical deduction leads to the same results (Schlögl, 1980).

Thermodynamic Analogies in Statistical Theory of Dynamic Systems

A. A. Platonov

From a macroscopic point of view non-equilibrium thermodynamic systems that take into account fluctuations are only a specific class among systems considered in statistical theories of optimal control and filtration (also called statistical theory of dynamic systems).

Specificity in thermodynamic systems is conditioned by association of their macroscopic description with the characteristics of micro-motion and this manifests itself in a rather limited regulation of the properties of macro-processes developed in such systems. Due to this fact in any phenomenological description of thermodynamic systems on the level of Langevin-type relaxation equations (Landau, Lifshitz, 1969; Stratonovich, 1967, 1969, 1970, 1980) there exist rigid connections between fluctuation and dissipation (relaxation) characteristics established by the fluctuation-dissipation theorems and the reciprocity relations. In the theory of dynamic systems such rigours are not common.

The class of stochastic systems considered in the theory of optimal control and filtration is much wider than that considered in statistical thermodynamics (e. g. Krasovsky, 1974). These theories develop practically independently, without interaction with the problems and apparatus of statistical physics and thermodynamics.

However, in various cases, it is possible to discover, though quite formally, some analogies between the behaviour and description of dynamic and thermodynamic systems.

Thermodynamics and Regulation of Biological and Processes

The developed formal apparatus in this branch of science, as well as a large number of independent results make the study of the possibility of extending certain analogies, methods and results from one theory to the other very tempting both in theory and application.

Physical conceptions were most widely used by Kraskovsky (1974) in the study of the general laws of statistical dynamics in Markov's non-linear dynamic systems and in addition, some of these aspects were considered in our papers (Platonov, 1979, 1980). Here we continue discussion of the possibilities of using the apparatus of statistical thermodynamics, mainly on the basis of its Markov version (Stratonovich, 1967, 1970, 1971, 1980) to study the behaviour of non-linear dynamic systems with noise. We also consider some analogies between the criteria of minimum and maximum entropy production, widely used in optimal control theory.

Some thermodynamic analogies in the stochastic approach

Following the formulation of the problem accepted in the paper by Krasovsky (1974), we consider the Markov type system given by stochastic differential equations of the form

$$\dot{x}_i = f_i(x, t) + \xi_i(t), \quad i = 1, \ldots, n \tag{1}$$

where $x = x(t) = \{x_1(t), \ldots, x_n(t)\}$ is the vector of phase coordinates (state-variables); $\xi(t) = \{\xi_1(t), \ldots, \xi_n(t)\}$ is the vector of random delta-correlated Gaussian processes (noises) with the (n x n) intensity matrix D independent of x and t; as to the functions f(x, t), we assume that they can be differentiated as many times as necessary at any point of the phase space.

Owing to the Markov character of the process x(t) the probability distribution density p(x, t) of phase coordinates at a given time moment satisfies the Fokker-Planck-Kolmogoroff ('FPK') equation

$$\frac{\partial p(x,t)}{\partial t} = -\frac{\partial}{\partial x_i}\{f_i(x,t)\,p(x,t)\} + \frac{1}{2}D_{ij}\frac{\partial^2 p(x,t)}{\partial x_i \partial x_j} \qquad (2)$$

here and below we imply summation from 1 to n over the pairs of identical indices.

Determining the entropy of the system in the traditional way as a value

$$S(t) = -\int_{-\infty}^{\infty} p(x,t)\ln p(x,t)\,dx \qquad (3)$$

(we imply integration over all the components of the vector x in the entire phase space) by means of (2), Krasovsky (1974) derived the following relation

$$\begin{aligned}\dot{S}(t) &= \int_{-\infty}^{\infty}\left(\frac{\partial f_i(x,t)}{\partial x_i}\right)p(x,t)\,dx + \\ &+ \frac{1}{2}D_{ij}\int_{-\infty}^{\infty}\left(\frac{\partial \ln p(x,t)}{\partial x_i}\,\frac{\partial \ln p(x,t)}{\partial x_j}\right)p(x,t)\,dx = \\ &= \left\langle\frac{\partial f_i(x,t)}{\partial x_i}\right\rangle + \frac{1}{2}D_{ij}\left\langle\frac{\partial \ln p(x,t)}{\partial x_i}\,\frac{\partial \ln p(x,t)}{\partial x_j}\right\rangle .\end{aligned} \qquad (4)$$

The brackets $\langle\ldots\rangle$ mean the averaging procedure

$$\langle\varphi(x,t)\rangle = \int_{-\infty}^{\infty}\varphi(x,t)\,p(x,t)\,dx.$$

The absence of special limitations upon the form of the functions $f_i(x, t)$ and upon the matrix of fluctuation intensities $\|D\|$ makes it possible to indicate, on the basis of (4), the classes of the dynamic systems having specific properties. In particular, in the absence of noises ($\|D\| = 0$) for generalised conservative systems, i. e. for the systems in which

$$\operatorname{div} f(x,t) = \frac{\partial f_i(x,t)}{\partial x_i} = 0 \qquad (5)$$

at any phase space point the entropy value is shown to be constant, since, for this case, according to (4), $\dot{S} = 0$.

In the presence of noises the entropy of a generalised conservative system always grows. This is readily seen when we notice that the second summand in the right-hand side of (4) can be presented in the form

$$D_{ij}\left\langle \frac{\partial \ln p(x,t)}{\partial x_i}\frac{\partial \ln p(x,t)}{\partial x_j}\right\rangle = \left\langle \left\{\eta_i \frac{\partial \ln p(x,t)}{\partial x_i}\right\}^2\right\rangle, \quad (6)$$

where η_i are random quantities whose moments are equal to $\langle \eta_i \eta_j \rangle = D_{ij}$.

The statement is proved by the fact that quantity (6) is non-negative.

We present without discussion (Krasovsky, 1974) another relation where the entropy equilibrium of a dynamic system is possible. This is the condition

$$\left\langle \frac{\partial f_i(x,t)}{\partial x_i}\right\rangle = -\frac{1}{2}D_{ij}\left\langle \frac{\partial \ln p(x,t)}{\partial x_i}\frac{\partial \ln p(x,t)}{\partial x_j}\right\rangle. \quad (7)$$

This relation appears essential to the theory since it indicates the weakest limitations, i. e. the most general conditions under which entropy equilibrium in the Markov dynamic system in question is possible.

In effect, this relation has a character close to the relations of Markov's thermodynamics (Stratonovich, 1967-1970) which describe the connections between general fluctuation and dissipation characteristics in thermodynamic systems. Constancy of entropy is, in its turn, a necessary condition for the existence of stationary distributions in thermodynamics. These points confirm the logic of the search for deeper analogies between non-equilibrium thermodynamics and statistical theory of optimal control.

The fundamental principle in statistical thermodynamics is universality in the description of equilibrium fluctuations in any thermodynamic system. Obedience of micro-processes to chemical equations makes it possible independently to identify the form of the thermodynamic parameter distribution of Gibbs, Boltzmann and others (Landau, Lifshitz, 1969) and their subsequent application in the analysis of processes in the system.

In the theory of dynamic systems, in the general case, it is impossible to construct, a priori, such a type of distribution. For certain classes this problem can however be solved (see Krasovsky, 1974) by seeking for first integrals.

Notice the following important circumstance: the existence of an equilibrium distribution is possible neither in thermodynamic nor in dynamic (closed) systems without the obligatory presence, simultaneously, of two processes, namely, dissipation and fluctuation.

Fluctuating stationary actions cause some energy flux in the system containing no other energy source, a closed system. In the absence of dissipative effects the latter would lead to continuous non-stationary evolution and to the impossibility for stationary states to appear. In the absence of fluctuating disturbances, dissipation would lead, in its turn, to degeneration of the distribution to the δ -function with zero arguments at a stationary point.

What has been said above is described by relation (7). It establishes relations among the dissipation characteristics ($f_i(x, t)$, noise intensities D_{ij} and moments of fluctuations of phase coordinates at which the entropy of the system becomes its integral of motion.

From relation (7) one can deduce further useful consequences. To this end, we introduce thermodynamic forces conjugated to phase coordinates (Landau, Lifshitz, 1969; Stratonovich, 1967-1971)

$$X_j = X(x) = -\frac{\partial \ln p_0(x)}{\partial x_j} = \frac{1}{\Theta} \cdot \frac{\partial F(x)}{\partial x_j}; \tag{8}$$

where

$$p_0(x) = e^{\frac{F - F(x)}{\Theta}}$$

and where F(x) is the generalised free energy of the system.

For the quantities $X_j = X_j(x)$ in the Gaussian approximation of (8) there hold the known relations

$$\langle x_i X_j \rangle = \delta_{ij} \tag{9}$$

$$\langle x_i x_j \rangle = \Theta R^{-1}_{ij} \quad \left(R_{ij} = \frac{\partial^2 F(x)}{\partial x_i \partial x_j}\bigg|_{x=0} \right) \tag{10}$$

$$\langle X_i X_j \rangle = \Theta^{-1} R_{ij} \, . \tag{11}$$

In the stationary case, with account taken of (8) and (11), then (7) takes the form

$$D_{ij} R_{ij} = -2\Theta \left\langle \frac{\partial f_i(x)}{\partial x_i} \right\rangle . \tag{12}$$

Expanding the functions f_i into a Maclaurin series and noting that moments of odd order are equal to zero, we obtain for (12) the expansion

$$\begin{aligned} D_{ij} R_{ij} &= -2\Theta \frac{\partial f_i(0)}{\partial x_i} - \Theta^2 \frac{\partial^3 f_i(0)}{\partial x_i \partial x_k \partial x_l} R^{-1}_{kl} + \ldots = \\ &= -2\Theta \frac{\partial f_i(0)}{\partial x_i} + O(\Theta^2) \, . \end{aligned} \tag{13}$$

Thus at small Θ a non-linear dynamic system behaves, practically, as a linear one, and there holds the equality

$$\mathrm{Sp}\,(DR) = -2\Theta \, \mathrm{Sp}A \tag{14}$$

where SpA, equivalent to Sp(DR), is the sum of diagonal elements of the matrices $\|A\| = \left\| \frac{\partial f_i}{\partial x_j} \right\|_{x=0}$; $\|DR\|$. On the other hand, the latter relation is a consequence of more rigid conditions i. e. of the existence of a stationary distribution. Substituting the Gaussian distribution into (2), we obtain the relation

$$- \Theta^2 \frac{\partial f_i(x)}{\partial x_i} - \frac{\Theta}{2} D_{ij} R_{ij} + \Theta f_i(x) R_{ij} x_j + \\ + \frac{1}{2} D_{ij} R_{ik} R_{jl} x_k x_l = 0, \tag{15}$$

which must hold at all X. This holds if the following matrix relations are valid

$$D_{ij} R_{ij} = -2\Theta \frac{\partial f_i(x)}{\partial x_i}\Big|_{x=0} = -2\Theta \frac{\partial f_i(0)}{\partial x_i} \tag{16}$$

$$D_{ij} R_{ik} R_{jl} = -\Theta \left(R_{ik} \frac{\partial f_i(0)}{\partial x_l} + R_{il} \frac{\partial f_i(0)}{\partial x_k} \right) + \Theta^2 \frac{\partial^3 f_i(0)}{\partial x_i \partial x_k \partial x_l} \tag{17}$$

$$\frac{\partial^2 f_i(0)}{\partial x_i \partial x_k} = 0. \tag{18}$$

Relation (16) mirrors (14) and (17) leads to the following

$$D_{ij} = -\Theta \left\{ \frac{\partial f_i(0)}{\partial x_k} R^{-1}_{kj} + \frac{\partial f_j(0)}{\partial x_k} R^{-1}_{ki} \right\} + O(\Theta^2). \tag{19}$$

Krasovsky (1974) gives examples of the apllication of relations (28) to (31) in the analysis of the statistical stability of dynamic systems.

Relations of the type (16) to (19) are also useful both in solving identification problems and in investigating the synthesis of optimal control systems through noise.

These relations are advantageous since they help simplify the corresponding Fokker-Planck-Kolmogoroff equation (because they are used to decrease the number of independent coefficients in the equations).

Relations (14) to (19), analogous to those presented in the paper by Krasovsky (1974) turn out to be close in many respects to the relations describing the fluctuation-dissipation theorem and to the reciprocity relations in Markov version (Stratonovich, 1967-1971).

However, these relations are obtained here from the stationary conditions, an in non-equilibrium thermodynamics they are derived under condition of time reversibility of the processes in a system.

Time reversibility and thermodynamic type relations for Markov dynamic systems

Time reversibility in dynamic systems is a little-studied problem and is not used practically in solving theoretical and applied problems in the theory of dynamic systems.

At the same time, on the basis of time reversibility principle a whole number of important theoretical results with a wide range of application (the Onsager reciprocity relations, the fluctuation-dissipation theorem, their non-linear generalisation) have been obtained in non-equilibrium thermodynamics.

Not all dynamic systems possess time reversibility, but it still takes place in a large number of cases. For example, in systems with Gaussian fluctuations of state variables. Obtaining thermodynamic type relations for such systems (close to the relations obtained by Stratonovich, 1967-1971) would make it possible to simplify the solution of many problems of optimal synthesis and control, and, in addition, to widen the analogies with the results of fluctuation theory and transport phenomena in statistical control theory.

The relations obtained would also permit establishment of necessary conditions where the processes in Markov dynamic systems are reversible. This makes it possible to synthesize and study the behaviour of a wide range of dynamic systems which are time reversible.

Some difficulties in the study are due to the absence of a definition of time reversibility for Markov non-stationary processes. Earlier studies (Platonov, 1978, 1979, 1981) lead to the following general and constructive definition:

The Markov process, $x(t) = \{x_1(t), \ldots, x_n(t)\}$, is called time reversible at the time moment t, if, for any τ_1 and τ_2, at a given t, there holds the relation

$$p(x_2, t+\tau_2 | x_1, t+\tau_1) = p(\varepsilon x_2, t-\tau_2 | \varepsilon x_1, t-\tau_1). \quad (20)$$

Here $p(x_2, t_2, x_1, t_1)$ is the probability density of a transition from the values x_1 of the process at the moment t_1 to the values x_2 at a moment t_2; $\varepsilon_i = +1$ for time-even components of the Markov process, and $\varepsilon_i = -1$ for time-odd components ($i = 1, \ldots, n$).

If condition (20) holds for any time moment, then we call the process x(t) reversible.

According to the latter definition, perfect time reversibility of the process is possible only in the case of its stationary rate.

In the publications mentioned above we obtained FPK type equations, but for the transition probability density $p(x, t | x_1, t_1)$, $(t \leq t_1)$ we have

$$\frac{\partial p(x,t|x_1,t_1)}{\partial t} = -\sum_{n=1}^{\infty} \frac{(-1)^n}{n!} \frac{\partial}{\partial x_{\alpha_1}} \cdots \frac{\partial}{\partial x_{\alpha_n}} \{Q_{\alpha_1 \ldots \alpha_n}(x,t) p(x,t|x_1,t_1)\} \quad (21)$$

$$\frac{\partial p(x,t|x_1,t_1)}{\partial t} = \sum_{n=1}^{\infty} \frac{1}{n!} Q_{\alpha_1 \ldots \alpha_n}(x_1,t_1) \frac{\partial}{\partial x^{(1)}_{\alpha_1}} \cdots \frac{\partial}{\partial x^{(1)}_{\alpha_n}} p(x,t|x_1,t_1)\} \quad (22)$$

which differ from the usual kinetic equations of Marxov theory (see e. g. Stratonovich, 1970, 1971) in the form of the coefficients. The transformed kinetic coefficients are determined by the relations

$$Q_{\alpha_1 \ldots \alpha_n}(x,t) = \frac{1}{p(x,t)} \sum_{m=0}^{\infty} \frac{(-1)^{m+n}}{m!} \frac{\partial}{\partial x_{\beta_1}} \cdots \frac{\partial}{\partial x_{\beta_m}} \{K_{\alpha_1 \ldots \alpha_n \beta_1 \ldots \beta_m}(x,t) p(x,t)\} \quad (23)$$

where $K_{\alpha_1 \dots \alpha_n}(x, t)$ are the usual intensity coefficients entering known equations of the type (2); $K_{\alpha}(x, t) = f_{\alpha}(x, t)$ are drift coefficients; $K_{\alpha_1 \alpha_2}(x, t) = D_{\alpha_1 \alpha_2}(x, t)$ are diffusion coefficients, etc.

From definition (20), using equations (21) and (22), we obtain the necessary and sufficient conditions for time reversibility of the process:

$$K_{\alpha_1 \dots \alpha_n}(x, t+\tau) = \varepsilon_{\alpha_1} \dots \varepsilon_{\alpha_n} Q_{\alpha_1 \dots \alpha_n}(\varepsilon x, t-\tau) \tag{24}$$

For complete time reversibility the relations (24) have a stationary character.

$$K_{\alpha_1 \dots \alpha_n}(x) = \varepsilon_{\alpha_1} \dots \varepsilon_{\alpha_n} Q_{\alpha_1 \dots \alpha_n}(\varepsilon x). \tag{25}$$

Relations (24) coincide with the analogous ones obtained by Stratonovich (1970), but, here, they are derived without the additional condition $p_0(x) = p_0(\varepsilon x)$, which is redundant. If (24) holds, the condition $p(x, t+\tau) = p(\varepsilon x, t-\tau)$ can be shown to hold too. (24) extends the time reversibility conditions to the non-stationary case.

The fluctuation-dissipation relations of Markov non-equilibrium thermodynamics are derived directly from (24) (Stratonovich, 1967-1971) by integrating with weight $p_0(x) \exp\{xX\}$ and a subsequent transition to asymptotic behaviour by taking into account the narrowness of the equilibrium distribution $p_0(x)$ (dispersion $\sigma^2 \sim kT = \Theta$). An essential difference of thermodynamic systems described by the Langevin type equations (1) from dynamic systems consists in the necessary appearance of non-Gaussian character in the random force fluctuation $\xi(t)$ and in non-linearity of the functions f(x).

For dynamic systems, in which the character of random effects is not regulated, $\xi(t)$ may be either Gaussian or non-Gaussian noise. In the Gaussian case the coefficients with the number of indices larger than two disappear. The remaining relations are generators for deriving exact and asymptotical 'thermodynamic type' relations for dynamic time-reversible systems.

Some analogues with regular theory of optimal control

Let us consider the analogies between non-equilibrium thermodynamics and the apparatus of optimal control theory.

First of all let us introduce some relations valid for every thermodynamic system.

As is known, in equilibrium states the fluctuations of internal thermodynamic parameters $\{x_1, \ldots, x_n\}$ are described by the distribution:

$$p_0(x) = \text{const} \exp\{S'(x)\} \tag{26}$$

where $S(x)$ is the entropy of a system, considered as a function of parameters x (see Landau, Lifshitz, 1976, formulae 110.1).

As is usually done in non-equilibrium thermodynamics, let us introduce the conjugated (to the sequent internal parameters x) 'thermodynamic forces' X by means of the following definition:

$$X_j = \frac{\partial \Delta S(x)}{\partial x_j} = X_j(x) \tag{27}$$

where $\Delta S(x) = S(x) - S_0$

S_0 represents the entropy in an equilibrium state.

Practically this means the replacement of variables and thus it allows us to rewrite the phenomenological relaxation equations for the thermodynamic parameters x(t):

$$\dot{x}_i = f_i(x) \qquad i = 1, 2, \ldots, n \tag{28}$$

in the form of fluxes

$$\dot{x}_i = f_i\{x(X)\} = J_i(X). \tag{29}$$

As is shown in the work of Stratonovich (1967-1970), relation (29) is an asymptotic being absolutely correct for the linear functions $f_i(x)$, it differs only little from $\dot{x}_i = \langle f_i(x) \rangle_X$ in the non-linear case.

We assume, that in the equilibrium state $\dot{x} = 0$, $x = 0$, and $X = 0$.

On movement away from the equilibrium state, under steady external conditions, the thermodynamic system turns back to the equilibrium state, which process is accompanied by a growth of entropy. This means that the entropy production, the function

$$\sigma(x) = \frac{d\Delta S(x)}{dt} = \sum_{k=1}^{n} X_k(x)\dot{x}_k \tag{30}$$

during the relaxation process, has a positive value and is equal to zero in the limit.

Note, that $\sigma(x)$ may be rewritten through the thermodynamic forces X

$$\sigma(X) = \sum_{k=1}^{n} X_k f_k\{x(X)\} = \sum_{k=1}^{n} X_k J_k(X). \tag{31}$$

Relations (27) and relaxation equations (28) provide a particular type of possible evolution of the conjugated parameters.

Indeed, if the symmetry of the matrix $\| \partial X_i / \partial x_j \|$ is taken into account, then

$$\dot{X}_j = -\sum_{k=1}^{n} X_k(x) \frac{\partial f_k(x)}{\partial x_j} + \frac{\partial \sigma(x)}{\partial x_j} \tag{32}$$

where $\sigma(x)$ is the entropy production function (30).

Relation (32) represents the connection between the 'force fluxes' X and the thermodynamic parameters x conjugated to X. In this sense it is close to relation (29).

If in (32) the variables x are replaced by X (by means of (27), then we get the relaxation equations for thermodynamic forces.

It is worth noting that the dependence X = X(x) and validity of (32) leads to the existence of (28).

Now let us assume, that in a thermodynamic system the real, steady external forces u are switched. Assume also that every force u_j impacts in the direction of the sequent internal thermodynamic parameter x_j.

As is shown by Stratonovich and Tolstopiatenko (1980), the relaxation equation in the presence of external forces

$$\dot{x}_i = f_i(x, u) \qquad i = 1, 2, \ldots, n \tag{33}$$

(which, like (28) can also be introduced phenomenologically), is only partially determined by the characteristics of (28).

The function f(x, u) necessarily satisfies the following conditions:

$$f_i(x, u)\big|_{u=0} = f_i(x) \tag{34}$$

where $f_i(x)$ is the right-hand side of (28) and

$$f\{x(X), u\}\big|_{X=u} = J(X, u)\big|_{X=u} = 0. \tag{35}$$

The last relation determines the necessary conditions for the existence of a steady state in the system under consideration.

For this reason, theoretically, it makes it possible to interprete the conjugated thermodynamic parameters X as definite kinds of forces, which are equal in a disturbed steady state to the real external forces, that affect the thermodynamic system and balance their influence.

The production of entropy in the case of switched external forces may be written in the following form:

$$\sigma(x, u) = \sum_{k=1}^{n} \{X_k(x) - u_k\} f_k(x, u) \tag{36}$$

or

$$\sigma(X,u) = \sum_{k=1}^{n} \{X_k - u_k\} J_k(X,u), \tag{37}$$

where X, is connected as above, with x through relation (27).

Equation (32) is now transformed in the following way:

$$\dot{X}_j = -\sum_{k=1}^{n} \{X_k(x) - u_k\} \frac{\partial f_k(x,u)}{\partial x_j} + \frac{\partial \sigma(x,u)}{\partial x_j} . \tag{38}$$

Further let us consider one of the fundamental principles in optimal control theory, the Pontryagin principle of maximum.

This determines the necessary conditions for optimisation of u(t), i. e. u(t) chosen among the 'possible') develops the evolution of phase coordinates x(t) of the dynamic system

$$\dot{x}_i = f_i(x,u) \tag{39}$$

in such a way, that the 'quality functional'

$$I = \int_{t_0}^{t_1} f_0(x(t), u(t))\,dt \tag{40}$$

reaches a minimum value, that this system passes over from point $x^{(0)} = x(t_0)$ to the point $x^{(1)} = x(t_1)$.

The solution of this problem (see Pontryagin et al., 1969) is based on consideration of the special Hamiltonian form:

$$H(x,\psi,u) = \sum_{k=0}^{n} \psi_k f_k(x,u) \tag{41}$$

which allows us to write canonical equations for the variables x, ψ :

$$\dot{x}_i = \frac{\partial H}{\partial \psi_i} = f_i(x,u) \tag{42}$$

$$\dot{\psi}_j = -\frac{\partial H}{\partial x_j} = -\sum \psi_k \frac{\partial f_k(x,u)}{\partial x_j} \qquad i,j = 1,2,\ldots,n \tag{43}$$

and

$$\begin{aligned} \dot{x}_0 &= \frac{\partial H}{\partial \psi_0} = f_0(x,u) \\ \dot{\psi}_0 &= -\frac{\partial H}{\partial x_0} = 0 . \end{aligned} \tag{44}$$

The principle of maximum declares that, for u(t) and the trajectory x(t) to be optimal, the vector-function $\psi(t)$, connected with functions u(t) and x(t), must satisfy the conditions:

1. At any $t \in [t_0, t_1]$ the Hamiltonian $H(x(t), \psi(t), u)$ considered as a function uf u reaches the maximum H_{max} (x(t), (t)) at the point u = u(t).
2. In the end of the time interval t_1

$$\psi_0(t_1) \leq 0; \; H_{max}(x(t_1), \psi(t_1)) = 0 \tag{45}$$

   should necessarily exist.

It happens, that if $\psi(t)$, x(t), u(t) obey equations (42) and (43), then $\psi_0(t)$ and H_{max} are constants, and the validity of (45) can be verified at any moment of time (not only at t_1).

The comparison between the equations adopted in optimal control and thermodynamic equations introduced above allows us to discover the following analogies:

1. There exists a formal equality of the dynamic (42) and relaxation (33) equations.
2. If the Hamiltonian (41) is determined as:

$$H(x,\psi,u) = \psi_0 \sigma(x,u) + \sum_{k=1}^{n} \psi_k f_k(x,u) , \tag{46}$$

then equations (42) to (44) take the form:

$$\dot{x}_i = f_i(x, u) \tag{47}$$

$$\dot{\psi}_i = -\sum_{k=1}^{n} \psi_k \frac{\partial f_k(x,u)}{\partial x_j} - \psi_0 \frac{\partial \sigma(x,u)}{\partial x_j} \tag{48}$$

$$\dot{x}_0 = \sigma(x, u)\,; \quad \psi_0 = \text{const} \quad . \tag{49}$$

3. Let vector u in (46) to (49) be fixed. And let us determine the functions ψ_j in the following way:

$$\psi_j = \frac{\partial \Delta S(x)}{\partial x_j} - u_j = X_j(x) - u_j \qquad j = 1, 2, .., n \,. \tag{50}$$

Then when $\psi_0 = -1$ equations (48) coincides with equations (38).

But the parameters X(x) satisfy equations (38) for every thermodynamical system. This means that in the case of a thermodynamic system under fixed forces u, functions ψ_j, introduced according to (50), automatically satisfy (48).

4. The introduction of $\psi_j = X_j - u_j\,;\ \psi_0 = -1$ into the Hamiltonian (46) leads to the following result

$$H(x, \psi, u) = -\sigma(x, u) + \sum_{k=1}^{n} \{X_k - u_k\} f_k(x, u) = 0 \,. \tag{51}$$

This shows that the time evolutions of processes in thermodynamic systems always develop in such a way that the Hamiltonian (46) reaches its maximum value $H_{max} = 0$ at every moment of time. This full entropy production along the trajectory

$$I = \int_{t_0}^{t_1} \sigma(x(t), u)\, dt$$

during such evolution (according to equation ((33), under fixed external forces u) reaches its minimal value.

The above allows us to consider the Pontryagin principle of maximum as a natural and convenient instrument for investigations in the field of optimal control, and of stabilisation in thermodynamic systems and others.

Alternatively, thermodynamics may represent new models, and encompass the ability to interpret methods and results, and allow for the development of investigation in the theory of dynamic systems

Stochastic Criterion in Dynamical Systems

E. V. Presnov

Theoretical models of physical, biological, and chemical phenomena often have the form of systems of ordinary differential equations

$$\dot{x} = v(x). \quad (1)$$

It is therefore of importance to know the qualitative behaviour of phase trajectories of such dynamic systems. Not so long ago it became clear (Lorenz, 1963) that already in the case of three equations with quadratic non-linearity in a corresponding three-dimensional phase space there existed an attracting invariant set, of measure zero, restricted and connected, which is different from a point or a limit cycle. This set is referred to (Ruelle, Takens, 1971) as a strange attractor. Locally it is organised as a direct product of a two-dimensional manifold on Cantor's set. Moreover, inside a strange attractor the fate of a mapping point cannot be predicted and is very sensitive to variation of the initial conditions. However, there exists no mathematical theory which would explain such a chaotic behaviour of the dynamic system. Neither has the strange attractor been exactly defined. The phenomenon of stochastisation even in the Hamilton systems, which cannot have strange attractors because the Liouville theorem is valid for them (Ruelle, 1981), has been studied insufficiently. It is possible that the question of irregular behaviour of the dynamic system (1) is associated with the absence in them of generalised Lyapunov functions $\mathcal{L}(x) \not\equiv 0$:

$$L_v \mathcal{L} \triangleq \frac{\partial \mathcal{L}}{\partial x_1} v_1 + \ldots + \frac{\partial \mathcal{L}}{\partial x_n} v_n \geq 0 .$$

The Lie derivative (the derivative of the function in the vector field direction) $L_v \mathcal{L}$ can be interpreted here (Presnov, Malyghin, 1982) as entropy production in a corresponding thermodynamic representation of the

Thermodynamics and Regulation of Biological and Processes

dynamic system (1). Indeed, the functions $X_i = \partial \mathcal{L}/\partial x_i$ may be considered quite formally as generalised thermodynamic forces. The vector field components $J_i = v_i$ will, in their turn, be generalised thermodynamic fluxes. Hence, the representation $L_v \mathcal{L} = \sum J_i X_i$ has the form of entropy production in a dissipative system. Thus, there exists a certain connection between irreversible thermodynamics and the theory of dynamic systems. And it is just the dynamic systems with global non-negative entropy production that have a sufficiently simple topological structure of phase space. Complicated behaviour of dynamic systems, on the contrary, is evidently connected with the absence in them of a non-negative entropy production defined in the entire phase space. But in the general case there is no constructive algorithm for dynamic systems which would enable their Lyapunov functions to be found. As concerns the Lyapunov functions, there exists a number of hypotheses (see, e. g. Shub, 1974). The above thermodynamic considerations make it possible to state the following:

> The dynamic system (1) does not have an attracting set which is not a manifold if, and only if, there exists for it a generalised Lyapunov function $\mathcal{L}(x) \not\equiv 0$; $L_v \mathcal{L} \geq 0$.

To substantiate this statement, we consider examples of dynamic systems which admit to generalised Lyapunov functions defined in the entire phase space.

I. For the linear dynamic system (1) there exists a generalised Lyapunov function. Indeed, let the linear operator $A: \mathbb{R}^n \to \mathbb{R}^n$, generalising the dynamical system $\dot{x} = Ax$, have k different eigenvalues $\lambda_1, \dots, \lambda_k$. Then the space $\mathbb{R}^n$ breaks up into a direct sum of subspaces invariant under A:

$$\mathbb{R}^n = N_{\lambda_1}^{(p_1)} + \dots + N_{\lambda_k}^{(p_k)}$$

($N_{\lambda_i}^{(p_i)}$ corresponds to the eigenvalue λ_i of p_i fold, $\dim N_{\lambda_i}^{(p_i)} = p_i$). Thus, the space $\mathbb{R}^n$ breaks up into a direct sum of three subspaces $\mathbb{R}^n = \mathbb{R}^{m_+} + \mathbb{R}^{m_-} + \mathbb{R}^{m_0}$ invariant under A, so that all contractions of the eigenvalue A on $\mathbb{R}^{m_+}$ have positive real parts, on $\mathbb{R}^{m_-}$ -negative parts, and on $\mathbb{R}^{m_0}$ we have $\mathrm{Re}\,\lambda_i = 0$. But for the linear system with $\mathrm{Re}\,\lambda_i > 0$ there exists a Lyapunov function $r^2_{m_+}$ (see e. g. Arnold, 1971). And for a system with $\mathrm{Re}\,\lambda_i < 0$ we have the function $-r^2_{m_-}$ which differs only in sign from the Lyapunov function $r^2_{m_-}$ of the system $\dot{x} = -Ax$ with $\mathrm{Re}\,\tilde{\lambda}_i > 0$. Finally, on the invariant subspace $\mathbb{R}^{m_0}$ we take the Hamiltonian H, and then the Lyapunov function of a linear system of general form is

$$\mathcal{L} = \frac{1}{2}(r^2_{m_+} - r^2_{m_-} + H), \quad L_{Ax}\mathcal{L} \geq 0.$$

On the other hand, the linear dynamic system is known to have no attracting sets which are not manifolds.

II. For the one-dimensional dynamic system (I), we take as a generalised Lyapunov function

$$\mathcal{L} = \int_{x_0}^{x} v(y)\,dy, \qquad L_v \mathcal{L} = v^2 \geqslant 0.$$

The Lyapunov function is written in a similar manner also for multidimensional systems with separating variables $(\dot{x}_i = v_i(x_i),\ \mathcal{L} = \sum \mathcal{L}_i)$.

III. To find the generalised Lyapunov functions of dynamic systems of general form, we shall use the following vector field expansion (Presnov, Malyghin, 1982; Bataille et al., 1982)

$$v(x) = \mathrm{grad}_{\nabla G}\,\Phi(\nabla G) + \mathcal{U}(\nabla G), \quad (\nabla G, \mathcal{U}(\nabla G)) = 0$$

(for expansions not equivalent to this one see Hendes, Duarte, 1981; see also the Hodge expansion theorem, e. g. from the paper by Ratiu, 1977). Then, for the thermodynamic representation of a dynamic system of the form

$$\dot{x} = \mathcal{U}(\nabla G), \quad (\nabla G, \mathcal{U}(\nabla G)) = 0$$

as a generalised Lyapunov function we take the potential $G(x)$

$$L_v G = (\nabla G, \mathcal{U}(\nabla G)) = 0 .$$

In a particular case for the Hamiltonian systems

$$\dot{x} = I \nabla H, \quad I = \begin{pmatrix} 0 & -E_n \\ E_n & 0 \end{pmatrix}$$

a generalised Lyapunov function is the Hamiltonian $H(x)$

$$L_v H = (\nabla H, I \nabla H) = 0 .$$

If a dynamic system is of the gradient type, i. e. has a purely dissipative representation with the potential $G = \frac{1}{2}(x, x)$: $\nabla G = x$

$$\dot{x} = grad\, \Phi(x)$$

a generalised Lyapunov function is the potential $\Phi(x)$

$$L_v \Phi = (\nabla \Phi, \nabla \Phi) \geqslant 0 .$$

IV. Finally, if for the dynamic system (1) there exists a time change which preserves phase trajectories, i. e.

$$\dot{x} = \lambda(x)\, v(x), \qquad \lambda(x) > 0 \tag{2}$$

such that $\lambda(x)\, v(x) = \nabla S$, there exists a generalised Lyapunov function for the dynamic system (2) (see case III). On the other hand, gradient systems do not possess strange attractors, and time change for such systems does not lead to their appearance.

Phases with Internal Degrees of Freedom and Their Heterogeneous Transformations

E. N. Eliseev

The appearance in phases under non-equilibrium conditions of non-uniformities in composition, structure, morphology, and properties is promoted by many non-equilibrium physico-chemical and structural transformations (Eliseev, 1967; Nicolis, Prigogine, 1977) which are clearly determined in the studies of regularities of their distribution in phases, particularly in the phases of a substance. On this basis the concept of a phase with internal degrees of freedom becomes most important in the analysis of products of non-equilibrium reactions.

Gyarmati (1971), when formulating non-equilibrium thermodynamics in the language of field theory, opened the way for the solution of problems in linear approximation. A change in the phase (or a system) is considered here as a one-parametric family of deformations (the parameter of the change is time). However, when physically non-realistic motion, which does not obey the continuity axiom, is excluded from field theory (and simultaneously from Gyarmati theory), many of the most important phase (system) transformations, which take place in the region of stability and occur in the development of each living organism, are excluded too.

Consideration of wide-spread heterogeneous phase transformation offers opportunities for the creation of another model of non-equilibrium thermodynamics based on involvement of various heterogeneous (including wave) scale transformation, which refer both to linear and non-linear phenomena. The wave model of thermodynamics of irreversible processes may provide an insight into these transformations - e. g. scale transformations proceeding in the process of development of living organisms. On the other hand, detection of heterogeneous wave scale transformations disproves the

Thermodynamics and Regulation of Biological and Processes

idea that wave propagation 'does not correspond to any stable physical process' (Glansdorff, Prigogine, 1971).

Scale heterogeneous transformations of phases and systems

For the description of scale heterogeneous transformations of phases (and systems) a research technique is proposed based on the methods of differential geometry, models of compact sets and methods of variational physico-chemical analysis (Eliseev, 1971, 1975, 1980, 1982).

As a model generalising all possible changes in the states, composition, and properties of real physico-chemical phases, we consider the state space. The main topological properties of this space are as follows

- connectedness (continuity) of a set of states (variational methods make it possible to establish the trajectory of phase and system variation)
- degeneration of processes resulting in the appearance of special invariant states (stability, indistinguishability)
- convergence of state successions near a set of trajectories of an irreversible process (the model of concrete variations of the state, composition, and properties are cluster diagrams, for example, the model of crystallisation differentiation in centered expansion and compression Riemann waves)
- extent of transformations under which the surroundings of a point in the space change essentially from point to point (scale transformation under phase transition in the critical point itself, transformation of a homogeneous into a heterogeneous phase, cell division of a homogeneous liquid phase under supercritical conditions with the formation of Bénard convection cells, phase exfoliation, etc.).

In comparison of the properties of the state space with the well studied space of crystallographic systems (crystal lattice) their essential difference is observed. So, the most important axioms of the space of crystallographic systems are as follows

1. discreteness
2. cover (the requirement of correctness of crystallographic systems)
3. rationality (the most general form is polyhedron)
4. local equality (identical surrounding of similar points in a crystallographic system).

The most important properties of the state space (which generalises the models of dynamic and evolutional transformations) are different

1. connectedness - continuity (succession of stable and unstable states, variations in the direction)
2. degeneration of irreversible changes and non-equilibrium systems (trajectories of the changes and singular points)
3. convergence of successive states (divergent and convergent evolution which in models of the type of clusters cones, helicoids, etc., involves all the trajectory variations in an irreversible process)
4. scale transformations which accompany phase transitions, evolutional variations (variations near similar points of the state space).

A generalised image of the space of crystallographic systems is a crystal lattice based on translation. A generalised image of the state space is a set of states (mainly, compact ones).

Once in the physico-chemical analysis attention turns from the equilibrium (static) state to the processes of dynamic variation (and then to evolutionary processes), and one should introduce new methods into the mathematical analysis of the results of physico-chemical experiments and their geometric models (Eliseev, 1971, 1975).

As an obvious example of the properties of the state space, Fig. 2 illustrates the specifics of molecular order in a nematic phase, and Fig. 3 shows nonlinear scale transformations in the evolution of living organisms. Comparison of such examples establishes common features of scale transformations in micro- and macrosystems.

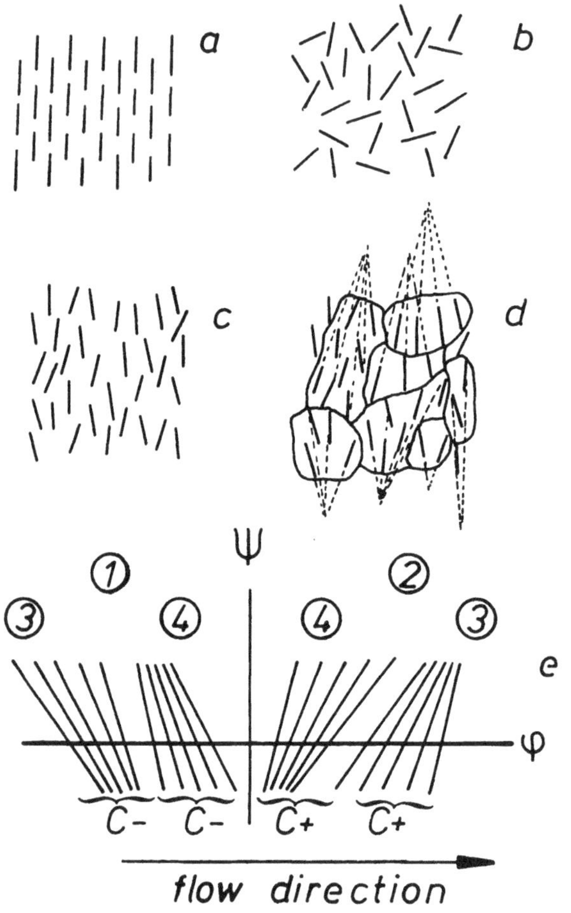

Fig. 2 Scheme of molecular organisation in crystalline (a), isotropic (b) and nematic (c, d) phases; bc: homogeneous-bundle orientation of molecule groups in nematic crystal; d: local regions characterised by common peculiarities convergence in the orientation of molecule groups; e: geometric criterion (Ovsyannikov, 1981) for distinguishing simple centred waves: compressions and discharges.
1: r-waves; 2: l-waves; 3: compression; 4: rarefaction; f.d.: flow direction.

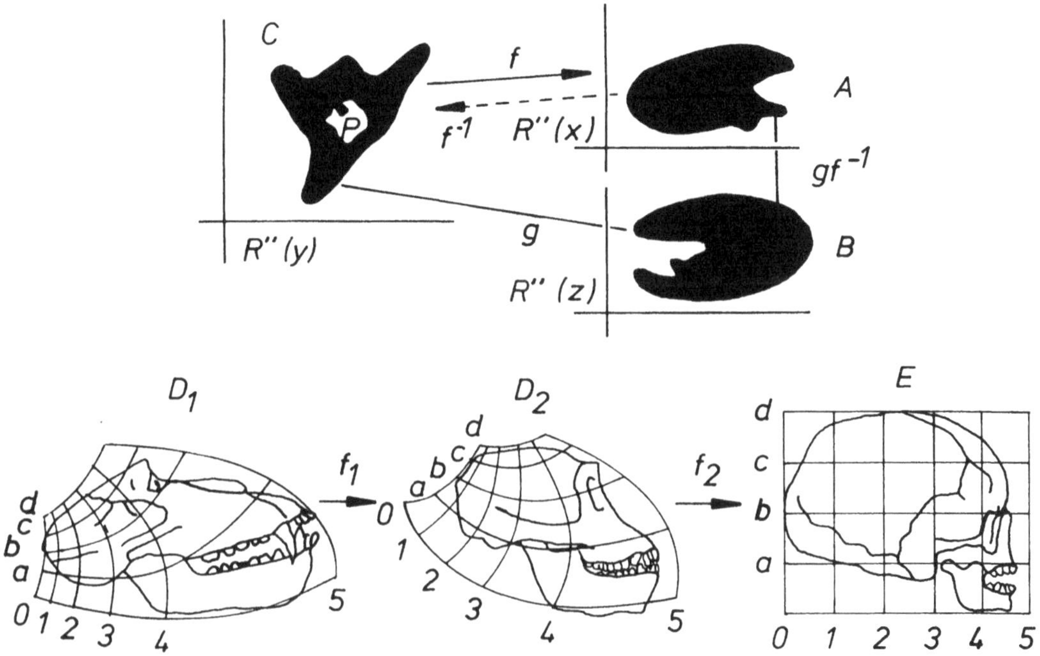

Fig. 3 Non-linear scale transformations. Smooth reflections of the regions A, B, and C manifested in the course of evolution of organic world: A-B, B-C; A-C: mutually single-valued, smooth mappings of the regions A, B, C in Euclid space (Mishenko, Fomenko, 1980). f and f^{-1} are mappings of region C on region A and vice versa; g: (C-B), $g \circ f^{-1}$: (A-B). If mapping f of region C on region A is set, establishing curvilinear coordinates in region C, mapping f^{-1} which maps region A onto region C may be looked upon as establishing curvilinear coordinates in region A (by means of Decart coordinates from region C). This note specifies transition from mapping f to inverse f^{-1}. D_1, D_2, E - scheme of scale transformations of coordinates on comparing crania of man (E), babuin (D_1) and chimpanzee (D_2) (Thompson, 1942).

The role of wave heterogeneous transformations in evolution of organisms

Fig. 4 presents schemes A, C, E. H showing the stages of coordinate grid deformation to which there correspond individual members of a phylogenetic line: to the hypothetical forms C and E (Fig. 4) there closely corresponds the ancestor of the present-day horse which really existed in the past (M_I) Miohippus, P - Protohippus, Miocene epoch.).

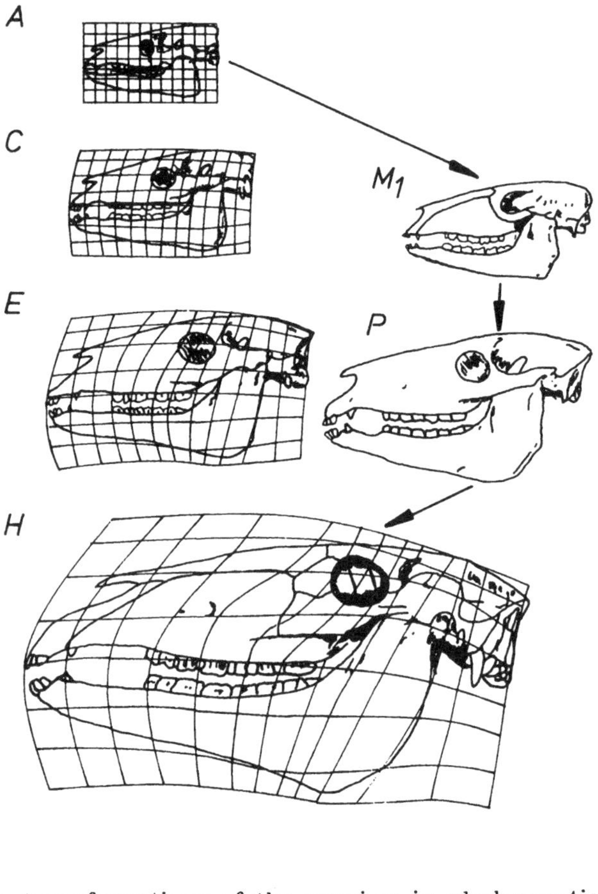

Fig. 4 Scale transformations of the cranium in phylogenetic sequence of ancestors of the horse (Thompson, 1942): gradual transition from linear coordinate system (A), to curvilinear system (with inserted outlines of intermediate fossil forms up to the outlines of the cranium of present day horse (H).

Fig. 5 analyses anisotropy in the development of skull contours in the phylogenetic line of horse ancestors presented in Fig. 4. Individual sectors of scale transformations are deformed each in its own way: different centered Riemann expansion or contraction waves are established for each sector.

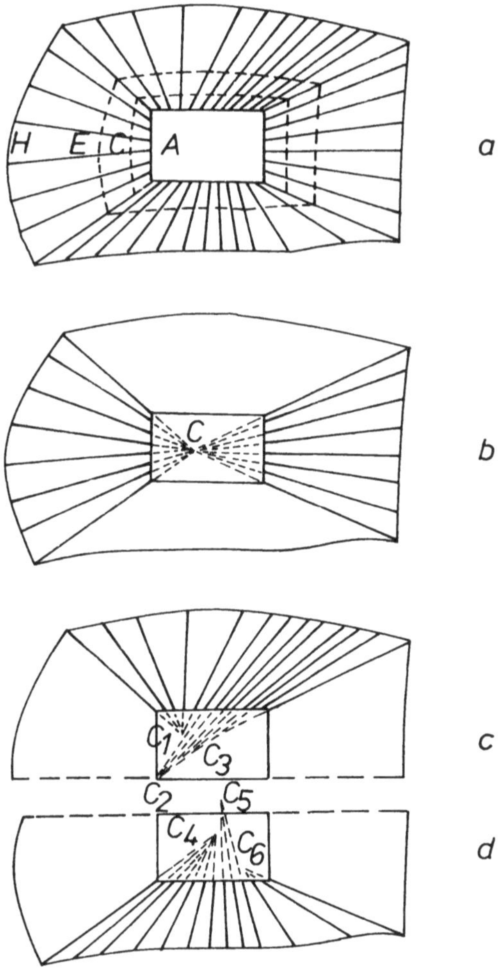

Fig. 5 Expansion to elements of scale transformations of horse cranium (see Fig. 4): a: scheme of compacts in scale transformations (initial scheme after Thompson, 1942); b, c, d: separate sectors of scale transformations isolated by us with the centres C, C_1, C_2, ..., C_6. These centres of compact sets appear as peculiar elements of wave transformations in the meaning of Riemann centred waves of compression and discharge.

To describe the dynamics of centered Riemann wave variation, one can apply models of wave deformation of the medium in the elastoplastic case (Fig. 6). The numerical method of Lagrangian cells is based on coordinates built in varying structures. The Lagrangian cells move along with the deformed medium (Fig. 6, b - g). In the absence of elasticity (Fig. 6, b, c) the plate, when loaded, vibrates and there arises a two-humped configuration of the upper plate surface (Butler, 1971). This phenomenon is the result of the fact that fluidity in the plate first occurs in the grid corners, whereas the middle domain has not yet lost elasticity. As a result, fluidity is observed in the entire plate, and wave points continue increasing b and c in Fig. 6).

Consideration of scale phase transformations (for example, those presented in Fig. 6) shows the exclusive role of wave heterogeneous transformations in the evolution of inorganic physico-chemical phases (especially under phase transitions) and living organisms.

Systematisation of elements of scale phase transformations

The most typical scale transformations are operations of degeneration (Table I), splitting and structure-formation (Table II), shift, twisting and change of morphology (Table III).

In Table I degeneration is considered via an example of isotopic models of the types of stationary states possible in a system with two degress of freedom (Ebeling, 1976): stable and unstable points, a stable centre. An unstable point-type regime, whose model is diverging vector beams, corresponds to the differentiation process. A quite stable point-type regime (a converging vector beam) corresponds to homogenisation.

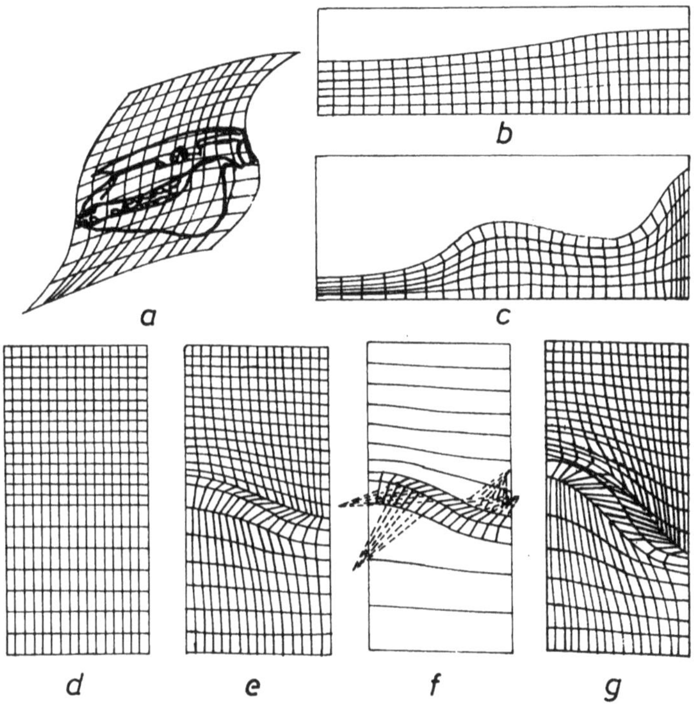

Fig. 6 Wave nature of deformation of square net of coordinates in the course of evolution from eocene ancestor (A) to the present day horse (II), analysed in Fig. 4 and 5: a: wavelike deformation of linear coordinates of horse ancestor cranium; b, c: results of numerical modelling of non-established uncompressed flow, in which there are free surfaces and interfaces (see Butler, 1973); d, c: initiation of cosinusoidal wave on the interface of two liquids under the effect of external perturbation: the horizontal plane (d) is deformed, passing into sinusoidal wave with an increasing amplitude. As a result depressed projection and rising swelling appear, which are characteristic of non-linear flows (Butler, 1973, p. 153). Fig. 6 e analyses location of Riemann centred waves of compression and rarity in non-linear wavelike flow (c, f), which serves as a model of the wave like scale transformation of morphology of the living organisms presented in Figs. 4, 5. a: according to Thompson (1943), b-e, g: according to Butler (1973), f: scheme of location of Riemann centred waves advanced by the author.

Table I *I. Operation of "degeneration" (⩾)*

Initial models of successive stationary states	*Transformational operation*	*Resulting model*
	⩾	• *point*
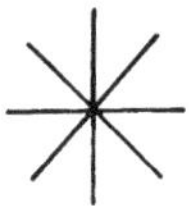	⩾	— *line*
	⩾	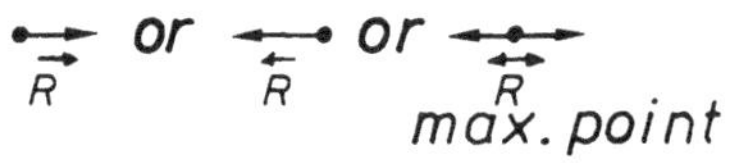
	⩾	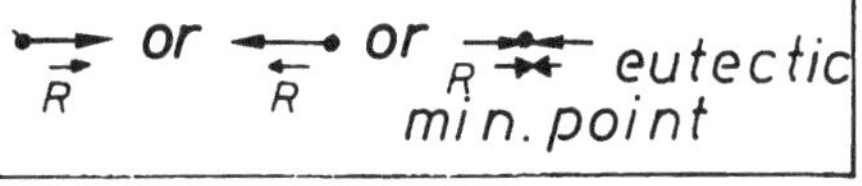

Table II *II. Operation of splitting and "structure-formation" (⩽)*

Initial model of elements of structure	*Transformational operation*	*Resulting model*
•	⩽	— *line*
—	⩽	 *linear parallelism (linearity) bundle*
	⩽	 *plane parallelism (streakiness)*
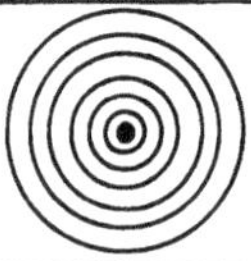	⩽	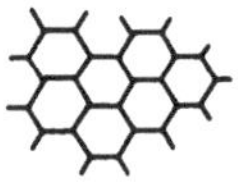 *cells (Bénard)*
	⩽	 *poly-stable state*

Table III

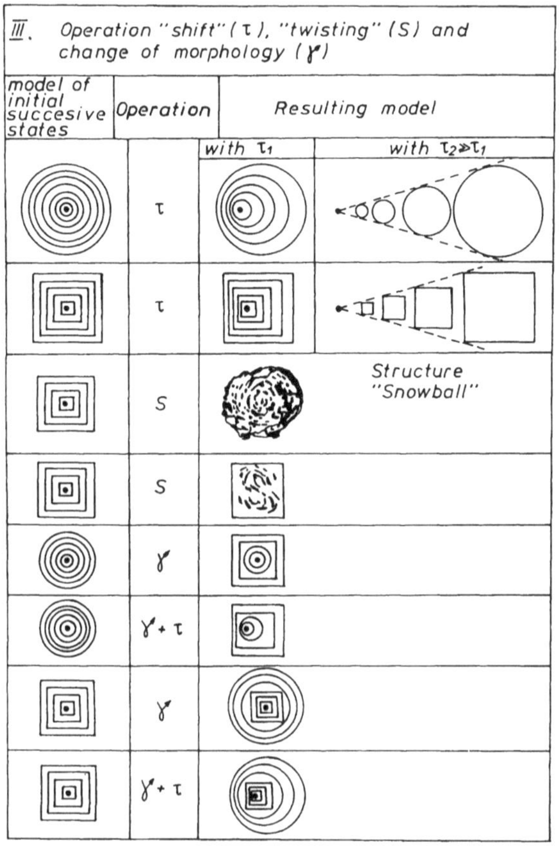

Summation of models scale transformation elements

The structure of the state space consists of combination of elements of scale transformations. Complicated elements of transformations result from summation of elementary ones (Tables I to III). Table IV presents the main operations of summation of models of elementary scale transformations. So, in summation of a stable point-type regime and a vector external action there occur different cases (a and b). The point-type regime and the mirror reflection plane form a real complicated transformation observed in liquid crystals. The case of summation of models of elementary transformations of the type a corresponds to the real example of liquid crystals (v) , and in the case of the type b to the example (u) . Summation of the vector transformation and the axes of symmetry (3, 4, 6) forms a trigonal, tetragonal, and hexagonal stable point (case d) or unstable points (case e). Table IV also presents other cases of summation of elementary scale transformations which result in the appearance of points of double descent (*f*), peritectic (*g*), spiral (*h*), bistable states (*i*), different saddles: triple 'monkey saddle' (*k*), tetragonal (*l*), hexagonal (*m*).

As an example of analysis of real complicated transformations we have chosen the schemes of molecule position near nuclei, the cases of relative position of molecules near disclinations in liquid crystals (Chistyakov, Pikin, 1981). n , Ψ , and Φ are parameters of the equation describing the molecular ordering in a liquid crystal.

Of the examples $n \div v$ only two cases (*p* and *s*) can be regarded as elementary scale transformations, the others are a summation of elementary transformations: n see a; o see k; r see h; t see l; u see b; v see b.

Table IV

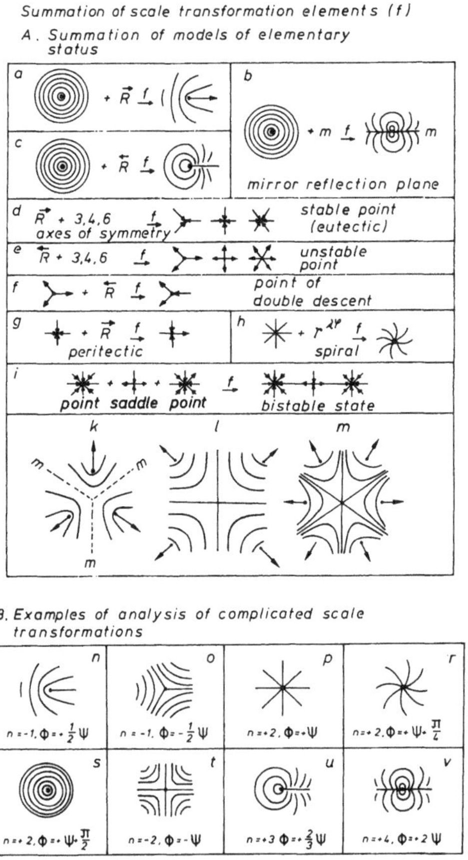

Essential difference in the concept 'phase' for equilibrium and non-equilibrium conditions

The definition of the concept 'phase' is based on different properties of phase space (in the equilibrium case phase homogeneity is stressed) and state space (continuous variation of the properties, composition, and structure within a phase, i. e. heterogeneity of non-equilibrium phases are typical of irreversible processes). This is just the meaning of 'single-phase medium' in the language of 'Heat transfer theory' (Petukhov, 1971). A single-phase medium is understood there as 'a continuous single- and multicomponent medium whose properties in the space can change only c o n t i n u o u s l y " (spacing by the author) (Petukhov, 1971, p. 9). As distinguished from a single-phase, a multiphase medium consists 'of a number of single-phase parts, on the boundary of which the physical properties undergo jump-like changes' (Petukhov, 1971, p. 9).

In irreversible non-equilibrium processes the changes and complication of the structure, composition distribution, and phase morphology are specific mechanisms of phase adaptation to non-equilibrium conditions. Heterogeneous transformation of phases help maintain their stability far from equilibrium. Phases with any kind of inhomogeneities are typical examples of phases with internal degrees of freedom which promote phase adaptation to non-equilibrium conditions.

Difference in the content 'phase' for equilibrium and non-equilibrium conditions is in agreement with the following conclusion by Semenchenko: 'The difficulty of theoretical thermodynamics consists in the fact that it is able to solve in the integral form not a single physical problem, even a very simple one, if only this problem is not referred to an ideal system although, in principle, it allows solution and solves practically any problem. This seeming contradiction is explained by the fact that thermodynamics itself cannot determine the form of thermodynamic functions and has either to seek for empirical formulae for them or to find these formulae by means of considerations taken from other fields of theoretical physics' (Semenchenko, 1950, p. 16-17).

Earlier (Eliseev, 1982) we considered phenomena of wave superstructural phase modulation with internal degrees of freedom, the wave character of phase composition variation under crystallizational fusion differentiation, specificity and systematization of concentration waves, which process was recently discovered in physics of periodic phenomena. The facts considered in (Eliseev, 1982) and partially those presented above make it possible to conclude that from general thermodynamic standpoints it is impossible to predetermine the form of the wave functions which modulate real physico-chemical phases. It is impossible to predetermine whether they will be cosine, saw-tooth, centered Riemann or some other waves. Thus, for the wave model of non-equilibrium thermodynamics Semenchenko's words remain valid that the formulae of the wave variations of media (phases and systems) should be sought for by appealing to the data and considerations from other fields of theoretical physics. To describe the wave phenomena observed in the organic world, observations and generalizations from experimental and theoretical fields of biology are needed.

Information Theory

Towards a Dynamic Information Theory

H. Haken

Introduction

The concept of information plays an important role in biology. Here the situation is complicated by the fact that the word 'information' is used with two different meanings. The first meaning stems from telecommunication theory, where a precise meaning is attached to the word 'information' and a formalism has been developed to determine the 'information content' (Shannon, 1948; Brillouin, 1962, 1964). We shall briefly remind the reader of this concept in the second part. There it will turn out that 'information' used in this sense is a measure for scarcity of a message, but it does not have a meaning such as 'useful' or 'useless' or 'meaningful' or 'meaningless'. In other words, semantics is absent in this approach. In addition it turns out that this concept is purely static. The second meaning of 'information' more or less coincides with its meaning in the language of our daily life. Unfortunately, both meanings are occasionally mixed up. For instance, sometimes biologists speak of the 'information' or 'information content' of DNA. At a first sight this seems to imply the first (telecommunication) meaning of information but a closer inspection reveals that this word is used in the sense of common language.

In our contribution we wish to show that these two interpretations of 'information' are quite different. In particular the scientific frame for the use of the word 'information' in the second sense seems still to be lacking and in this paper we wish to show how one may cast this concept into a rigorous mathematical form. It is thus hoped that a solid basis is laid for a further treatment of the problem.

Thermodynamics and Regulation of Biological and Processes

In order to explain our approach, let us briefly remind the reader of some simple facts relating to information used in ordinary language. To be more precise let us speak of messages which may consist of a string of signs or sounds. Whether such a message is useful or useless, meaningful or meaningless depends not only on the string of signs but also on the receiver, For instance, when we consider the t-RNA of a phage as a message and put it into a heap of sand, there will be no response. On the other hand, when we put this t-RNA into certain bacteria, new kinds of processes in these bacteria will be triggered by that message. Similarly when somebody approaches us in a language unknown to us we cannot react properly.

While these comments are, of course, somewhat trivial, casting these problems into mathematical form is far from being trivial. To come closer to its solution we shall proceed in three steps.

1. We shall remind the reader of some basic facts of information in the sense of information theory developed in the realm of telecommunications.
2. We cast the new concept of information proposed in this paper into mathematical form.
3. We draw some general conclusions.

A reminder of information theory

Let us start with some simple considerations (Shannon, 1948; Brillouin, 1962, 1964; Haken, 1978; Jaynes, 1957 a, b). Let us assume that we can transmit R different messages which can be received with equal a priori probability. Then the information content of a single message will be the bigger the scarcer the message is, or in other words, the bigger the number of possible messages is. In order to find an adequate meaning for the information content, we treat a special case in which the messages R can be decomposed into two independent kindes of messages, R_1 and R_2, so that the total number messages transmitted is given by

$$R = R_1 R_2 \tag{1}$$

We require that the information content belonging to (1) is additive, i. e.

$$I(R_1 R_2) = I(R_1) + I(R_2). \tag{2}$$

The only solution to this functional equation is

$$I = K \ln R \tag{3}$$

where K is an arbitrary constant which generally is assumed to take the form

$$K = 1/\ln 2 \tag{4}$$

This allows us, as is well known, to measure I directly in bits. Let us now apply (3) to calculate the information content I for a simplified morse alphabet which contains only dots and dashes but no intermissions. Let us consider a word of length N which contains N_1 dashes and N_2 dots with

$$N_1 + N_2 = N. \tag{5}$$

These dashes and dots can be arranged in different sequences. Their number can easily be shown to be

$$R = \frac{N!}{N_1! N_2!} \quad . \tag{6}$$

In the more familiar equations of information theory we determine the information content per symbol, i, which is defined by (7). Using (6) and (3) we evaluate

$$i \equiv \frac{I}{N} \quad . \tag{7}$$

In the limit that N becomes large but

$$p_j = \frac{N_j}{N} \qquad j = 1, 2 \tag{8}$$

i. e. the relative frequences of the occurrence of dashes or dots remains fixed, and using Stirling's formula we readily obtain after simple transformation

$$i = -K(p_1 \ln p_1 + p_2 \ln p_2) \tag{9}$$

When there are several symbols j = 1..., equation (9) can easily be extended to

$$i = -K \sum_j p_j \ln p_j \tag{10}$$

which is the fundamental equation of information theory. This equation describing the information content, or in short the information, can be used in two ways. Either (I) the relative frequencies of the symbols j are given, for instance, the relative frequencies of letters in a language, and then formula (10) allows us to determine, in this sense, the information of a language, or (II) application of formula (10) follows if one wishes to make 'unbiased estimates' on the frequency distribution p_j. As can be shown, in such a problem one has to choose the extremum of

$$-\sum p_j \ln p_j = \text{Extr.} \tag{11}$$

under certain constraints such as the normalization condition

$$\sum p_j = 1 \tag{12}$$

and possibly further conditions in the form

$$\sum p_j f_j^{(k)} = f_k \,. \tag{13}$$

In order to underline the importance of this approach we just remind the reader that by means of this formalism one may derive the fundamental equations of thermodynamics for entropy, free energy etc. (Jaynes, 1957 a, b; see also Haken, 1978). At the same time the drawback of this approach should be apparent. It is based on counting states or possible realizations but it does not contain any dynamics or semantics.

The new approach to dynamic information theory

Since we wish to introduce new concepts it may be advisable to start with rather concrete situations. Let us consider a set of messages which are specified by a string of numbers. The central problem consists in modelling the receiver. We do this by invoking modern concepts of dynamic theory or, still more generally, by concepts of synergetics (Haken, 1978, 1983). We model the receiver as a dynamic system whose behaviour is described by trajectories within a state space.

The point of each trajectory at time t is then determined by the vector $\tilde{q}(t)$. To be still more specific we assume that $\tilde{q}$ obeys a set of first order non-linear differential equations of the form

$$\dot{\tilde{q}} = \tilde{N}(\tilde{q}, \alpha) + \tilde{F}(t). \tag{14}$$

In it $\tilde{N}$ is a vector function which depends on $\tilde{q}$ and on certain control parameters α. Later on it will turn out that stochastic forces $\tilde{F}(t)$ play an important role. We assume that the receipt of a message by the system means that the parameters α and the initial value of $\tilde{q}_0$ are set by the message. For the time being we shall assume that these parameters are then uniquely fixed. An extension of the theory to an incomplete message is straightforward. We first ignore the role of fluctuations. We assume that before the message arrives the system has been in an attractor which we shall call the neutral state. The attractor may be a resting state, i. e. a point sink, but it could equally well be a limit cycle, a torus or a strange attractor, or a type of attractor still to be discovered by dynamic system theory.

We shall call this attractor $\tilde{q}_0$. After the message has been received and the parameters α and the intitial value $\tilde{q}$ are newly set, in principle two things may happen. Let us assume that we are allowed to wait for a certain measuring time so that the dynamic system can be assumed to be in one of its possible attractors. Then (I) either the message has left the system in the $\tilde{q}_0$ state and in such a case the message is evidently useless or meaningless.

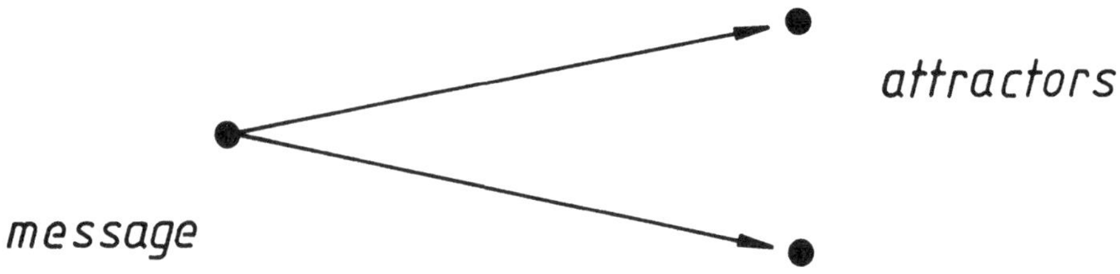

Fig. 7 Explanations in the text.

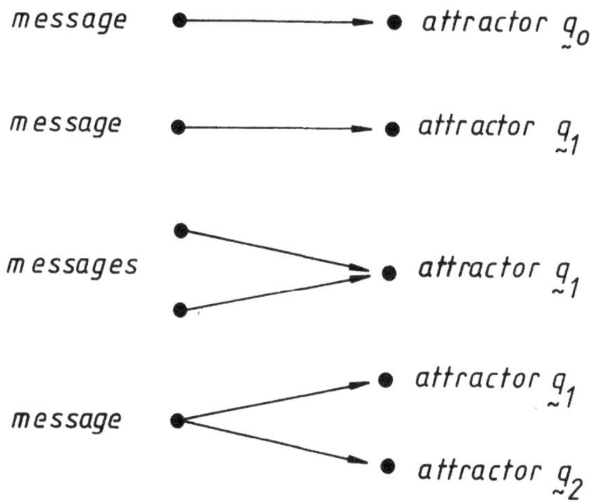

Fig. 8 Explanations in the text.

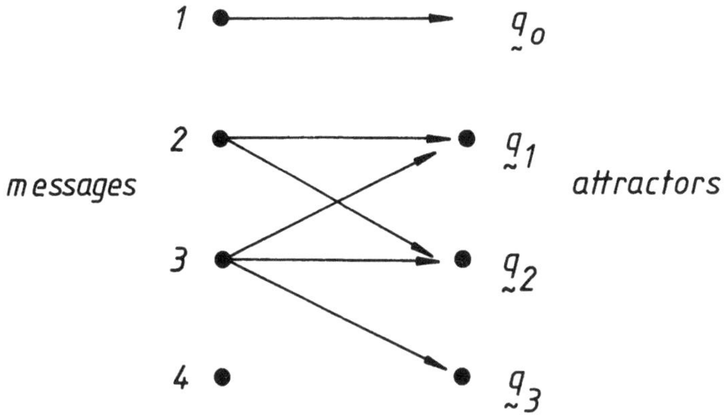

Fig. 9 Explanations in the text.

Or (II) the system goes into a new attractor. We first assume that this attractor is uniquely determined by the incident message. Clearly, different messages can give rise to the same attractor. In this case we will speak of redundancy in the messages.

Finally, especially in the realm of biology it has been a puzzle to know how information can be generated. This can easily be visualised however if we assume that the incident message produces the situation depicted in Fig. 7, which is clearly ambiguous. Two new stable points (or attractors) can be realised dependent on a fluctuation within the system itself. Here the incident message contains information, in the ordinary sense of the word, which is ambiguous and the ambiguity is resolved by a fluctuation of the system. Loosely speaking, the original information is doubled because now two attractors become available.

Taking all these different processes together we can consider the elementary schemes of Fig. 8. Of course, when we consider the effect of different messages, more complicated schemes such as those of Fig. 9 may evolve.

We shall now treat the question 'In what way can we attribute values to the incident messages?' or, more precisely, we want to define a 'relative importance of the messages'. To this end we have first to introduce a 'relative importance' for the individual attractors. In reality, the individual attractors will be the origin of new messages which are then put into a new dynamic system and we can continue this process ad infinitum. However, for practical purposes, we have to cut this hierarchical sequence at a certain level, and, at this level we have to attribute values of the relative importance to the individual attractors. Since this procedure can be clearly demonstrated for a one-step process, consider this process in detail.

Let us attribute a 'relative importance' to the individual attractors where the attractor 0 with $\underset{\sim}{q}_0$ has the value 0, while the other attractors may have values $0 \leq p_j' \leq 1$, which we normalise to $\sum_j p_j' = 1$.

The fixing of p'_j depends on the task the dynamic system has to fulfill. We may think of a specific task which can be fulfilled just by a specific attractor or we may think of an ensemble of tasks whose fulfillment is of a given relative importance. Clearly the relative importance of the messages p_j does not only depend on the dynamic system but also on the tasks it has to fulfill. The problem now is: 'What are the values p_j of the incident messages?' To this end we consider the links between a message and the attractor into which the dynamic system is driven after receipt of this message. If an attractor k (including the 0 attractor) is reached after receipt of the message j we attribute to this process the matrix element $M_{jk} = 1$ (or $= 0$). If we allow for internal fluctuations of a system, a single message can drive the system, via fluctuations, into several different attractors which may occur with branching rates M_{jk} with $\sum_k M_{jk} = 1$. We define the 'relative importance p_j" by

$$p_j = \sum_k L_{jk} p'_k = \sum_k \frac{M_{jk}}{\sum_{j'} M_{j'k} + \varepsilon} p'_k \tag{15}$$

where we let $\varepsilon \to 0$. (This trick serves the purpose that the ratio remains determined in the case that denominator and nominator vanish simultaneously). We first assume that for any $p_k \neq 0$ at least one $M_{jk} \neq 0$. p_j is normalised as can be shown by the following

$$\sum_j p_j = \sum_{kj} \frac{M_{jk}}{\sum_{j'} M_{j'k} + \varepsilon} p'_k = \tag{16}$$

$$= \sum_k \left(\sum_j \frac{M_{jk}}{\sum_{j'} M_{j'k} + \varepsilon} \right) p'_k = \tag{17}$$

$$= \sum_k p'_k \tag{18}$$

where the value of the terms in the bracket in (17) is 1. Now consider the case that some $M_{jk'} = 0$ for $p_{k'} \neq 0$ and all j. In this case in the sums over k in (16) and (17) some coefficients of $p'_k \neq 0$ vanish and, since $\sum p'_k = 1$, we obtain $\sum_j p_j < 1$. If this inequality holds, we shall term it an <u>information deficiency</u>.

In a more abstract way we may adopt the left hand side of (15) as a basic definition where we assume

$$\sum_j L_{jk} \leq 1 \; ; \tag{19}$$

the equality sign holds in the case of absence of an information deficiency.

We note that the requirement $M_{jk} = 1$ in the case of a single final attractor for an incident message can be generalised as

$$0 < M_{jk} \leq 1 \, . \tag{20}$$

Equation (15) allows us immediately to write down the equations when several systems are coupled one after the other. For instance, in the two step process we immediately obtain

$$p_j = \sum_k L^{(1)}_{jk} \, p'_k = \sum_{kk'} L^{(1)}_{jk} \, L^{(2)}_{kk'} \, p''_{k'} \tag{21}$$

where it can be shown that $\sum p_j = 1$ provided $\sum p_{k'} = 1$ and $\sum_j L_{jk} = 1$. The individual steps are

$$\sum p_j = \sum_j \sum_{kk'} L_{jk} \, L_{kk'} \, p''_{k'} = \tag{22}$$

$$= \sum_{kk'} \Big(\underbrace{\sum_j L_{jk}}_{=1} \Big) L_{kk'} \, p''_{k'}$$

$$= \sum_{k'} \underbrace{\sum_k L_{kk'}}_{=1} \, p''_{k'} = 1 \, . \tag{23}$$

We may define

$$L'_{jk'} = \sum_k L^{(1)}_{jk} \, L^{(2)}_{kk'} \, . \tag{24}$$

Because the L's are positive we find that

$$L'_{jk'} \geq 0 \tag{25}$$

and, because of the normalisation properties, (in case of absence of information deficiency)

$$\sum_j L'_{jk'} = \sum_k \sum_j L^{(1)}_{jk} \, L^{(2)}_{kk'} = \sum_k L^{(2)}_{kk'} = 1 \tag{26}$$

we readily obtain

$$L'_{jk} \leq 1 \tag{27}$$

so that L_{jk} obeys the inequality

$$0 \leq L'_{jk} \leq 1 . \tag{28}$$

We note that the recursion from p" or some higher order $p^{(n)}$ to p may depend on the paths.

Some conclusions

This approach not only introduces the new concept of the relative importance of a message but it also provides an algorithm to determine p, which has some conceptual and practical consequences. For a given task or ensemble of tasks this algorithm allows selection of the message to be sent, namely the one with the biggest p_j. If there are several p_j of the same size it does not matter which message is sent. From the conceptual point of view we may decide, whether by dynamic processes information is annihilated, conserved or generated. To this end we make use of the concept of information outlined in the second part, namely in the sense of conventional information theory. But instead of the information content caused by the relative frequency of symbols we use the relative importance within a set of messages, i. e. we introduce the quantities

$$S^{(0)} = -\sum p_j \ln p_j \tag{29}$$

$$S^{(1)} = -\sum p'_k \ln p'_k \tag{30}$$

where p_j, p'_k have the meanings introduced in the third part. If $\sum p'_k = 1$, as is assumed here, and $\sum p_j < 1$, an information deficiency is present. In the case $\sum p_j = 1$ annihilation of information occurs if

$$S^{(1)} < S^{(0)} \; , \tag{31}$$

conservation of information if

$$S^{(1)} = S^{(0)} ; \tag{32}$$

and generation of information if

$$S^{(1)} > S^{(0)} \tag{33}$$

The meaning of this definition quickly becomes clear when special cases are treated. If, for instance, two messages lead to the same attractor there is a redundancy in the system and the information content (in the traditional technical sense of the word) becomes smaller. It is reduced from

$$S^{(0)} = -K\left(\tfrac{1}{2}\ln\left(\tfrac{1}{2}\right) + \tfrac{1}{2}\ln\left(\tfrac{1}{2}\right)\right) = K\ln 2 \tag{34}$$

to

$$S^{(1)} = -K \cdot 1 \cdot \ln 1 = 0 . \tag{35}$$

In the case of a one-to-one mapping of p_j onto p_k' we find the transfer of $\{p_j\}$ into the same set $\{p_k'\}$, may be performed except for the permutation of indices, i. e. for different numeration of states. In such a case clearly (32) holds. For the case where equation (33) holds, for instance where one $p_j = 1$ and all other $= 0$ and transferred into e. g. $p' = p'' = 1/2$, all others equal to 0. Then $S^{(0)} = -K \cdot 1 \cdot \ln 1 = 0$ is enlarged to

$$S^{(1)} = -K\left(\tfrac{1}{2}\ln\left(\tfrac{1}{2}\right) + \tfrac{1}{2}\ln\left(\tfrac{1}{2}\right)\right) = K\ln 2 . \tag{36}$$

Of course these examples are not meant to prove the definitions (31) to (33) but rather to illustrate their meaning.

The approach based on synergetics (Haken, 1978, 1983) has some further features. Semantics has become a problem in the study of the response (attractors) of the dynamic system. The system may be error-correcting (or supplement missing information). If the incident message does not set the initial state $\underset{\sim}{q}$ on the attractor (i. e. not correctly), it may set the initial state $\underset{\sim}{q}$ within the basin of the attractor and the system pulls the state vector into the attractor corresponding to that basin, i. e. into the correct state. An interesting problem will it be to determine the minimum number of bits required to realise a given attractor (or to realise a given value of 'relative importance'). Within the present scheme, the learning of a system can also be modelled. A system can be 'sensitised' or 'desensitised' with respect to message j e. g. by letting more or fewer parameters react on specific messages.

It is hoped that the concepts outlined above may serve as a starting point for the development of a dynamic information theory. Clearly there are a number of obvious generalisations, for instance to time dependent $M_{jk}(t)$ or to a permanent influx of messages which continuously controls the development of the system. Contact with control theory can be made at this point and other aspects, e. g. to find optimal choices for p_j, may be linked with game theory.

On the Problem of the Valuability of Information

R. L. Stratonovich

The interest of biologists and biophysicists in questions connected with entropy and information, particularly, with the value of information has increased recently (see e. g. Schmalhausen, 1968; Volkenstein, 1980, 1981). In this connection it is reasonable to discuss more or less thoroughly the definition and determination of the value of information. We shall interpret this concept in line with that adapted in a recent book (Stratonovich, 1975).

Information is always needed for some purpose. For a living organism three types of information are essential. First, there exists information arising from the organs of sense, information about the external medium, about the conditions under which the organism exists, about the location of food, enemies, etc.. Second, there is information arising from the internal nerve receptors, information about facts of vital activity of the organism. This information is needed for correct regulation of the processes of vital activity, for promoting satisfactory functioning of the internal organs. Finally, there is hereditary information stored in DNA. This is needed for proper development of the organism, for genus continuation and heredity transfer.

To make it possible to apply precise mathematical methods, it is desirable to characterize, quantitatively, the use to which the information is put. Therefore, it is necessary, for example, to introduce a quantitative measure - the value of a catch, the value of the correct regulation of some processes, etc.. This arbitrarily introduced numerical function can be called the function of welfare. After fixing this function, one can introduce the value of information, i. e. a number quantitatively describing the maximum use of information.

Thermodynamics and Regulation of Biological and Processes

In biology the use of information cannot, of course, be easily characterized quantitatively. This is particularly difficult in the case of hereditary information. This means that the possibilities of applying exact mathematical methods in biology are limited.

In this introduction it is not out of place to define the concepts under considerations. The quantity of information is equal to the quantity of removed uncertainty, i. e. is equal to the uncertainty before the appearance of information (a priori uncertainty) minus the uncertainty after the appearance of information (a posteriori uncertainty). Maximization of the transmitted or recorded amount of information on a fixed transmitter or recorder gives the information capacity.

This definition has much in common with the one of the value of information. The benefit of information is equal to the increase of welfare, i. e. to the welfare after the reception of information minus the welfare before the reception. Maximization of the information use, in the described sense, in different actions of the recipient, gives the value of information. Thus, the determination of the information value is correlated with the behaviour of the recipient. Such a necessity is intuitively clear. Indeed, what is the value of information if the recipient misunderstands or ignores it?

After this introduction we pass over to a quantitative determination of these concepts.

Quantity of information and the value of individual communication

We begin with a simple example. Suppose that in a forest of 10 hectares a purse with 100 roubles is lost. If a man is looking for it, within a certain time let him be able only to search through 0.5 hec.. Assume the probability density for purse location to be uniform. Then the probability that the man will find the purse is equal to $0.5/10 = 0.05$, and the mean profit from the search will be 100 roub. x $0.05 = 5$ roubles.

Suppose that each hectare (square) is marked with digits 1, 2, ..., 10. If the man learns that the purse is in one of the squares 3, 4, or 5, his search will be limited to these squares, and the probability of finding the purse will be equal to 0.5/3 = 0.167. Accordingly, the mean profit from the search will be 100 roub. x 0.167 = 16 roub. 70 kop. Subtracting from this sum the guaranteed profit equal to 5 roub., we deduce that the piece of information has given a profit of 11 roub. 70 kop. This is the value of this communication.

Besides the value, one can also indicate the quantity of information contained in a concrete communication. It is equal to the logarithm of the probability that the purse is in the region indicated in the communication:

$$J_1 = -\log_2 p_1 = -\log_2 \frac{3}{10} = 1.74 \text{ bits} .$$

We should stress that the value of the communication will be realized only provided that the behaviour of the man is optimal, i. e. he understands the communication, looks for the thing in the indicated region and not in another place, etc.. This optimal behaviour is typical of problems associated with information value. If a person cannot, or does not want, to read, the value of information contained, say, in a guide is equal to zero. The information value should always be understood as a maximum value which can be obtained under optimal conditions and optimal behaviour. Otherwise, the concept of information value loses its definition.

Mean value and mean quantity of information contained in a statistical ensemble of communications

A statistical ensemble arises if a situation repeats many times. In our example the purse will, each time, be in another place according to the uniform probability distribution. What will then be an ensemble of communications? Different cases are possible.

Case 1. It is decided in advance that the reporter says whether or not the purse has got into the region of squares 3, 4, 5. If not, the probability of finding it is equal to 0.5/7 = 0.071. In this case the mean profit of the search is 7 roub. 10 kop., so that the value of this information is 2 roub. 10 kop.. The quantity of information in this piece is equal to the logarithm (with an opposite sign) of the probability of getting into the region of squares 1, 2, 6, 8, 9, 10, where the purse is.

$$J_2 = -\log_2 p_2 = -\log_2 \frac{7}{10} = 0.514 \text{ bits}.$$

Now let us find the mean value and the mean quantity of information. The purse is in the region of squares 3, 4, 5 with a probability of 0.3 and in the remaining regions with a probability of 0.7. To obtain the mean value, one should average the individual values with the indicated probabilities, i. e.

$$V = p_1 \cdot 11.7 + p_2 \cdot 2.1 = 0.3 \cdot 11.7 + 0.7 \cdot 2.1 = 4.98 \approx 5 \text{ roubles}.$$

The quantities of information obtained above are averaged in a similar way. This gives the following quantity of information

$$J = -p_1 \log_2 p_1 - p_2 \log p_2 = 0.882 \text{ bits}.$$

Case 2. We may decide differently. The purse can be reported to be either in the region of squares 3, 4, 5, or in the region of squares 1, 2, or in the region of squares 6, 7, 8, 9, 10. If it is in the region of squares 1, 2, the probability of finding is 0.5/2 = 0.25. The profit of the information is 25 - 5 = 20 roub. The quantity of information here is given by

$$J_2 = -\log_2 p_2 = -\log_2 \frac{2}{10} = 2.28 \text{ bits}.$$

Now assume the purse to be reported to be in the region of squares 6, 7, 8, 9, 10. The probability of finding it in this case is 0.5/5 = 0.1, and the value of the information is equal to 10 - 5 = 5 roub., and the quantity of information is as follows

$$J_3 = -\log_2 p_3 = -\log_2 \frac{5}{10} = 1 \text{ bit}.$$

By averaging the indicated values of 11 roub. 70 kop., 20 roub., 5 roub., and the informations J_1, J_2, J_3 with the corresponding probabilities 0.3, 0.2, 0.5 of being in this or that region, we obtain the value and the quantity of information in this case as

$$V = 0.3 \cdot 11.7 + 0.2 \cdot 20 + 0.5 \cdot 5 \quad 10 \text{ roub.}$$

$$J = -p_1 \log_2 p_1 - p_2 \log_2 p_2 - p_3 \log_2 p_3 = 1.478 \text{ bits.}$$

Other cases are of course possible too.

In the first case the reporter says 'yes' or 'no', i. e. whether or not the purse is in the region of squares 3, 4, 5. He can, in principle, (in the optimal case) send one bit of information. So, the information capacity of the message is equal to one bit, whereas actually he sends a smaller quantity. This shows an insufficiently good stipulation. The reporter had better indicated the region of five squares where the purse is. Then the information capacity would be completely used, and the quantity of information would be 1 bit. In the second case, to make the quantity of information equal to the information capacity, i. e. $\log_2 3 = 1.58$ bit, one would divide the forest into three equal parts and indicate the one where the purse is.

So, we see that the information capacity and the quantity of information actually transmitted or recorded is not the same. It is very easy to calculate the information capacity of a book or a DNA molecule, but it is not so easy with the quantity of the recorded information in a fixed situation. To this end, the situation must be clearly described, and its statistical model built.

Going back to our example, we can state that both for the indicated non-optimal division of the forest into regions and for the optimal division, for which it is not difficult to estimate numerically the value and the quantity of information, the following regularity is observed: as the information increases, its general value increases too. Of course, there exists no direct proportionality, so one can only approximately measure the value of the unit information irrespective of the total amount of information.

The quantity and value of information in the presence of erroneous messages

All the messages so far described considered by the recipient to be correct. Now a message will be assumed correct or erroneous with certain (known) probabilities. Let us go back to the example considered. Let, say, a message y come (which can be erroneous) that the purse is, or is not in the region of squares 3, 4, 5. On the basis of this message one should choose a posteriori the distribution of the purse in different places of the forest. Remember that the a priori probability distribution p(x), where x is the point of purse location, was uniform: $p(x) = 0.1\ \text{hectare}^{-1}$. In addition, one should assume the probability of a correct or an erroneous message. In the general case this may depend on the purse location. Let the message that the purse is in the region of squares 3, 4, 6 be y = 1, and for the opposite message y = 0. The reliability of the messages is characterised by the function p(y/x), i. e. by the probability of the message y under the condition that the purse is at the point x. In this case the probability of the erroneous message as a function of the purse location is as follows

$$P_{err} = \begin{cases} 0 \text{ at } x \text{ in the region of squares } (3,\ 4,\ 5) \\ 1 \text{ at } x \text{ outside the region of squares } (3,\ 4,\ 5) \end{cases}$$

The probability of the true message is equal to $P_{corr} = 1 - P_{err}$.

The product p(y/x)p(x) is none other than a joint distribution of the purse location and the message. The a posteriori distribution of the purse location corresponding to the message y is determined by Bayes' formula

$$p(x|y) = p(y|x)p(x) / \int p(y|x)p(x)\,dx. \tag{1}$$

If the man looking for the purse has got the message y = 1, he must go to the place in the forest, where the function

$$p(x|1) = p(1|x).0.1 / \int p(1|x')\,0.1\,dx'$$

is maximum. If he has got the message y = 0, he must go where the function

$$p(x|0) = p(0|x) / \int p(0|x')\,dx'$$

is maximum.

Under the condition that the indicated functions are constant in the neighbourhood (on an area not less than 0.5 hec.) of the point where the functions have maximum values, the probability of finding the purse is equal to

$$0.5 \max_x p(y|x) / \int p(y|x')\,dx'.$$

Multiplying this quantity by 100 roub. and subtracting 5 roub., we obtain the value of information corresponding to the fixed message

$$V(y) = \left(50 \max_x \frac{p(y|x)}{\int p(y|x')\,dx'} - 5 \right) \text{ roub.}$$

This method can also be applied in the case where the a priori distribution is non-uniform.

Let us now clarify the concept of the quantity of information, assuming for generality, p(x) to be non-uniform. Suppose that some concrete message y was sent. The measure of uncertainty of the purse location before the message has been received is the following integral

$$H_{pr} = -\int p(x) \log_2 p(x)\,dx.$$

The measure of uncertainty remaining after the message has been received is a similar integral, but the a priori distribution should be replaced in it by the a posteriori one. Taking into account (1), we have

$$H_{po}(y) = -\int p(x|y) \log_2 p(x|y)\, dx =$$
$$= -\int \frac{p(y|x)p(x)}{p(y)} \log_2 \frac{p(y|x)p(x)}{p(y)}$$

where $p(y) = \int p(y|x)p(x)\,dx$.

The measure of the quantity of information here is the decrease of uncertainty

$$J(y) = H_{pr} - H_{po}(y) = -\int p(x) \log_2 p(x)\, dx + \int \frac{p(y|x)p(x)}{p(y)} \log_2 \frac{p(y|x)p(x)}{p(y)}\, dx.$$

Remember that the concrete message y is fixed here. To obtain the quantity of information averaged in y, one should carry out averaging in (2) with the weight

$$p(y) = \int p(y|x)p(x)\,dx.$$

This gives the following quantity of information

$$J = \int J(y)p(y)\,dy = -\int p(x)\log_2 p(x)\,dx + \sum_y \int p(x,y) \log_2 \frac{p(x,y)}{p(y)}\,dx$$
$$= \sum_y \int p(x,y) \log_2 \frac{p(x,y)}{p(x)p(y)}$$

where $p(x,y) = p(y|x)p(x)$.

As is known, this measure of the quantity of information was introduced by Shannon, who obtained for it a number of mathematical results.

The above methods are also applicable in the case where the message has an arbitrary degree of complexity. In this case, corresponding to messages with errors, the described regularity is also observed: the larger the quantity of information, the larger its benefit and its value.

If for one and the same problem and for one and the same message the value and the quantity of information is calculated for the case of the presence or absence of erroneous messages, then for possible erroneous messages the value and the quantity of information will of course be much less. To increase these values, one should use excessive information by doubling messages. As a consequence of the independence of violation of the truth in doubling messages, the resulting probability of violation decreases.

Value of information in the general case

In the general case it is reasonable to introduce some function which describes the benefit of information. In different cases it may have a different meaning: the meaning of welfare, pleasure, the absence of pain, the absence of a fine, profit, etc.. It must depend on the value of the variable (or variables), on the coordinates of the object about which one has to get information. Designate this variable or variables by x. Further more the welfare function must depend on the behaviour of the subject who tries to increase welfare by using information. We call these actions regulating and designate them by u. So, one should introduce the function C(x, u) of at least partially unknown variables x that affect the welfare and the regulating actions of the subject.

If there exists no information except a priori, the maximum that the subject can do is maximise the mean welfare-calculated with the a priori distribution $p_{pr}(x)$. This gives the welfare equal to

$$\max_{u} \int C(x, u) p_{pr}(x) dx .$$

Now suppose that a message y arrives concerning the meaning of x. If the message is true, y is a function of x (y = f(x)), which is, in the general case, degenerate. If the message can be erroneous, y is a random quantity correlated with x and described by the conditional probabilities p(y/x), i. e. by the distribution with respect to y at different fixed x. On obtaining this or that message y, one can calculate the a posteriori distribution

$$p(x|y) = \frac{p(y|x)\,p_{pr}(x)}{p(y)}, \quad \left(p(y) = \int p(y|x)\,p_{pr}(x)\,dx\right)$$

and the corresponding mean welfare

$$\int C(x, u)\,p(x|y)\,dx.$$

The best thing that the subject can do now is undertake something which could maximise this function, i. e.

$$\max_{u} \int C(x, u)\,\frac{p(y|x)\,p_{pr}(x)}{p(y)}\,dx.$$

In this case the value of the message y is equal to

$$V(y) = \max_{u} \int C(x, u)\,p(y|x)\,p_{pr}(x)\,[p(y)]^{-1}\,dx - \max_{u} \int C(x, u)\,p_{pr}(x)\,dx.$$

The mean value of the information can be obtained by averaging the latter quantity over y with weight p(y), which gives

$$V = \int \left\{ \max_{u} \int C(x, u)\,p(y, x)\,p_{pr}(x)\,dx \right\} dy - \max_{u} \int C(x, u)\,p_{pr}(x)\,dx.$$

This allows a general determination of information value at a fixed a priori distribution $p_{pr}(x)$.

Particular case 1. Assume x and u to be points of one and the same space and $c(x,u)$ to have the form

$$c(x,u) = \begin{cases} c_0 & |x-u| \leq b \\ 0 & |x-u| > 0 . \end{cases}$$

Here $|x-u|$ is the distance between the points x and u. Then the mean

$$\int c(x,u)p(x)dx$$

is expressed by

$$c_0 = \int_{|x-u|<B} p(x)dx = c_0 p\{|x-u| \leq b\},$$

where $p\{|x-u| \leq B\}$ is the probability of the fact that x will be at a distance b from u or closer. It is natural to interpret the function c(u,x) as follows. The purpose is 'to find', 'to get', 'to see' some catch. This can be done under the condition $|x-u| \leq b$. In case of success the benefit is C_0. The use of the information must provide an increase in the probability of achieving the object and the mean expected catch. Thus, we are dealing with a problem of the type considered in earlier sections. In this case the information value is equal to the expression

$$V = C_0 (\langle P_{ps} \rangle - P_{pr}), \tag{3}$$

where P_{pr} is an a priori probability of achieving the goal, P_{ps} is an a posteriori probability.

Particular case 2. Let now the goal be a regulation of some process, namely, stabilisation of the variable x_1 near the value b. If we take the mean square error criterion, then the goal will be maximisation of the quantity $-(x_1-b)^2$. Suppose that x_1 is the known function of the value x of the variable considered at the previous stage and of the regulating action u, the region of the possible values of which can be limited. Substituting $x_1 = f(x,u)$ into $-(x_1-b)^2$, we obtain the benefit function

$$c(x,u) = -[f(x,u) - b]^2 .$$

More complicated formalisations of the stabilisation problem are, of course, possible. The problem may be sequential, and the benefit function may be a time integral. Note that on introducing a benefit function, one can look for optimal regulation at fixed observation by applying methods of the optimal control theory. The number of particular cases, where it is reasonable to introduce the benefit function to increase the mean profit of information use, is also very large.

Why is it useful to know the information value?

It is desirable to know the value of information of this or that message when the information capacity of the reception channel or the memory capacity (if information is to be memorised) is limited because of their construction. Then it is only the most valuable information that must be received or memorised. Take, for example, the case considered above. Suppose that the capacity of the reception channel (or of the memory) is only 2 bits. Then one should take the most valuable information of just this quantity. Such information is as follows:
indication of one of the four equal regions of 2.5 hec. where the purse is. The value of this information is calculated to be equal to 15 roub. Any other piece of the same information capacity, corresponding to a division of the forest into four unequal regions is of smaller value. In this example the conclusion is, of course, somewhat trivial. This is because of the simplicity of the problem. In the case of complicated benefit functions the choice of the most valuable information at a given information capacity is non-trivial.

Another situation is also possible where the capacity of the reception channel or of the memory are not a priori limited in their construction, but an increase of memorised or gained information is associated with an increase in some expenditure, in particular, energy. In this case no information should be recorded or gained. If the value of information exceeds the expenditure, such information should be recorded. Expenses in recording (or gaining) information should be expressed in the same units as the information value.

Energy expenditure in memorising information is a wellknown fact. The information memorising time t is associated with the depth of the potential well Δu , i. e. with expenditure of energy per bit of information, by the formula

$$\ln t = \frac{\Delta E}{kT} - C ,$$

where T is the absolute temperature of the recording device, k is the Boltzmann constant, C ist the constant typical of a given recording device. If the time t, required to memorise the information, is fixed and T and C are known, one should only set up the cost of energy in the same units as the information value. After this, one can compare the benefit of the information with expenditure for recording and decide what information should be recorded.

Other determinations of the value and benefit of information

The above interpretation of information value was first proposed by Stratonovich (1965). Earlier publications (Kharkevich, 1963; Bongard, 1963) considered other approaches to the problem of the value and use of information. Now we shall briefly discuss the corresponding ideas.

Kharkevich (1963) suggests that the value of information should be determined by the formula

$$V = \log_2 (P_{ps} / P_{pr}) , \qquad (4)$$

where P_{pr} is the a priori, and P_{ps} the a posteriori probability of achieving the goal. This formula should be compared with equation (3) of our theory, or, if the value, C_o, of achieving the goal is assumed equal to unity, with the formula $V = P_{ps} - P_{pr}$. For the example presented in the first sections the logarithmic measure of the value of information proposed by Kharkevich is seen to be irrelevant. In another example consider the goal be to get 100 roub.. Let, without any information, the probability P_{pr} be equal to zero, and, after receiving information become 1/2. The intuitive estimate of such information is 50 roub., whereas, using (4), we calculate this value to be equal to infinity, which is nonsense.

Now, let us turn to Bongard's conception (1963) which considered as absolute some probability distribution p_i^o of the discrete random quantity x. This distribution is thought of as 'correct'. The quantity of useful information leading to replacement of the a priori by the a posteriori distribution is determined by

$$J_n = -\sum p_i^{(pr)} \log p_i^{(pr)} + \sum p_i^o \log p_i^{(ps)}. \qquad (5)$$

The quantity of useful information can be negative, which corresponds to misinformation. It can be easily verified that at fixed p_i^o and $p_i^{(pr)}$ the measure (5) becomes maximum when $p_i^{(ps)} = p_i^o$. Thus, according to this concept, the 'use' of information consists in making the a posteriori distribution as close to the absolute distribution p_i^o as possible. The measure (5) correctly characterises the quality of the calculating algorithm described in section 1 of the paper by Bongard (1963), but it becomes irrelevant when applied not to an algorithm, but to normal problems. In normal problems, for example, in the one considered above, there is no absolute distribution p_i^o which one should approach. The goal is to determine the coordinates of the unknown quantity. In equation (5) the goal is replaced: the goal of determining a random quantity is replaced by another goal of approaching an absolute distribution.

We present here the example considered by Bongard himself in section 3 of his paper. We read there: 'A certain A wants to find, in an office open from 10 h to 18 h a certain B, who is known to be there for two hours every day (from 10 h, from 12 h, from 14 h or from 16 h). To specify the information, A adressed a friend of his working in the same office. The friend said that B comes to the office after 14 h twice as often as before 14 h. After this A called to the secretary. In reply to his question he heard: '... the time-table for the next month has not yet been made, but B always comes 5 times a week from 12 h and once a week he is here from 14 h to 16 h. Oh, sorry, I am mistaken. Once a week from 12 h to 14 h and 5 times from 14 h to 16 h. The latter statement was correct.'

Bongard (1963) takes the probability distribution (0, 1/6, 5/6, 0) in two-hour intervals, corresponding to the correct statement of the secretary, as absolute and calculates, using formula (5), how close this or that statement approaches the a posteriori distribution.

> In this problem the knowledge of the secretary is not the end in itself. The end goal is to find comrade B in the office and for this purpose to learn his time-table.

Let us ourselves continue describing what happened next to comrade A. After the conversation with the secretary, which could not, of course, satisfy him, A called B at home. He was out. A spoke to B's wife, who said: '... my husband's time-table has already been sanctioned by his chief. He will work Monday, Tuesday, Wednesday, and Thursday from 14 h to 16 h. But I do not know about Friday and Saturday: he will either work on Friday from 14 h to 16 h and on Saturday from 12 h to 14 h, or vice versa. The latter statement was correct.'

If we take as absolute the knowledge of the secretary, in line with Bongard (1963), the wife's statement appears to be misinformation, although it is true. May be it is better to take as absolute the statement of the wife? Then, eliminating the last uncertainty, B himself will give 'misinformation'.

We see that there exists no absolute distribution (in the latter version we mean distributions in different time-tables). Information given by the friend, the secretary, the wife are equal successive specification. In these pieces of information the quantity should be calculated not by (5), but by

$$J = -\sum_i p_i^{(pr)} \log p_i^{(pr)} + \sum p_i^{(ps)} \log p_i^{(ps)}$$

as is normal in information theory. This is undoubtedly the quantity of useful information since all the three pieces mentioned above are useful. The determination of the value (5) is seen to be needless in the ordinary situation.

Relativistic Information and Biology

G. Jumarie

Introduction

The problem of applying information theory to questions of theoretical biology is one of the most controversial in the present state-of-art. If one believes Thom (1975), the information theory initiated by Shannon (Weaver, Shannon, 1949), (see also Brillouin (1978) who gave an excellent review of its prospects in physics) refers mainly to fields of probability and has valuable application only to technical communication problems, which was its principal purpose. Among the questions which cannot be fruitfully considered by this approach, Thom cites the following: How can one define the amount of information which is contained in a form?

Nevertheless the results which have been obtained through use of this theory in biology are somewhat amazing, and stimulating e. g. the theory of ageing by Gatlin (1972), the analysis of the transformation process DNA RNA Proteins; the information theoretical approach to biological systems by Atlan (1972). These results, even though they are not in final forms, are sufficiently important (in our opinion) to justify further investigations.

In fact, the literature shows the necessity for modification of standard information theory to meet the needs of the biological framework, and a recent publication which considered this question (Ryan, 1980), suggests the new concept of *information-entropy interfaces* and speaks in terms of *relative information about a topic.*

Thermodynamics and Regulation of Biological and Processes

My contention is that all these are tentative efforts, even though they are intuitively justified, and somewhat 'ad hoc' in the way that they attempt to adapt a theory which is now soundly established in a technical framework. I think that it is the concept of information itself which requires new approaches and new definitions in order to be applied fruitfully to such areas as biology, linguistics, sociology, etc.; and in this way we have been led to propose a so-called relativistic information theory which explicitly deals with syntax and semantics. This theory has been built up mainly to account for the subjectivity which is involved by communication via natural languages, so that, at first glance, its applications to biology would not appear to be relevant. Why not? It has been done with the standard theory?

In the present paper, I shall examine how this relativistic information can be utilised to analyse biological systems, and we shall see that the results so obtained are quite meaningful. Therefore there is interest in identifying the biological senses of the various parameters which define the theoretical model. For the convenience of the reader, I shall first sketch the background on relativistic information, and then I shall consider the ageing problem.

Relative Information, Relative Entropy

The purpose of this section is to provide the reader with the main elements of a new approach to modelling information entropy which attempts first, to introduce explicitly the relativistic features of information, and second, to emphasise the discrepancy between information and indeterminacy. To this end, I return to the primary framework of information, which is the setting of communication between human observers.

Natural language

Natural language (NL) is the basic 'device' by means of which information is conveyed among people. Formally, a NL is defined by a space of symbols, Ω , the syntactic space; and a space of meanings, Ω' , the semantic space which is the universe of discourse onto which the NL makes sense. Thus Ω can be viewed as composed of words, while Ω' is composed of concrete objects. The internal structure of Ω is defined by some organising rules which are referred to as the syntax of the NL, and likewise, Ω' can be considered as structured by a field of possibility which discards any combination of the elements, which does not make sense. A given word α may have several meanings a, and one defines a *lexem* as being a pair (α , a), clearly a word with a given meaning (Fig. 10).

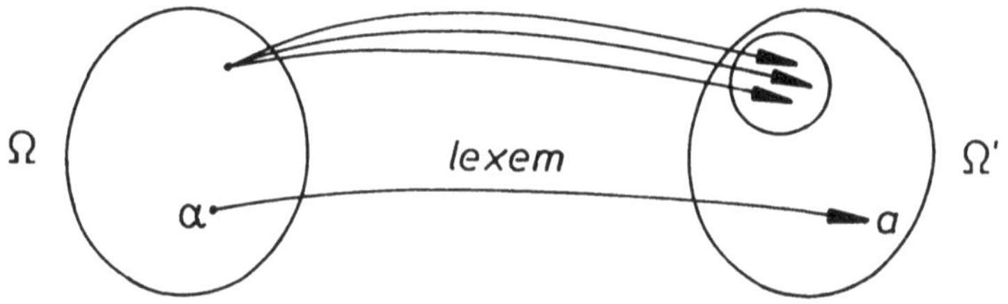

Fig. 10 Natural language.

Exactly like in Shannon's theory, I assume that Ω is described by the field of probability associated with the occurrences of the different words, and therefore I define the entropy of Ω , H(α). Likewise we shall define the entropy of Ω' as H(a). So, in the Shannon theoretical framework, H(α) is the amount of information involved with Ω alone, and likewise for H(a) and Ω' .

My argument is that there is a coupling effect between Ω and Ω' which, as a result, directly infers a relativistic feature for the entropy, and a discrepancy between information and uncertainty; and the problem is now to devise a model for this interaction.

The model

In human systems, I shall consider, as a result of subjectivity, the possibility that any observer has, of interpreting and recovering (or not recovering!) the exact content of the message he is receiving in such a way that the meaning which is given to the message is not the meaning which is observed as such, but is rather the meaning the observer thinks it should be. This behaviour could be generated by directly referring to prior internal models in the mind of the observer. It seems that the following principle is valid in a large number of cases.

Principle. Subjectivity works in such a way that indeterminacy decreases when information increases and conversely, indeterminacy increases when information decreases.

This principle may be applied as follows:

I. The total amount of information involved in Ω and Ω' can be defined by the expression H(α) + H(a) i. e. α and a are independent, and thus

$$H(\alpha a) = H(\alpha) + H(a).$$

There is no inconsistency in this definition. Since a message is composed of the pair (Ω , Ω'); and the concept of stochastic dependence appears when two messages are created i. e. (Ω , Ω') and (Σ , Σ'). When the message (Ω , Ω') is considered we work at a level of internal dependence, and it is exactly the latter that we are trying to represent.

II. The amount of total indeterminacy involved by the message (Ω , Ω') is H(a) - H(α).

Indeed, if H(a) = H(α), then on the average each word has only one lexem, and there is no relative indeterminacy on Ω' with respect to Ω . Moreover, despite the fact that several different words may have the same meaning (synonyms) H(a) $\geqslant$ H(α), so that H(a) - H(α) $\geqslant$ 0.

III. Subjectivity works at the level of the pair (Ω , Ω'). In the absence of any subjectivity, (Ω , Ω') is characterised by the entropies H(α) and H(a). When an observer R considers (Ω , Ω') with subjectivity, these entropies become H(α /R) and H(a/R). One way to represent mathematically this principle, is to state that the product '(information) · (indeterminacy)' is not affected by the presence of subjectivity viz:

$$H^2(a/R) - H^2(\alpha/R) = const \; . \tag{1}$$

Subjective Entropy

Equation (1) defines the geodesic of the special relativity. As a result, there exists a constant $u_\alpha(R)$ which depends upon both α and R, such that the following equations are satisfied:

$$H(\alpha/R) = \rho(u_\alpha)[H(\alpha) + u_\alpha(R)H(a)] \tag{2}$$

$$H(a/R) = \rho(u_\alpha)[H(a) + u_\alpha(R)H(\alpha)] \tag{3}$$

$$\rho(u_\alpha) := (1 - u_\alpha^2(R))^{-1/2} \; . \tag{4}$$

The symbol := means that the left-side is defined by the right side, while =: means that the right side is defined by the left side.

Two observers R and R' are referred to as independent observers whenever the subjectivity of either of them does not affect the subjectivity of the other.

Let (α ,a) and (β ,b) denote two messages which involve the subjectivities u_α and u_β to the independent observers R and R' respectively. Then the compound observer RR' will observe the message ($\alpha\beta$, ab) with the subjectivity $u_{\alpha\beta}$ such that

$$u_{\alpha\beta} = (u_\alpha + u_\beta)/(1 + u_\alpha u_\beta) \tag{5}$$

and the corresponding expression of $H(\alpha\beta/RR')$ and $H(ab/RR')$ can be obtained from (2), (3) and (4).

Transinformation, transuncertainty

The *subjective transinformation* provided by (α /R) about (β /R') is

$$I[(\beta/R')/(\alpha/R)] := H(\beta/R') - H[(\beta/R')/(\alpha/R)] \quad (6)$$

where H $[$ (β /R')/(α /R)$]$ is defined by the equation

$$H(\alpha\beta/RR') := H(\alpha/R) + H[(\beta/R')/(\alpha/R)] \quad (7)$$

The *subjective transuncertainty* about (β /R') as caused by (α /R) is

$$\mathcal{U}[(\beta/R')/(\alpha/R)] := K(\beta/R') - K[(\beta/R')/(\alpha/R)] \quad (8)$$

where K(·) and K(·/·) are defined by the expressions

$$K(\alpha/R) := \varrho(u)[H(\alpha) - uH(a)] \quad (9)$$

$$K(\alpha\beta/RR') := K(\alpha/R) + K[(\beta/R')/(\alpha/R)]. \quad (10)$$

The *balance condition* in communication is as follows. If (α /R) and (β /R') provided that the following conditions are satisfied:

$$H(\alpha/R) \geq K(\beta/R') \quad (11)$$

$$H[(\alpha/R)/(\beta/R')] = 0\ . \quad (12)$$

For further information about this approach see Jumarie (1978; 1979, 1980 a, b, 1981, 1982 a, b, 1983).

Application to Genetic Information

As pointed out in the introduction I shall apply relativistic information to biology in a formal way only and show later whether the results so obtained justify further investigation.

Notation. 'dna' represents the generic element which describes the bases A, C, G and T of the DNA. Likewise, 'rna' is the generic element for the bases A, G, G and U of the RNA; and the generic element for amino-acids is referred to as 'aa'.

Model A

In this model, we assume that the pair (DNA, amino-acids) define a (natural) language with the corresponding spaces Ω = (bases) and Ω'=(amino-acids); therefore H(α) = H(dna) and H(a) = H(aa). The observer R is the ribosomal center which decodes the message. Application of relativistic information results in

$$H(dna/R) = \varrho(u)[H(dna) + uH(aa)] \tag{13}$$

where the biological meaning of u remains to be understood. Let p be defined by the equation

$$H(aa) = p\,H(dna) \tag{14}$$

then (13) becomes

$$H(dna/R) = k(p,u)H(dna)$$

with

$$k(p,u) := (1 + pu)/(1 - u^2)^{1/2}. \tag{15}$$

Model B

In this model, assume that the natural language of the process is the pair (DNA, RNA) with the syntactic space $\Omega := \{dna\}$ and the semantic space $\Omega' := \{rna\}$. Equation (2) then yields

$$H(dna/R) = \rho(u)[H(dna) + uH(rna) . \tag{16}$$

But,

$$H(dna) = H(rna)$$

so that the result from (16) is

$$H(dna/R) = k(1, u)H(dna) \tag{17}$$

where k(1, u) is defined by (15) for p = 1.

A few comments

In model A, u is related to the properties of RNA, and more explicitly, any transmission error which could be caused by faulty production of the RNA molecules would be described by u. In Model B, u is also related to the production of RNA, but it is more apparent that it could be dependent upon the chemical properties of the various chemical species in the cell.

When u = 0, then H(dna/R) = H(dna) and the production of proteins is regular. When u < 0, then H(dna/R) < H(dna), and there is underproduction of proteins; for u > 0, H(dna/R) > H(dna) and there would be overproduction of proteins. Therefore the following

Assumption. According to this relativistic information modelling, there is the possibility that some chemical substances may alter the dynamics of the biological cell in such a manner that there is overproduction of proteins.

Dynamics of Information Transmission

The main idea

Consider the ageing process as being directly associated with the transfer of information in the duplication process of the cell. Ageing would be caused by loss of information from the cell to its replica, so that the duplication process would stop at some instant.

Before examining the ageing problem itself, it is approriate to consider models of information transfer in cascade processes.

Multi-level subjective model

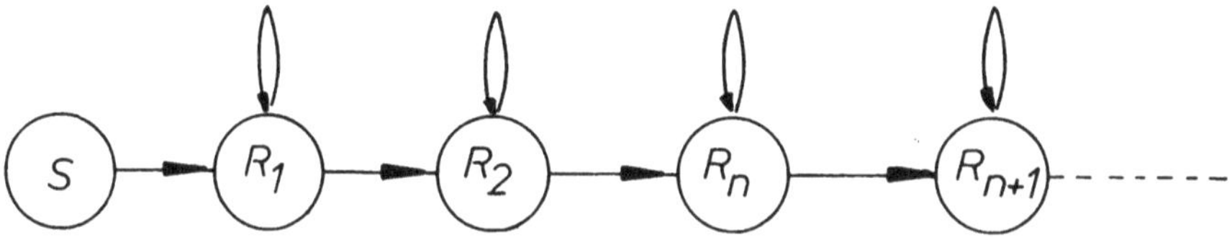

Fig. 11 Information transfer in cascade systems.

Figure 11 shows a scheme for information transfer in a cascade system. The source S which contains amount $H(\alpha)$ of information is observed by R_1 with a subjectivity v and thus receives $H(\alpha/R_1)$. This information is transferred to R_2 with a subjectivity v, and thus observes S with the subjectivity u_1 and so receives the information $H(\alpha/R_1/R_2)$. And so on. R_{n+1} with subjectivity v, observes S via , ,..., R_n with subjectivity u_{n+1} and so receives $H(\alpha/R_1/../R_n/R_{n+1})$.

Result. Assume that $v < 0$, then

$$H(\alpha/.../R_{n+1}) < H(\alpha/../R_n) \qquad (18)$$

and

$$H(\alpha/.../R_n) \downarrow 0 \text{ as } n \uparrow \infty . \qquad (19)$$

On the contrary, for $v > 0$,

$$H(\alpha/.../R_{n+1}) > H(\alpha/.../R_n) \qquad (20)$$

and

$$H(\alpha/\ldots/R_n) \uparrow \infty \text{ as } n \uparrow \infty . \tag{21}$$

Indeed, as a consequence of the composition law for subjectivity,

$$u_{n+1} = (u_n + v)/(1 + u_n v) \tag{22}$$

which provides $u_n \to +1$ for $v > 0$ and $u_n > -1$ for $v < 0$. But

$$H(\alpha/\ldots/R_n) = k(1, u_n) H(\alpha) \tag{23}$$

where $k(1, u_n)$ is defined by (15) and with $p = 1$, which yields (21).

One-level subjective model

This model is quite similar to the preceding one, except that the subjective process works at one level only. Namely, R_{n+1} observes R_n with the subjectivity v, but does not have any prior modelling of $R_1, \ldots, R_{n-1}$ so that the latter do not affect his total subjectivity with respect to S.

In such a case, one has

$$H(\alpha/R_1/\ldots/R_n/R_{n+1}) = H(\alpha/R_n/R_{n+1}) = k^{n+1}(1, v) H(\alpha) \tag{24}$$

therefore the following

Result

$$H(\alpha/R_n/R_{n+1}) \uparrow \infty \text{ as } n \uparrow \infty, \text{ for } v > 0$$

$$H(\alpha/R_n/R_{n+1}) \downarrow 0 \text{ as } n \uparrow \infty, \text{ for } v < 0 .$$

The Ageing Process

The basic assumptions

The duplication model of a cell is given in Figure 12

$$\text{DNA} \xrightarrow{\text{RNA}} \text{ribosome} \longrightarrow \text{proteins}$$

Fig. 12 Faulty transmission in cell duplication

The information of the DNA is conveyed to the ribosome via the RNA, and the ribosome decodes this information to produce proteins. The following is assumed:

(H1) The first source of error in the production of RNA is that the RNA molecules would not be exact replicas of the DNA molecules, but would involve some errors with respect to the latter.

(H2) The second source of error in the production of proteins is that the ribosome does not produce the proteins defined by the RNA, but rather the ribosome reads the RNA, making additional mistakes, and, as a result produces wrong proteins.

(H3) The parameter u of the relativistic model is directly dependent upon the cytoplasmic medium.

With these assumptions, the models A and B apply directly, and simulate life and death as follows. Consider Model B (given that the reasoning is the same for both A and B) and assume that the information $H(\alpha)$ of S is used to encode a message whose entropy is $H(\beta)$, while the task of each observer R_n is to decode this message.

I. Assume that the encoding process is meaningful when $u = 0$. Then have

$$H(\alpha/R_n/R_{n+1}) = H(\alpha)$$

and the encoding feasibility condition is

$$H(\alpha) \geq H(\beta). \tag{25}$$

II. Assume now that $v < 0$. The encoding feasibility condition at the level of R_n is

$$H(\alpha/R_{n+1}/R_n) \geq H(\beta) \tag{26}$$

that is to say, by virtue of (24)

$$k^n(1, v) \geq H(\beta)/H(\alpha). \tag{27}$$

Let N be defined by the equation

$$N \ln k = \ln [H(\beta)/H(\alpha)] \tag{28}$$

then the encoding process works for $n \leq N$, while it does not apply when $n > N$ (Fig. 13).

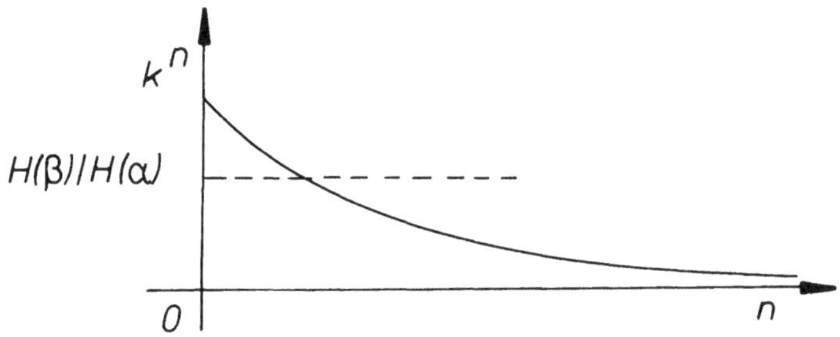

Fig. 13 Encoding limit in cascade transfer.

In the duplication process of biological cells, the age t of the process is $t = T_n$, where T denotes the time required for a cell to duplicate. At $t = NT$, the duplication process stops.

A Numerical example

Consider 4 bases and 20 amino-acids; in such a case

$$H(\alpha) = 3 \ln 4$$

and

$$H(\beta) = \ln 20 .$$

Moreover, for small v,

$$\ln k(1,v) \cong v$$

and (28) yields approximately

$$N \cong \frac{1}{v} \frac{H(\beta)}{H(\alpha)} .$$

Assume that $v = -2.10^{-6}$; then $N = 0.175.10^{6}$, and if $T \cong 5$ hours, the duration of the duplication process is $5(0.175.10^{6})$ hours, in other words the process ceases after 99 years.

Relativistic Information and Thermodynamic Entropy

Shannon defined the entropy of the random experiment $\alpha := \{(A_1, A_2, \ldots, A_n), p(A_i) := p_i \text{ for every } i\}$ by the expression

$$H_c(\alpha) = -\sum_{i=1}^{n} p_i \ln p_i$$

Later, the entropy has been mathematically derived in the form

$$H(\alpha) = Q H_c(\alpha)$$

where Q denotes an integration constant. Boltzmann introduced the thermodynamic entropy $S(\alpha)$ defined by the expression

$$S(\alpha) := k H(\alpha)$$

where k is the so-called 'Boltzmann-constant'.

Is the similarity between $H(\alpha)$ and $S(\alpha)$ merely formal and superficial, or is there a significant relationship between them?

If this analogy is not a pure coincidence, is $H(\alpha)$ completely identical to $S(\alpha)$, in which case k would be merely a scaling factor?

It appears that answers to these questions have not yet been found.

Assumption. The new preposition in an attempt to solve these problems, is to consider $S(\alpha)$ in the form

$$S(\alpha) = k(1, u)\, H_c(\alpha)$$

where u is a parameter which would characterise the medium under consideration. k(1, u) would not have the same value for a thermodynamic medium and a biological one, and it would be related to the meaning of the information relative to the medium.

With this assumption, consider the duplication process of cells, and consider the cell, C_n, at the n-th stage of duplication. Its thermodynamic entropy would be

$$S(C_n) = k^n(1, u)\, H(\alpha)$$

and

$$S(C_n) \downarrow 0 \quad \text{as } n \uparrow \infty .$$

This appears to contradict the Second Law of Thermodynamics. In fact, this inconsistency is only apparent, because the n-th cell in question is not a closed system. The closed system to be considered at the n-th stage of duplication is composed of cells C_1, C_2, ..., C_n, and the corresponding thermodynamic entropy is

$$S(C_1 + C_2 + \ldots + C_n) = H_c(\alpha) \sum_{i=0}^{n} k^i = \frac{1 - k^{n+1}}{1 - k} H_c(\alpha)$$

and where $k < 1$,

$$S(C_1 + \ldots + C_n) \uparrow \infty \quad \text{as } n \uparrow \infty$$

so that both these evolutionary processes are consistent with the Second Law of Thermodynamics.

Conclusions

One of the main criticisms, and probably the only one, against the utilisation of standard information theory in biology is that it will yield only in conclusive results, since it does not deal with semantics, that is to say the meaning of the information with respect to the medium under consideration. So, the main purpose of relativistic information is to explicitly introduce syntax and semantics into the formulation. It was then quite appropriate to examine what kind of results it could provide when applied to biology, and it appears that it is also quite meaningful in biology and it leads to conclusions which are quite consistent with the well known features of living organisms. Furthermore, and this is one of the objectives of any theoretical modelling, relativistic information suggests new assumptions about the physical nature of living organisms.

In view of these results, the task now is to identify clearly the origin, and the meaning of, the parameter u in a biological framework.

Thermodynamics of Information Processes

Entropy and Information
The Supernatural Actions of Maxwell's Demon

I. Lamprecht

'Herr, die Not ist groß!
Die ich rief, die Geister,
Werd ich nun nicht los.'

(Goethe, Der Zauberlehrling)

Introduction

Since the dawn of mankind human life has been accompanied by gods, angels and demons and all the creatures between them. They governed life and death, determined the present and future and distributed good and bad luck among men. Especially demons - entities between gods and men - played an essential role in many or nearly all natural philosophies. Though originally not necessarily of bad intention they later developed into creatures which were responsible for the disadvantages and misfortune that affected the life of the community or the individual. Superhuman powers were essential to enable them to perform things impossible for the mortal contemporaries.

With growth of natural sciences and the dawn of enlightenment more and more demons left our daily life making it more predictable and less colourful. So it is amusing that science has brought back a demon, who now - for more than 100 years - has haunted physicists and philosophers. 50 years ago it seemed that Szilard (1925, 1929) exorcised him, and 20 years later Brillouin (1951) showed which tricks he was working on, but it became obvious in recent years that he is still alive, and at work in many of the natural sciences, such as biology, but that he was also successfully

Thermodynamics and Regulation of Biological and Processes

conquered new fields like masers or reactor techniques: Maxwell's demon, though once exorcised seems to be immortal.

Demons developed from good to bad, and the question arises as to which category Maxwell's demon fits into. His creator never talked about a 'demon', he just wanted a 'finite being' and asked Tait (see e. g. Heimann, 1970) to 'call him no more a demon but a valve', because he felt the negative and too anthropomorphic implications. Although he haunted sciences for more than a century, he was perhaps the most stimulating creature of human imagination, the most active homunculus ever produced, a chimera from which science derived considerable progress in thermodynamics, information theory and philosophy.

The last century confronted men not only with an exact definition of energy, the First Law of Thermodynamics and the impossibility of a perpetuum mobile of the first kind, but also confronted them with an inevitable dictum of ever increasing entropy and the heat death of the universe. Often man experiences the unidirectional nature of life, of catastrophic irreversibility; many poems and proverbs from all over the world express this feeling (Angrist, 1967, 1967). On the other hand living systems seem to be the only systems which could oppose the Second Law of Thermodynamics; could be a 'deceiver' in the play of entropy. Maxwell's demon with his 'so sharpened faculties' performed the work men wanted to do: to counteract the increase of entropy and to avoid final death by complete disorder or thermal equilibrium. In this sense the demon was an image of god (Daub, 1970), a superman, 'the first man-made creature (that) knowingly could obviate the results prescribed by the second law, ...' (Huneeus, 1976). He corresponded to the wish of mortal man, to find - at least in microscopic regions - an entity, that can escape the dictum of entropy. He could perform the actions men desired to perform, 'only we can't, not being clever enough' (Maxwell).

And Loschmidt expressed the hope:

> 'Thereby the terrifying nimbus of the second law, by which it was made to appear as a principle annihilating the total life of the universe, would also be destroyed; and mankind

could take comfort in the disclosure that humanity was not solely dependent upon coal or the sun in order to transform heat into work, but would have in inexhaustible supply of transformable heat at hand in all ages.'
(cited after Daub, 1970)

Maxwell's Demon - His Birth and Faculties

The First Law of Thermodynamics was easily accepted by the scientific world - the sad fate of the would-be inventors of the perpetuum mobile was the strongest proof of its truth (M. Planck). It was different with the Second Law - there seemed to be two domains which could shirk its negative prognosis of ever-increasing entropy and disorder: animate matter and microscopic systems. Lord Kelvin (W. Thomson) expressis verbis excluded living organisms in his statement of the Second Law; and many scientists followed this lead. In an article about 'Gestalt-Theory' (Huneeus (1976) writes:

> 'Of the many problems posed by biological entities to the physical sciences, their obvious differentiation within the growth span of the organism, is a flagrant violation of the second law and hence, they, as a whole, escape the realm of thermodynamics. Only living organisms can go against the second law.'

On the other hand it had become clear that the Second Law was a statistical law, and Maxwell wanted to point out the differences between individual molecules and the immense number of particles in a macroscopic vessel. He conjured up his 'neat-fingered' being as a way of giving life to the difficult idea that the Second Law of Thermodynamics is a statistical truth and not a dynamic certainty (Klein, 1970).

Maxwell's thought-child was born in 1867 in a letter to Tait, in a discussion of a system of two vessels A and B:

> 'Now conceive a finite being who knows the paths and velocities of all the molecules by simple inspection but who can

> do no work except open and close a hole in the diaphragm by means of a slide without mass.
>
> Let him first observe the molecules in A and when he sees one coming the square of whose velocity is less then the mean square velocity of the molecules in B let him open the hole and let it go into B. Next let him watch for a molecule of B, the square of whose velocity is greater than the mean square velocity in A, and when it comes to the hole let him draw the slide and let it go into A, keeping the slide shut for all other molecules'.
> (cited after Heimann, 1970)

He continued to describe him as a 'very observant and neat-fingered being', a 'pointman for flying molecules', a 'doorkeeper ... exceedingly quick'. He discussed this quaint creature in further letters to Tait, Thomson and Rayleigh and wrote about him in his Theory of Heat to represent a 'being whose faculties are so sharpened that he can follow every molecule in its course'.

It was William Thomson who baptised this tiny being as 'intelligent' or 'the sorting demon of Maxwell'. Thomson used the word 'demon' in the sense of superman without ascribing any evil intentions to the demon. In his paper 'The Sorting Demon of Maxwell' he wrote:

> 'He is a being with no preternatural qualities, and differs from real living animals only in extreme smallness and agility. He can at pleasure stop, or strike, or push, or pull away any atom of matter, and so moderate its natural course of motion. Endowed equally with arms and hands - two hands and ten fingers suffice - he can do as much for atoms as a pianoforte player can do for the keys of the piano - just a little more, he can push or pull each atom in any direction'.
> (cited after Daub, 1970).

Thomsons's definition of the demon gives him the following faculties:

- animation
- atomic dimensions
- intelligence

and the question was posed are these three conditions necessary and sufficient for him to work against the Second Law of Thermodynamics (Ehrenberg, 1967)? Maxwell was not content with Thomson's choice of the name 'demon' and suggested to Tait that he should 'call him no more a demon but a valve'. In spite of this objection the demon kept his anthropomorphic shape, and 45 years after his birth the 'pointsman for flying molecules' was transformed to *Homo sapiens* by von Smoluchowski in his lecture about fluctuation phenomena, Brownian motion and the validity of the Second Law of Thermodynamics. Von Smoluchowski (1912) showed that violations of nearly all the usual statements of the Second Law could be found if sufficiently small systems are investigated.

Fig. 14 Action of Maxwell's demon (explained in the text).

What is done by the demon to a system of two vessels A and B which are connected by a hole closed by a slide, a shutter or a trap-door? The classical experiment is depicted in Figure 14 a and 14 b. Starting with an even distribution of molecules ('hot' and 'cold' or 'rectangular/white' and 'round/black') in both compartments, the sorting demon keeps the cold white particles on the left and opens the trap-door only when a hot black one approaches from left to right. After a while the separation is perfect, the left hand side becomes cooler, the right hand side hotter, and the demon can look with pride on the battle won against the Second Law. The border between cold and hot could only be crossed by the hot molecules, and this is why Poincarè compared the demon with a custom-officer.

Of course, modifications to this scheme have been proposed. The simplest device was to create a 'pressure demon' instead of Maxwell's 'temperature demon'. At the end of the action a pressure difference between vessel A and B exists which may be used to drive a piston and to gain useful work from the system. In this case the demon could easily be replaced by a spring door which only opens on one side and remains closed to the other.

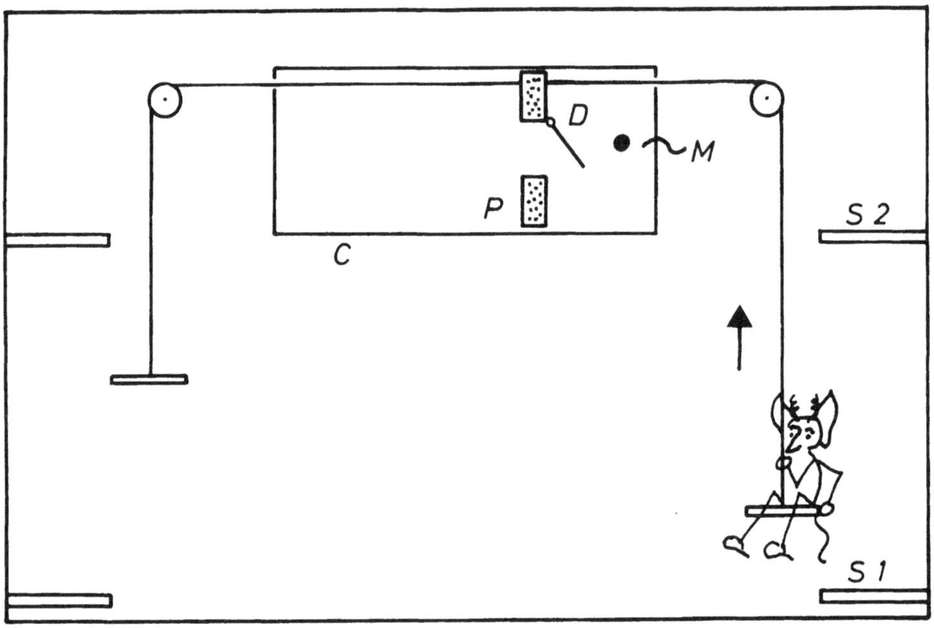

Fig. 15 Pearce's device acting with the help of Maxwell's demon (explained in the text).

Pierce (1961) proposed the different arrangements shown in Fig. 15 - a frictionless imaginary machine - which would certainly not work in a practical sense but which helps to understand the thought-experiment. Just one molecule, M, is enclosed, in a cylinder, C, by a piston, P, and fitted with a door, D. As the whole system is at temperature T, the molecule has an average energy proportional to T. As the piston operates without friction, it needs no energy to shift it around in the cylinder, to the left, or the right and to center it with the door closed. The thought-experiment requires that the position of the molecule be determined and then that a little weight (or even a baby demon of real mass) be placed on the pan and the molecule allowed to push the piston, by one, or several impacts, to the other side. Simultaneously the weight (or the demon) is raised and one receives organised work by means of disorganised heat energy.

If the cylinder has a half length of L, the mean force, F, of the molecule on the piston can be calculated as

$$F = kT / L . \qquad (1)$$

The maximum work W is gained when the distance L is doubled, so that W amounts to

$$W = \ln 2 \cdot kT = 0.693\, kT. \qquad (2)$$

Other imaginary machines were proposed of the ratched and pawl type (see e. g. Feynman et al., 1963), with spring loaded doors, with ciliates or with valves, just as Maxwell proposed to his friend Tait. All these devices had one aim: to pick a hole in the Second Law. A patent was even given to a laboratory instrument which was said to demonstrate what Maxwell's demon was supposed to do (Sussman, 1966).

Henry Adams, American historian and philosopher, who tried to find an analogy for the Second Law of Thermodynamics for his own theory of history, was so fascinated by the tiny intelligence that he wrote in an amusing letter to his brother Brooks (May 2, 1903): 'You and your brother Charles will be both be quite sure that Clerk Maxwell's demon who runs the Second Law of Thermodynamics ought to be made President. ... Your You ... will think me "very silly", but I have passed my life among the snorting demons, and the little cusses amuse me' (Cater, 1947).

Maxwell's demon - a life for sciences

William Thomson, the godfather of Maxwell's thought-child, played a double role in the future life of the demon. By his formulation of the Second Law of Thermodynamics

> 'It is impossible by means of inanimate material agency to derive mechanical effect from any portion of matter by cooling it below the temperature of the coldest of the surrounding objects.'

in his 'Theory of Heat' (1853) he stimulated speculation that animate nature could escape the dictum of entropy. Helmholtz supposed that in the case of single molecules the principle of entropy increase could fail, e. g. in the case of life processes which are based on finest biological structures, and that one might finally find a graduated criterion for vital effects: the invalidity of the entropy principle. Lewis, in his 'The Anatomy of Science' (1926) wrote that living organisms are

> '... cheats in the game of entropy', which 'alone seem able to breast the great stream of apparently irreversible processes. These processes tear down, living things build up. While the rest of the world seems to move towards a dead level of uniformity, the living organism is evolving new substances and more and more intricate forms'.
> (cited after Butler, 1946)

Auerbach stated in his book 'Die Weltherrin und ihr Schatten' (The queen of the world and her shadow, 1902), talking about the 'queen' energy and her 'shadow' entropy that animate material was created to bridle the entropic activities or even to overcome them. The energy of living entities should have a really 'ektropic' efficiency, a term which corresponds to the later introduced concept of negentropy.

Besides animate systems those containing a small number of particles were supposed to offer a chance of circumventing the Second Law. Starting from Glasstone's comment that

> '... the second law of thermodynamics is a statistical law which can only be applied to a system consisting of a large number of particles, as for example, an assemblage of molecules. If it were possible to work with systems of single, or a few, molecules the law might well fail.'

McClare (1971) states that in living organisms single molecules can be highly important and such small microbes as *Mycoplasma laidlawii* with a diameter of $\sim$200 nm contain only a few ribosomes and single molecules of the rarer enzymes and coenzymes. Nevertheless, in no case has there been any reason to believe in a violation or circumvention of the Second Law.

With the growth of cellular, and especially of molecular, biology membranes penetrated more and more into the domain of Maxwell's demon (see e. g. Patlak, 1956, 1957; Rashevsky, 1960). Their permeability to one type of molecule and their impermeability to another seemed to ordain them as a field for his activities. Hill wrote in 1924 - but full of doubts:

> 'A one-sided, selective permeability, maintained by an active, "purposeful", but molecular mechanism in the cell, something like Maxwell's demon with his trapdoor, might enable the cell to evade the statistical rules which govern larger systems. Philosophically, perhaps even practically, the discovery of such a mechanism, if ever made, will be the greatest step taken by science: if life and matter are co-equal and co-eternal, as some believe, it may be that life is maintaining the entropy at a reasonable constant level ...'

He continues that there is no evidence that the ordinary laws of physics and chemistry - including those of thermodynamics - do not apply to the living cell and animal and that both, the First and the Second Law, hold true in all experiments. And then he adds:

> 'The First Law of Thermodynamics - that of the Conservation of Energy - has been shown, directly and indirectly, to apply to living cells and bodies, with an accuracy equal to that of any experiments which may be made to test it. The Second Law, that which deals with the availability of energy, is more difficult to test, but there is no evidence of any value that it also does not hold.'

N. Wiener (1961), father of cybernetics, considers the existence of metastable states of matter and that they might correspond to some sort of biologically active demons:

> 'There is no reason to suppose that metastable demons do not in fact exist; indeed, it may well be that enzymes are metastable Maxwell demons, decreasing entropy, perhaps not by the separation between fast and slow particles but by some other equivalent process. We may well regard living organisms, such as Man himself, in this light. Certainly the enzyme and the living organism are alike metastable: the stable state of an enzyme is to be deconditioned, and the stable state of a living organism is to be dead. All catalysts are ultimately poisoned: they change rates of reaction but not true equilibrium. Nevertheless, catalysts and Man alike have sufficiently definite states of metastability to deserve the recognition of these states as relatively permanent conditions.'

R. B. Dean (1941) reflects about the sodium and potassium concentrations inside and outside of muscle fibers, being led to the conclusion that some sort of a pump, located possibly in the fiber membranes must be active in pumping sodium out and potassium in. He had no idea of the underlying mechanisms but clearly saw the thermodynamic consequences:

> 'A differentially permeable membrane will produce a concentration difference at the expense of the kinetic energy of the diffusing molecules. This violates the second law of thermodynamics and is in effect a Maxwellian demon.'

and concluded:

> '... we are not justified in assuming that it (the second law) is violated in biology without very convincing evidence. It is

> safer to assume that there is a pump of unknown mechanism which is doing work at a constant rate excreting the sodium as fast as it diffuses into the fiber.'

It was left to later experiments to establish the existence of energy driven pumps in membranes, but to yet later experiments to find intercellular membrane channels with asymmetric permeabilities which would support the modern reappearance of the demon (Flagg-Newton, Loewenstein, 1980).

Nearly simultaneously with Maxwell's views Loschmidt (1869) raised the question of constructing a mechanical apparatus, which provokes, without employment of work, a spatial separation of spheres in two compartments of volume V, so that in one compartment accumulate all the spheres whose mean velocity is larger than the mean velocity of the whole, the rest remains in the second compartment. And he continues:

> 'Obviously it is quite possible, and in the more general case too, that where the spheres are assembled into larger groups, molecules, that a law other than that of elastic spheres governs the collisions. ... It is, therefore, theoretically not impossible to transform the gas from a lower temperature to a higher one, or, to increase its density without employment of work or other compensation.'

Loschmidt did not use a demon for his separation of spheres, but he expressed the idea in the same way as Maxwell

> '... an opening in the place of σ in the separating wall, which opens and closes at the calculated moments.'

and established by this a violation of the Second Law.

Other leading scientists of that time followed such trains of thought and accepted the importance of the entropy principle but supposed that violations might exist in the domain of the small or the large. Svante Arrhenius, whilst admitting the universality of the Second Law, thought that rare exceptions might occur leading to the rebuilding of worlds - perhaps by some conscious Maxwellian demon. And M. G. Bruhat came to the conclusion that 'the second law loses its value when phenomena are investigated whose dimensions are near to the molecular dimension' (see Demers, 1944).

Demers (1944) analysed thought-experiments such as, how to draw work from a single heat reservoir by means of fluctuations. He arrived at the conclusion that the useful energy obtained is always smaller than the energy originally supplied and presented some new formulations for the Second Law of Thermodynamics:

- A system with only one temperature can produce no work.
- It is unlikely that a system with only one temperature can produce useful work.
- A system with only one temperature cannot produce useful work.

In a following paper Demers (1945) proved the validity of the Second Law as a consequence of quantum theory when processes like the activities of Maxwell's demon are investigated.

Bridgman (1953) gave his attention to these fluctuation phenomena and their connection with the First and Second Law, and posed the question 'Is it possible, by ingeneous enough construction, to violate the second law of thermodynamics on an industrially profitable scale?' Bearing in mind the many doubts Maxwell's demon have provoked, the denial of such a prospect cannot be based on 'the completely random character of the small-scale fluctuations, for it is possible on a larger scale to extract useful energy from the completely random motion of the waves of the sea.' His argument continues in the field of statistical mechanics and probabilities. 25 years earlier Bridgman (1928) came to similar conclusions in a quite different way. He showed that the 'principle of detailed balancing' (known too as: law of entire equilibrium, principle of microscopic reversibility or the hypothesis of the unit mechanism) is a consequence of the Second Law. It demands that there be no net changes in time in any part of the system, if it is in equilibrium. For unidirectional processes in the system would enable us to introduce a small energy extracting instrument, use such processes to render energy, remove the instrument and allow the system to return to equilibrium again. This of course would be a violation of the Second Law.*

---

*'An interesting converse use of the principle is as a criterion of the physical reality of processes invented for the purposes of analysis' (Bridgman, 1928)

Following the papers of von Smoluchowski (1912) and Szilard (1929) Brillouin made clear how Maxwell's demon deals with information to decrease the entropy in the system. He established close connections with Shannon's discussion of telecommunication, which can be stated in the following progressive form:

information $\rightarrow$ telegram $\rightarrow$ negentropy on the cable $\rightarrow$
reception of telegram $\rightarrow$ reception of information.

Negentropy is not a term used by Shannon; he compares information with positive entropy, whilst Wiener makes it parallel to negentropy. As the exorcism of the demon is very much associated with the transformation of information into negentropy, this aspect is treated separately below.

Laing (1974) has developed a quite different aspect of demonology. He opposes a 'computing automaton' against the classical, the normal demon (the 'demon proper', as he calls him) and shows that such a generalised Maxwellian demon system can carry out Turing computations. His computing demon is composed of two finited capacity counters, a register chamber and a reservoir chamber, storing positive integers or zero and being increased or decreased one unit at a time. The demon proper acts as a finite state control unit and changes the counter of the 'fast' molecules to a maximum value, and the 'slow' molecule counter to zero. When this final state is reached the demon proper signalises his successful action to an outside observer. This basic Maxwellian demon system was generalised by Laing in further steps to more complicated devices incorporating more, and infinitely large chambers, and numbers of particles, enlarged memory of the demon and an increased observational scope for the demon. Such systems can then simulate the computation of an arbitrary Turing machine.

To complete this chapter mention should be made of a demonstration device which was developed by Sussman (1966; 1974; see also Stong, 1973): his 'Maxwell Demon Bottle' which allows the demonstration of the usual statistical behaviour, and action, of a Maxwell demon. A sealed bottle with an extended neck contains two sorts of spheres which will roll via random distribution into the neck in a columnar array when the bottle is inverted. When suitable spheres are used and the bottle swirled with a circular, wrist motion one type of sphere will drop into the bottle neck first because it loses momentum more rapidly. In this case the 'demon' produces a real separation of the two sorts of spheres.

Maxwell's demon - His friends and relatives

Although perhaps the most popular, Maxwell's demon was not the only thought-child that haunted science and philosophy. Even he had a double existence or - perhaps better - had a twin-brother. Maxwell's neat-fingered being sorted the molecules in a system in such a way that one compartment became hotter and the other cooler, thus creating a temperature difference which could be used for useful work. Several authors have pointed out (see e. g. Pierce, 1961; Ehrenberg, 1967; Poplavskii, 1979, 1981) that it would be easier for the demon to produce a pressure difference between the two vessels of the system. When Maxwell proposed to call him (the temperature demon) no more a demon but a 'valve' this twin-brother should not be named a pressure demon but more a 'spring door'.

A hundred years earlier Laplace invented another famous demon who was supposed to know exactly the positions and velocities of all the particles in the universe. The mechanical interpretation of the world in that period meant that this demon could predict the course of the world for all eternity. The next chapter will prove that the fate of Laplace's demon is the same as that of Maxwell's tiny door-keeper - elimination from the notebooks of physics.

As pointed out above, Loschmidt (1869) in his wish to pick a hole in the Second Law discussed a thought-experiment very similar to that of Maxwell. Although he did not talk about a demon or an intelligent being he was made father of the 'Loschmidt demon' who was as active as his English cousin and likewise returned to modern physics. This 'non-demon', as Daub (1970) calls him, had to change his profession later. He should be able to reverse the dynamic behaviour of a system and to retrace a seemingly irreversible path. The 'Loschmidt demon differs from the familiar Maxwell demon in that he accomplishes the reversal by changing the sign of the system Hamiltonian rather than by exercising control over microscopic variables' (Rhim et al., 1971). One could perhaps say that Maxwell's demon is the mechanic of both and Loschmidt's the mathematician.

Another near relative of Maxwell's demon, the 'crossroads policeman', was invented by Shotter (1966). His father sketches briefly 'the nature of a mechanical procedure whereby the presence of an elementary class of Maxwell demons can be detected and machines constructed automatically to mimic their behaviour'. The problem posed by Shotter is the following. Two main one-way roads cross at a special point with no allowance for left or right turns and ideal traffic flows, without any crashes, overtaking and spontaneous creation of cars. The question posed was 'Is it possible by mere observation of the cars entering and leaving the crossroad to decide whether or not a policeman directs the traffic at that point?' For this end four observing 'clerks' are placed at the four arms of the crossroad documenting their observations. If it is possible to draw from these documents the rules by which the policeman controls the cars then the desired machine could be constructed which could mimic the behaviour of the policeman-demon.

Sir Arthur Eddington (1955) troubled a geographical type of demon to explain the differences between real curved geography and the flat world of a (Mercator) map to simple minded travellers. In his example Greenland is horribly distorted when Central Europe is correct, making all real distances in Greenland shorter than those apparent from the map: a positive demon is active to 'help travellers on their way', while a poor Greenlander would suggest a negatively acting demon in Greece. At the end the experience could be that 'the demon is never, where you are; it is always the other fellow who is haunted by him'.

A 'pandemonium' followed from the model constructed by Selfridge (1959) of a process which should be able to 'adaptively improve itself to handle certain pattern recognition problems which cannot be adequately specified in advance'. At increasingly more complex levels different types of 'demons' are active. In the last level in a machine for recognition of Morse code these demons are, from bottom level to top level: data or image demons; computational demons together with some subdemons; cognitive demons; and finally the decision making demon himself, ruling this pandemonium.

This ends the 'demonic' survey! Of course, there are more demons or thought-children of demonic origin active in science and philosophy, but none as important and stimulating as Maxwell's - and perhaps Loschmidt's - creatures.

Maxwell's demon - His exorcism and death

'One cannot get something for nothing, not even an observation' and 'We cannot get something for nothing, not even an observation, far less a law of nature'. These two sentences of Gabor (1961) will be the tread running through the chapter on the exorcism of Maxwell's demon and his colleagues.

Of course, it would be much easier to deny the existence of Maxwellian demons in nature and thus have no cause to deal with them. Following Chambadal (1963), one could reject Maxwell's trap-door in the same way as the light ether, as such a door must be completely impermeable, without mass, friction and heat capacity, a hybrid chimera, material and non-material. But - as many demonologists have pointed out - one would miss interesting insights into the character of the Second Law of Thermodynamics and its application to physical, chemical and biological systems.

The exorcism started in 1912 with a lecture of M. von Smoluchowski on 'experimentally provable, the usual thermodynamics contradicting molecular phenomena'. The author pointed out that there really exist phenomena showing deviations from ideal thermodynamic equilibrium and that it would be not necessary to engage a demon to perform a separation of particles. Every suitable valve, perhaps an opening protected by small elastic hairs or eyelashes could do this job. But at the same time he expressed doubts: if statistical mechanics is correct, the construction of a microscopically acting rachet is impossible. Such a ratchet must exhibit fluctuations around a position of minimum energy. Brownian motion of the trap door would result in random opening and closing of the door. This would make a long range operation of an automatic device like a spring valve, a ratchet or a pawl ineffective. A *perpetuum mobile* could only be produced with a completely new type of valve, which are excluded not, by technical reasons but from principles. In the discussion following the lecture, von Smoluchowski conceded that an intelligent being looking through the physical laws could create processes which could contradict the Second Law of Thermodynamics. But the intelligence would be from outside the framework of physics, and, possibly even here, compensations would be necessary.

Szilard (1925, 1929) took up Smoluchowski's idea of the action of intelligent beings and pointed out that a *perpetuum mobile* would operate only if the experimentalist were a kind of 'deus ex machina' and operated without expenditure of energy. But there is always coupling between different parameters of the system, and Szilard was able to show that inanimate machines may achieve essentially the same results as intelligent beings. He concluded that 'in this example we followed the 'life processes' of an inanimate machine and were able to see that they produce exactly that amount of entropy which is demanded by thermodynamics'.

From this first step of coupling entropy decrease to information by Szilard, Brillouin (1951, 1959, 1963) continued to the final exorcism of the demon. At Maxwell's time little was known about black body radiation so that it was not included in the demon's action scheme. Brillouin (1951) showed that the complete cycle is

$$\text{negentropy} \longrightarrow \text{information} \longrightarrow \text{negentropy}$$

and that the demon was just a transforming agent between information and negentropy.

Negentropy is an interesting quantity introduced by Schroedinger (1944) in his book 'What is life?' as the statement that organisms are feeding on negative entropy or 'sucking up' order from their environment. Brillouin (1951) shorted it to 'negentropy'. It corresponds to 'ektropy' as mentioned above (Auerbach, 1902) or to the 'grade' of energy in the discussions of the 'degradation of energy' by Lord Kelvin.

Through these investigations it became clear that 'it is a fundamental and inescapable fact that an observation always create a perturbation. ... An observation necessarily requires a certain temporary coupling between the system and the measuring instrument; this coupling results in a perturbation of the observed system' (Brillouin, 1963). Thus an observation creates a certain amount of information, but at the same time inevitably increases the entropy of the measuring device. Or to turn it the other way round: The information obtained is paid for in negentropy.

After these more general statements take a closer look at the actions of the demon, following Brillouin's (1963) arguments. In doing this one should bear in mind that in his closed system of uniform temperature T_o, the demon could see nothing unless he imported a torch into his system.* The filament of the torch is heated up to a temperature $T_1 > T_o$ so that visible quanta of an energy

$$h\nu_1 \gg kT_o \tag{3}$$

are emitted; where h is Planck's constant, k Boltzmann's constant and ν the frequency of the quanta. The battery yields a total energy, E, during the experiment, and the emission of light is associated with entropy production, S_ν, where

$$S_\nu = E/T_1\ . \tag{4}$$

When this energy is absorbed by the gas at the lower temperature, T_o, a positive entropy increase of

$$S = E/T_o \tag{5}$$

appears in the system. To distinguish the molecules, the demon must at

---

* As Chambadal (1963) comments it would be easier for the demon to use the electrical energy of the torch to increase the temperature of one compartment.

least absorb one quantum scattered by a molecule. As his eyes are at the ambient temperature T_0, this absorption increases the entropy by

$$S_d = h\nu_1/T_0 = k\,h\nu_1/kT_0 = kb\,. \qquad (6)$$

Because of (3) the ratio $h\nu_1/kT_0 \gg 1$. The entropy of the closed system prior to the action of the demon can be written as

$$S_0 = k \ln P_0 \qquad (7)$$

where the total number of configuration is P_0. Because of the information which the demon received the number of possible configurations is decreased by a small amount $p \ll P_0$. Thus, the entropy of the system decreases too by an amount ΔS_i

$$\begin{aligned} \Delta S_i &= S - S_0 = k\{\ln(P_0 - p) - \ln P_0\} \\ &= k \ln\{(P_0 - p)/P_0\} = k \ln\left(\frac{p}{P_0} - 1\right). \end{aligned} \qquad (8)$$

Expanding as a Taylor series and taking only the first term, because $p/P_0 \ll 1$, yields

$$\Delta S_i = -kp/P_0\,. \qquad (9)$$

In total, therefore, the entropy of the system changes by

$$\Delta S_d + \Delta S_i = kb - kp/P_0 = k\,(b - p/P_0) > 0, \qquad (10)$$

since $b = h\nu_1/kT_0 \gg 1$ and $p/P_0 \ll 1$. Equation (10) proves that the sorting demon acts in full agreement with the Second Law of Thermodynamics and that the demon could only reduce the entropy increase by a small amount.

This first complete exorcism of the demon was intensively discussed and examined from different standpoints. Poplavsky (1975) wrote:

> 'It was shown that the important thing in Maxwell's paradox is not that the 'demon' is a thinking being (he might be called a robot), but that information is used for purposeful control, leading to a decrease of entropy.'

Angrist and Hepler (1967 a, b) investigating entropy and information in different fields of human life as language, poetry or sciences state that a poet is transmitting a coded message to his reader and that he, in breaking the code, has to pay a price for the information. They continue:

> 'The observer, whether he be natural philosopher, demon, or poetry reader, requires sources of high-quality energy to pay for the information he receives.'

Looking just at the scientific aspects of the problem they find that only measurements can create information, but that no measurements are possible in closed systems. Thus they reach a rigorous form of the Second Law of Thermodynamics relevant to information:

> 'Our information about an isolated system can never increase. Reversible processes conserve, irreversible ones lose information.'

These statements are close in sense to a comment by Einstein that, in an equilibrium system, no intellectual mechanism seems to be able to work.

Returning to the demonic device of Pierce (1961) illustrated in Fig. 15, an exorcism of the demon could pursue the following path: When the piston is placed in the center and the door locked it is necessary to know on which side to put the weight. One needs one bit of information to decide in which half of the cylinder is the molecule enclosed. The minimum energy to transmit 1 bit of information at a temperature T is 0.693 kT Joule which is just equal to the maximum work obtained in the device:

> 'Thus, we use up all the output of the machine in transmitting enough information to make the machine run.'*

The informational approach was not the only one directed towards the exorcism of the demon. This tiny intelligence has to know the very position and velocity of the considered molecule in order to open the trapdoor at the appropriate time. Position and velocity - or its product with mass, its momentum - are a pair of complementary quantities whose exact determination is limited by Heisenberg's uncertainty principle. Thus, the demon violates this fundamental law of modern physics which was, of course, un-

*It can be shown in a simple case that the limitation imposed by thermal noise on the transmission of information results in the failure of machine designed to convert the chaotic energy of heat into the organised energy of a lifted weight (Pierce, 1961).

known to Maxwell.* Following a calculation by Demers (1944) the kinetic energy of a molecule with mass m and a mean thermal velocity, V, at temperature, T, is

$$mV^2/2 = 3/2\, kT \tag{11}$$

and its momentum, p,

$$p = mV . \tag{12}$$

With a molecular weight P where

$$P = m . N \tag{13}$$

(N being Avogadro's number) and the well-known relation

$$R = k . N \tag{14}$$

(R the gas constant), p can be shown to be

$$p = mV = (3\, R . T . P)^{1/2} / N \tag{15}$$

and with R = $8.31 \cdot 10^7$ erg deg $^{-1}$mol^{-1}, N = $6.02 \cdot 10^{23}$ mol^{-1}

$$p = 2.61 . 10^{-20}\, (T . P)^{1/2} . \tag{16}$$

Heisenberg's uncertainty principle demands that the product of the uncertainties in the measurement of the position, Δ x, and the momentum, Δ p, obeys the inequality

$$\Delta x . \Delta p \geq h / 2\pi \tag{17}$$

where h = $6.63 \cdot 10^{-27}$ erg.s is Planck's constant. Combining (12) and (17) one arrives at

$$\begin{aligned} \Delta p / p &\geq h / (2\pi m V \Delta x) \\ &\geq 0.37 . 10^{-7}\, (T . P)^{-1/2} / \Delta x \end{aligned} \tag{18}$$

---

*It was shown that no violation takes place when one deals with heavy molecules or low pressures (Demers, 1944; Brillouin, 1963).

and with usual experimental conditions for water molecules (P = 18) at T = 273 °K

$$\Delta p/p \geq 5.27 \cdot 10^{-10}/\Delta x . \qquad (19)$$

Under normal pressure one molecule occupies a volume of 22,400/6.06.10^{23} 22,400/6.06 · 10^{23} cm^3 corresponding to a cube of 3 · 7 · 10^{-20}cm^3 and with sides of 3.33 nm. If the position should were to be known to, at least, a tenth of such an elementary cube, the value of (19) amounts to Δ p/p $\geq$ 1.58 · 10^{-2}, a measurement that only could be performed with 'light' of a corresponding wavelength, i. e. with X-rays. When the demon uses visible light with a wavelength of 500 nm, the uncertainty Δ x of the molecule's position cannot be better than this value, so that (19) becomes

$$\Delta p/p \geq 10^{-5} .$$

But in a cube with sides of 500 nm there are, under normal pressure,

$$n = \frac{(5 \cdot 10^{-5})^3}{22 \cdot 4 \cdot 10^3} \cdot 6 \cdot 06 \cdot 10^{23} = 3 \cdot 38 \cdot 10^6$$

molecules so that no definition is possible. To make it possible the pressure has to be reduced to less than $1/3.38 \cdot 10^6\ p_0 \approx 3 \cdot 10^{-4}$ bar which corresponds to a moderately high vacuum.

Before leaving this chapter let us assess the results and the comments on the different approaches to exorcism. Chambadal (1963) pointed out that Maxwell would have supplied his thought-child with a sixth sense if he had known, or foreseen, Planck's theory of black body radiation. For it was essential for him that the demon performed his measurements without expenditure of energy; that 'he got something for nothing.'

Ilse Walker (1976) showed that Maxwell explicitly separated the two concepts of 'knowledge' and 'energy' and anticipated information theory. More important is her statement that Maxwell committed the heresy of placing the observer inside his own system and included his experience in the statements on the physical behaviour of the system. In classical physics exists a tacit agreement that there is no interference between the experimenter and his subject, that the latter is independent of the measurement. In this respect, Maxwell's child is a modern, a quantum mechanical demon. Similar ideas were published by Kosloff (1981) and also by Bhandari (1976) who regretted that the existence of a non-classical observer has received so little attention. He believes that 'there is no logical basis for excluding the possibility of such an observer and that it should be considered as the modern analogue of Maxwell's demon'.

Problems such as those discussed above always appear when one leaves the range of 'anthropomorphic' physics and deals with very large quantities (velocity of light, e. g.) or very small ones such as Planck's constant, h. If, in classical physics h is taken as infinitely small, frequency, v, must tend towards infinity because of the energy relation $E = h \cdot v$ and the wave character of material is lost. Or, E becomes zero together with h, and the quantum aspect of waves vanishes. Starting with such reflections Costa de Beauregard (1960) asked if, for the quantity '$k \cdot \ln 2$', which relates entropy S and information I through

$$\Delta S = k \cdot \ln 2 \cdot \Delta I \qquad (20)$$

the same holds true. In the pre-cybernetic period of physics $k \cdot \ln 2 = 1.1 \cdot 10^{-23}$ J. deg^{-1}mol^{-1} was taken as zero and the transition: negentropy $\longrightarrow$ information (the observation) supposed to be gratis, then the reciprocal transition: information $\longrightarrow$ negentropy (the free action) would be impossible and just a psychological illusion. In his opinion the word 'information' has a double meaning

- in the transition: negentropy $\rightarrow$ information it means the gathering of knowledge
- in the transition: information $\rightarrow$ negentropy it means the faculty of organisation.

If one postulates the reciprocal transition between negentropy and information, one accepts the two meanings of information. Beauregard continues that this reciprocity reflects a real symmetry between the two processes of observing and acting a symmetry which is normally masked by the small quantity of $k \cdot \ln 2$ in anthropomorphic units and denied when put equal to zero in (20).

McClare (1971) pointed out that the discussion as to whether or not the Second Law of Thermodynamics was merely statistical goes wrong, was because 'the second law must be generally true, or it is not true at all'. He puts the proof in the following simple form

> 'If the second law is statistical, then Maxwell's demon is possible, then the second law is false. But the second law is true, therefore the demon is impossible. Therefore the second law is not statistical.'

This means that the statement 'the second law of thermodynamics is statistical' implies the existence of demon-like machines working on one molecule (see Pierce, 1961). As these machines are falsified, so the statement must be false, too.

When the exorcism of Maxwell's demon is performed via Heisenberg's uncertainty principle it is evident that this exorcism must also hold for Laplace's demon. If he does not know by nature the physical parameters of all the particles in the universe then he is bound to observations, as his Maxwellian colleague, and to the same restrictions of quantum mechanical interference when making measurements. Thus, it seems, that together with Maxwell's tiny intelligence we have extinguished the whole demonic family.

Maxwell's demon - His resurrection and future life

It was stated above that biological systems were thought of as 'cheats in the game of entropy' (Lewis); to act as Maxwell's demon, because it was obvious that living systems created negentropy instead of disorder, that they slowed down the general decrease of order in the world. In arguing like this no attention was given to the fact that living beings are *per se* open systems engaged in a steady exchange of energy and negentropy with their environment. It is evident that Maxwell's demon tried his comeback in this field of sciences, and, when one follows recent publications it seems that his resurrection was perfect. Of course, sometimes an ordering mechanism is called a Maxwell demon even if it is clear that it uses energy in different forms for its activities.

In this sense Patlak (1956) talks about 'positive' Maxwelllian demons, hypothetical mechanisms which allow a particle to cross a potential barrier, by itself, more readily in one direction than in the opposite one. Even 'negative' demons exist which hinder the crossing of the particle. They resemble those mechanisms envisaged by Dean (1941) in his active transport model, being different only in the fact that those did not use energy for their job. Patlak supposes that the demons could be active by increasing the particle's velocity or by decreasing the barrier resistance for short periods through chemical reactions and thus facilitate the crossing of the barrier. In a second paper on active transport (Patlak, 1957) the author no longer uses the expression 'demons' as they were thought not to violate the Second Law of Thermodynamics; being bound to the energy input for successful actions.

In the same way Wang and Wang (1976, 1979) discuss energy conservation and entropy increase during active transport and the activities of demons of the Maxwell type. These demons use - in contrast to their classical brothers - short distance forces (such as van der Waals forces) for obtaining information and do not detect the molecules long before they reach the membrane. The detailed calculations of the reactions involved show that there is no contradiction of the Second Law and that entropy increase and energy conservation are rigorously obeyed.

The kidney is a very active sorting device in the human body, being closely analogous to a Maxwell demon. It clearly differentiates between sodium and non-sodium ions and thus needs information about the molecules which shall be excreted. For secure handling of information the signal to noise ratio should at least be 4, so that at a body temperature, T, with a thermal noise of $kT = 0.027$ eV the signal must amount to 0.1 eV. For an excretion of 10^{22} ions/min a signal energy of $0.1 \cdot 10^{21}$ eV/min = 2.7 W is necessary. With an energy consumption of ca. 7 W by the human kidney an efficiency of 40 % is calculated which is not bad but very distant from demonic (Johnson and Knudsen, 1965).

Gol'tsman (1978) employs a 'Maxwell quasi-demon' acting in the ATP splitting system of muscles, a demon who is of course not violating the Second Law. Real demons were envisaged by Primavesi (1979) who admired their unique combination of intelligence and minute size. He wondered if molecular sieves had the proper size for Maxwell demons and would it be possible to incorporate a skewed symmetry into such a molecular sieve. With particles in the range of golf balls this would be no problem - but in molecular dimensions? With such a device it would be easier to construct a pressure difference than a temperature difference (see above). He himself is not sure about the realisation and the efficiency of such a machine ('... it could work but produce only one mouse-power per km^2, or it could, just possibly, provide us with an inexhaustible source of power'), but in some way he answers his question by stating:

> 'Indeed, if there is a biological way of using a pressure difference to power an organism, there is probably some lowly bacterium already doing this.'

But such a bacterium has never been detected.

Nearly simultaneously Flagg-Newton and Loewenstein (1980) published an article about 'Asymmetrically permeable membrane channels in cell junction'. Such junctions connect cells in tissues and cultures, provide a path for direct diffusion from cell to cell and are symmetric for small molecules under normal conditions. Larger molecules which show an interaction between permeant and channel experience a skewed symmetry of permeability as described by Primavesi (1979). These authors nowhere talk about demons acting, but give a special hint as to the validity of the First Law of Thermodynamics.

This paper was included in a recent discussion of the circumvention of the Second Law by means of membranes of asymmetric permeability (Deutsch, 1981). The author claims that the Second Law of Thermodynamics is applicable in two compartments which are separated by a membrane, but not in the membrane itself, and that it should be possible to incorporate some skewed symmetry into the pores of membranes. He proposes some macroscopic examples for asymmetric pores and their action, by which a concentration gradient would be created and heat converted to useful work in a cyclic manner, but it is not clear how these ideas could be transposed to microstate membrane pores. This article was the subject of much controversary; summarised by Greco (1983). He comes to the conclusion that there are no biological membranes with the proposed properties and that 'the microphysical domain may escape confinement by an equation of state but not the long arm of the second law'. Only the last sentence can give hope to some of the natural philosophers:

> 'But an answer can readily be given to the question 'Has the second law of thermodynamics been circumvented?' Not yet.'

Maxwell's demon would be astonished to discover the different fields in which his help is asked for. Of course, one of the most severe problems of modern mankind is public abuse of the environment. Sugawara (1977) published a paper about such abuse and entropy increase and showed that there are two types of Maxwell demons working in this area of daily life:

- active people in a community who are strongly committed to prevention of public abuse of the environment
- microorganisms purifying polluted water in sewage treatment plants.

Both, of course, are dependent on energy (and information) uptake and are not Maxwell demons *sensu strictu*.

In another very modern field, demons are asked for help, as a 'Maxwellian demon improves reactor efficiency' (Stengel, 1976). Because of radioactive poisoning real demons are not employed but suitable potential barriers are established which, in the case of charged particles, could be either an electrostatic double ramp barrier, or, a suitably shaped electromagnetic field. High energy particles climb up the ramp, and, by dropping down at the other side they regain their original energy in a conservative manner, while low-energy particles are trapped. In the opinion of the author, no energy, entropy or information is involved in the process of sorting and nevertheless a net decrease in entropy results. This idea is incorporated into a mirror-type plasma bottle from which only high-energy particles can escape axially while the others turn to the side (Maxwell-splitter).

When producing 'Doppler-free spectra' by means of laser irradiation of gas atoms, the monochromatic laser light selectively interacts with just one 'velocity group' of particles, as the natural linewidth for atoms is far smaller than the Doppler linewidth in a gas: the realisation of Maxwell's demon (Pritchard, 1977). Correspondingly Poplavsky (1979) talks about Maxwell's demon in connection with the ordering of molecules in a two-level maser. To create the desired population density inversion, which makes the maser active, the excited molecules are physically separated from the unexcited ones by a nonhomogeneous electric field, the ordering demon.

In a paper about time-reversal experiments performed in systems of dipolar-coupled nuclear spins, Rhim and coworkers (1971) reactivated Loschmidt's demon to reverse the dynamic behaviour of the system. They had in mind no violation of the Second Law of Thermodynamics when they wrote that their systems behave irreversible on a (large) thermodynamic scale, but dynamically reversible in a microscopic one. They continued:

> 'We shall find it useful to introduce the concept of a 'Loschmidt demon' who is able to reverse the dynamical behaviour of a system thereby retracing a path which seemed to be irreversible.'

Their demon became active in various spin-echo experiments managing to reverse the apparent sign of the dipole-dipole interaction and forcing the system to retrace its own history.

A fairly odd member of the demonic family will end this chapter. In his article about 'Gestalt-Theory' and 'Gestalt-Therapy' in psychotherapy Huneeus (1976) commented that Maxwell's demon was the first man-made creature that, knowingly, could violate the Second Law and then compared him with the modern psychotherapist in stating:

> 'In all schools of psychotherapy, the therapist can be considered as a Maxwell demon. ... Gestalt Therapy with its strong emphasis on the 'awareness of the ongoing proces' relies heavily on all organismic functions as the means of producing information relevant to the patient.'

Here, only the informational aspect of the demon is considered, so that there is no violation of the Second Law of Thermodynamics from an energetic point of view.

Epilogue

When one studies the papers which have appeared since the 'final' exorcism of Maxwell's demon and his companions by Brillouin (1951) one cannot be convinced that they really were expelled from science. Maxwell's thought-child still enjoys the attention of many scientists and inventors. For two reasons:

Jeans stated in 1925 that 'Maxwell's sorting demon could effect in a very short time what would probably take a very long time to come about if left to the play of chance' (cited after Walker, 1976); and Hill (1924) mentioned that the discovery of a molecular mechanism in the cell, 'something like Maxwell's demon with his trapdoor, ... will be the greatest step taken by science.' It was and it is the hope of some contemporaries that such processes really do exist and that they - developed to an industrial scale - could solve the energy problems of mankind. The papers of Bridgman (1953), Primavesi (1979) and Deutsch (1981) point in this direction.

The second reason is a less practical and more theoretical one. Even if one is convinced that Maxwell's demon cannot exist and that the Second Law of Thermodynamics holds good, in macroscopic as well as in microscopic states, in physical as well as in biological systems, it is worthwile dealing with these thought-experiments to discover the limitations, and the correct formulations, of this fundamental law of physics. In this sense Wiener (1961) wrote that by simple denial of the existence of the demon 'we shall miss an admirable opportunity to learn something about entropy and about possible physical, chemical and biological systems'. And Ehrenberg (1967) feels that it is necessary to '... be grateful to the good old Maxwellian demon, even if he does not assist in providing power for a submarine. Perhaps he did something much more useful in helping us to understand the ways of nature and our ways of looking at it'. In the same sense Chambadal (1963) claims that it is not necessary to exorcise the demon, because it is hard to find a more evident illustration of the statistical meaning of the Second Law.

Daub (1970) gave a nice farewell to the tiny intelligence:

> 'Witnessing the demise of his creature, he (Maxwell) would not long mourn at the grave but rather be grateful for the exciting years his thought-child had enjoyed. Steeped in the knowledge and love of biblical imagery, Maxwell would doubtless take pleasure in the thought that by becoming mortal his doorkeeper had prepared the way for new life in science'.

After considering some of the papers cited in the last chapter it is not certain that the demon is dead; perhaps he is just sleeping or even quite vital and active in fields where, our senses not being sharp enough, we could not observe him until now. Keeping the question of demonic mortality open we want to close his *curriculum vitae* with a comment of Walker (1976):

> 'It is, after all, in the widely branched Trees of Life and Knowledge that the Demons dwell, since time immemorial'.

Thermodynamics and Information

R. P. Poplavsky

Association between classical equilibrium thermodynamics and information has two aspects, which should not be confused. On the one hand, according to Boltzmann, the determination of statistical entropy is, in practice, characterised by insufficient information about the microstates of a physical system which is in a given macrostate. Therefore, information on microstates is completely related to physical entropy by definition. On the other hand, for other kinds of information processes information about macrostates is essential, i. e. information as to the values of macroscopic parameters of a physical system. A primary source of such information is measurement. The natural consumers of such information are control systems. In all these and similar cases information is not complementary to entropy: all information processes are essentially irreversible. These processes are the object of the study of thermodynamics of information processes (Poplavsky, 1981).

The association between thermodynamics and information had been discussed long before the appearance of Shannon's information theory, and the first work in this area was the well-known paper by Szilard (1929). After information theory had appeared, the main contribution to this field was made by Brillouin (1960). Further development in this field is contained in a number of papers by the author which have been reviewed in a monograph (Poplavsky, 1981).

Here we define an elementary information interaction as interaction between two bodies in a thermostat, where a first body Y does some work R upon a second body X and transfers an energy U to it. When the value of the generalised coordinate x of the second body is established from the corresponding value of the generalised parameter (force y) of the first body,

Thermodynamics and Regulation of Biological and Processes

then this process constitutes an elementary active information process. All active measurement and control processes are composed of such elementary processes. These processes are characterised by the fact that thermodynamics and information are connected in them directly and obviously.

Indeed, at first the bodies Y and X are in the initial equilibrium state. Then, as a result of interaction, the bodies which were disconnected earlier (and their parameters y and x) prove to be connected, and now the connected system Y + X relaxes to a new equilibrium state. Connection of the parameters y and x, which were earlier independent, is some ordering in a system that can be characterised by a lowering of statistical entropy - a negentropic effect ΔN. The negentropic effect turns out to be exactly equal to the information quantity I (y, x) which appeared in the body X about the body Y. The information quantity $I = \Delta N$ depends logarithmically on the relative accuracy of the ratio $1/\sigma$, i. e. the ratio of the *a priori* root-mean-square error to the *a posteriori* root-mean-square error. It can be easily shown (Poplavsky, 1981) that

$$1/\sigma^2 = 2U/T \tag{1}$$

where U is energy transferred from the body Y to the body X, and T is the thermostat temperature. Here the temperature is measured in energy units, and the entropy in relative (dimensionless) units. Then (Poplavsky, 1981)

$$\begin{aligned} \Delta N = I(x, y) = \tfrac{1}{2} \ln (1/\sigma^2 + 1) \approx \\ \approx \ln 1/\sigma = \ln \sqrt{2U/T}, \quad \sigma \ll 1 . \end{aligned} \tag{2}$$

This negentropic effect and the information obtained, of course, are not free: there must be a thermostat entropy increase, ΔS_T, which, according to the Second Law of Thermodynamics, cannot be less than ΔN. This means that the usual assumption in energetic processes is not valid here, that arbitrarily by decelerating the transition process, one can direct the dissipation Q to zero ($R \longrightarrow U$). Since $\Delta S_T = Q/T \geq N$, the transition process velocity cannot be zero. A regulator realising deceleration also participates in heat motion and is a source of another component of dissipative losses Q_2, which are usually ignored. If the first component falls linearly as the velocity decreases, the second is inversely proportional to the velocity, so that the minimum in the summed losses is reached as soon as the transition process velocity becomes of the order of the heat velocity. The minimum resulting losses are as follows

$$Q_{\Sigma min} = 2\sqrt{\mathcal{U}T}, \quad \Delta S_{T min} = 2\sqrt{\mathcal{U}/T} = \sqrt{2}/\sigma \tag{3}$$

Without deceleration

$$Q_{max} = \mathcal{U}, \quad \Delta S_T = \mathcal{U}/T. \tag{4}$$

Thus, $\Delta N \ll \Delta S_T$ as $1/\sigma \gg 1$, i. e. the measurement (or control) process is essentially irreversible.

The monograph by Poplavsky (1981) also considers other control system ordering models which differ from an elementary information interaction (in particular, a model with a Maxwell demon) and which also give a low entropy efficiency η :

$$\eta_{max} = \Delta N/\Delta S_T \ll 1. \tag{5}$$

Thus, an association between thermodynamics and information has obvious physical meaning only when the produced quantity of information is directly characterised by the negentropic effect.

To establish such association in classical information transfer theory special model representations must be introduced. Indeed, a signal is a non-equilibrium system, whereas a receiver placed in a thermostat is an equilibrium system. After a signal-receiver interaction, the receiver acquires a new equilibrium state and, naturally, finally dissipates the signal energy, i. e. increases the thermostat entropy. On the other hand, if the quantity of information contained in the signal is taken as a measure of deviation from equilibrium (corresponding to the initial negentropy), then in the interaction with the receiver we again see only an entropy increase.

Another approach is possible. In the proof of the basic theorem Shannon (1963) uses the idea that the statistical properties of the 'signal-noise' ensemble are similar to the properties of noise. When considering a thermal noise, we are dealing with Gaussian distribution and, here, one can imagine that the 'signal + noise' ensemble is similar to noise with some effective temperature $T_{eff} = T + \Delta T$. For Gaussian noise this is equivalent to the fact that the ensemble of signals is also Gaussian with a temperature ΔT. Assuming that these ensembles possess the properties of a Boltzmann gas (with heat capacity per molecule c_v and number of molecules n), we find that in the receiver this gas is cooled from the initial temperature T_{eff} to the thermostat temperature T. The negentropic effect of such gas cooling is

$$\Delta N = n c_v \ln(1 + \Delta T / T). \tag{6}$$

On the other hand, due to energy dissipation the thermostat energy increases

$$\Delta S_T = \Delta Q / T = n c_v \Delta T / T. \tag{7}$$

If the Boltzmann gas is considered one dimensional (i. e. $c_v = 1/2$), equation (6) for ΔN is exactly equivalent to Shannon's formula for the quantity of information I in n independent samples (degrees of freedom of a message) and for a signal-noise ratio $\Delta T/T$.

With such an interpretation, from (6) and (7) it is apparent that

$$\Delta S_T > \Delta N = I \tag{8}$$

and that the lower boundary of the energy value of the natural unit of information is equal to T and is never reached ($\Delta S_T \rightarrow \Delta N = I$ only at $\Delta T \rightarrow 0$). It is seen that the smaller the signal-noise ratio, the closer to reversible is the transfer process.

It is only under such an interpretation, where the quantity of transferred information I proves equal to the negentropic effect of gas cooling, that Shannon's formula directly follows from thermodynamics, and the expression $\Delta S_T \gtrsim I$ from the Second Law.

In reality, the model representation described is, however, rather artificial: an ensemble of signals is not really similar to a Boltzmann gas since the latter involves, simultaneously (large n), all microstates, while in information transfer these microstates are realised successively in time. Hence, in the information transfer (as distinguished from the measurement considered above) there occurs no real negentropic effect (ordering in a physical system).

Thus, in spite of wide-spread opinion, associating between thermodynamics and classical information transfer theory are rather artificial and require, as has already been mentioned, special model representations.

In conclusion we briefly discuss specific features of thermodynamics of information processes which distinguish it from the usual equilibirum thermodynamics.

The first specific feature has already been mentioned and consists in the fact that all information processes are essentially irreversible. The minimum dissipation in active information processes (3), that leads (along with (2)) to (5), does not, in practice, affect the mechanical efficiency value

$$\eta_{MECH_{max}} \approx 1 - \sqrt{2T/\mathcal{U}} \approx 1 \quad (\mathcal{U} \gg T) \tag{9}$$

due to which idealisation of reversible processes in ordinary (energetic) thermodynamics is well justified.

The second specific feature consists in a decisive role of thermal fluctuations and in principle the impossibility of neglecting them. In equilibrium thermodynamics the factor $N \gg 1$ is always present (where N is the number of subsystems, degrees of freedom, molecules, etc.) which smoothes fluctuations and enables the mean values of macroscopic parameters to be considered as practically constant. In this case relative flutuations tend to zero as $N^{-1/2}$.

This is not the case with information processes. Information is contained only in those degrees of freedom in which the mean energy exceeds the value T/2. Therefore, in all information processes (in particular, in signal discovery) those degrees of freedom in which information is expected are investigated separately. Accordingly, in information processes the factor $N \gg 1$ is absent, and the decrease of thermal fluctuations can be reached only at the expense of energy increase. Thus, for all information processes an improvement of quality (accuracy and precision) is reached by means of energy increase. For active processes (of the type described above of elementary information interaction in measurement or control) an increase in accuracy requires an increase in the interaction energy U, and along with it, the energy dissipated Q. For passive processes (e. g. discovery) an increase in precision of the discovery requires an increase in the signal energy dissipated in the receiver.

In all cases, however, the energy factor (E/T) cannot, of course, be compared with $N \gg 1$, so it is impossible to neglect fluctuations in information processes.

Entropy of Information Theory and Thermodynamic Entropy

Z. A. Terentjeva

In respectful memory
of my mother
Terentjeva Marija Alexandrovna

Very many papers are written in search of associations between information theory and thermodynamics. And one cannot but agree with Jaynes that connection between these fields can be established 'only by finding a new point of view from which thermodynamic entropy and entropy of information theory represent one and the same conception' (Jaynes, 1957 a). The solution of the problem proposed by Jaynes and then developed by Tribus (Jaynes, 1957 a, b; Tribus, 1961 a, b) has, however, an essential subjective element which raises objections. The essence of this approach consists in the fact that from the known mean value of a parameter (e. g. energy) and from the requirement of a maximum entropy value $S = -k \sum_i p_i \ln p_i$, using the Lagrangian multiplier method, one finds the probability density as a function of energy, and from this entropy and all the parameters of a given system. The requirement of maximum entropy is explained by the desire to obtain the least subjective estimate of the probability distribution, which is interpreted as an expression of human ignorance. From this point of view, the methods of statistical mechanics are a particular case of the general methods of error prevention. For an objective researcher it will probably be easier to understand the efficiency of this approach if he notices that for one and the same object under investigation. proceeding from different sets of initial data, one can calculate all other parameters (if, of course, the regularities of the behaviour

Thermodynamics and Regulation of Biological and Processes

of this object are sufficiently known). From this point of view the requirement of entropy maximum is part of the initial data which are considered as known (for any reason).

Proposing another approach to the calculation of entropy of a thermodynamic system, where the apparatus of the information theory is also used, but without a subjective interpretation, and where the probability distribution is assumed to be known (in this point this approach contradicts the above-mentioned one) it is best of all to begin with a kind of digression on the history of the entropy concept in information theory.

Information theory started, as is known, with the work of Shannon, who understood entropy as a measure of information quantity obtained by an observer.

However in communication theory, entropy was first of all used for estimating disturbances and was an objective characteristic of the phenomenon (Shannon, 1948, 1949). In this sense it is used now, not only in communication theory (Krasovsky, 1963).

Bearing in mind the definition of entropy as a measure of uncertainty in a quantity or in a situation, Khinchin (1953) suggested that the theory of entropy should be regarded as a part of probability theory. He stressed that it was most widely used in information theory because the quantity of information given by the realisation of a certain scheme is considered to be equal to the entropy of the scheme. The huge number of papers written on this topic has evidently led to the fact that entropy calculated by Shannon's formula, in practice, is now used in the latter interpretation only. The entropy of a probability distribution was, of course, used in its initial sense, too, and its character was particularly demonstrated in situations with time-dependent uncertainty. A random process which exists irrespective of whether or not somebody knows about it is described by quite definite parameters which are also independent of the knowledge of the observer. The entropy of the process is one parameter (Kobozev, 1948; Shannon, 1949; Rozenblat-Rot, 1957). Thus, we would like to emphasise that the application of the mathematical techniques of information theory are not necessarily connected with the estimation of the quantity of information obtained by an observer although one can, of course, transfer to it at any moment.

Now, consider from this standpoint a random walk process for an individual molecule of an ideal gas (Terentjeva, Kobozev, 1976). The uncertainty of the situation is determined, in this case by the region of the most probable particle location and this increases with time. Assuming the probability distribution to be known and Gaussian, and, taking into account that when normalised to one particle it has the meaning of the probability for this particle to find itself in a time interval Δt in a certain region of coordinates $(q, q + \Delta q)$. For a one-dimensional case, and, with zero initial conditions ($t_o = 0$, $x_o = 0$), we obtain the probability density as equal to

$$\rho(x,t) = A\exp(-x^2/2\bar{x}^2), \qquad (1)$$

where $\bar{x}^2 = 2Dt$ (D is the diffusion coefficient), A depends on the integration region in the normalisation at

$$\text{at } |x| \leq \infty \qquad A = (2\pi\bar{x}^2)^{-1/2} = (4\pi Dt)^{-1/2} \qquad (2)$$

and

$$\text{at } 0 \leq x \leq x^o \quad A = 1/x^o. \qquad (3)$$

For each time moment t the uncertainty of the situation is estimated by the differential entropy

$$H(x,t) = -\int \rho(x,t)\ln\rho(x,t)\,dx, \qquad (4)$$

which is a function of time and is called the entropy of the process.

The uncertainty of the particle state may depend not only on the uncertainty in the coordinate, but also that in the momentum, as well as in other parameters. The entropy, as a measure of uncertainty of the particle state, may be assumed to be composed of the uncertainty values for each of the parameters

$$H = \sum_i H_i \, . \tag{5}$$

In the absence of uncertainty, i. e. at a definite value of some parameter, its probability $p_i = 1$ and corresponding $H_i = -1.\ln 1 = 0$.

Consider now a concrete system, $n = \nu N$ (ν is the number of moles) of particles of an ideal gas of volume V at a temperature T. For simplicity let it be a monatomic (lest we should take into account the uncertainty due to vibrational and rotational degrees of freedom) and inert (all the atoms are only in the ground state) gas. The uncertainty in the state of each particle of this system depends only on the coordinates and momentum. In the phase space of the particle on its random walk among the other particles of the system and connected with a random change in momentum, there is a corresponding walk of a point revealing the state. Taking into account the limited volume, i. e. (3), the calculation of the entropy (4) for each of the coordinates gives

$$H(q_i) = \ln q_i^{\circ} \qquad i = 1, 2, 3$$

The components of the momentum can take values from 0 to ∞. Therefore, the integration of (4) with account taken of (2) gives

$$H(p_i) = \ln (2\pi e \bar{p}_i^2)^{1/2} = \ln (2\pi e m k T)^{1/2} \quad i = 1, 2, 3 .$$

The total particle entropy is

$$H^{(1)} = \sum_{i=1}^{3} H(q_i) + \sum_{1}^{3} H(p_i) = \ln V + \ln (2\pi e m k T)^{3/2} = \ln V (2\pi e m k T)^{3/2} \tag{6}$$

where m is the mass of one particle, V is the volume of the whole system.

For a system of n identical particles of an ideal gas the state probability density is equal to the product of the probability densities of the particles with consideration of the fact that each state of the system characterised by a certain set p and q can be realised by n! methods

$$\varrho = (\vec{p}_1,\ldots,\vec{p}_n,\vec{q}_1,\ldots,\vec{q}_n,t) = \prod_{i=1}^{n} \varrho(\vec{p}_i,\vec{q}_i,t)\, n!$$

where p_i implies the set (p_x, p_y, p_z), q_i stands for the set of coordinates (x, y, z) of the i-the particle. The entropy of the system as a measure of uncertainty of the system state is equal to

$$H^{(n)} = -\int_p\ldots\int\int_q\ldots\int \varrho(\vec{p}_1,\ldots,\vec{p}_n,\vec{q}_1,\ldots,\vec{q}_n,t) \ln \varrho(\vec{p}_1,\ldots, \vec{p}_n,\vec{q}_1,\ldots,\vec{q}_n,t)\, d\vec{p}_1\ldots d\vec{p}_n d\vec{q}_1\ldots d\vec{q}_n = nH^{(1)} - \ln n! \quad (7)$$
$$= n \ln \frac{V(2\pi e m k T)^{3/2} e}{n} =$$
$$= \nu N \ln \frac{V^\circ(2\pi e m k T)^{3/2} e}{N},$$

where V° is the volume of one mole.

Thus, the entropy of a system of identical particles is smaller by ln n! than the sum of the entropies of the individual particles, and the entropy per particle is smaller than the individual entropy of the particle (6). For a system of quite different particles the entropy would be equal to the sum of the entropies of the particles and would be a function of the entire volume and, therefore, could not characterise the substance. This is obvious since a system of quite different particles cannot constitute any substance.

The quantities (6) and (7) are relative since they contain dimensionality under the logarithm sign. This is due to the use of differential entropy, which is a relative quantity, and of the classical approximation under which, to each state, there corresponds a point in the phase space. According to quantum mechanics, the least volume corresponding to a system state is equal to h^{3n}, and it is natural to choose this quantity as a measure of the phase space. Then, entropy becomes an absolute quantity

$$H^{(1)} = \ln \frac{V(2\pi e m k T)^{3/2}}{h^3} \quad (8)$$

$$H^{(n)} = \nu N \ln \frac{V^{\circ}(2\pi \ell m k T)^{3/2} \ell}{h^3 N} . \tag{9}$$

The latter formula unites the Boltzmann constant with the known formula for the entropy of translational degrees of freedom for a system of identical particles. But as has been noticed by many researchers, the Boltzmann constant results from the arbitrary choice of temperature units, and, if temperature were measured in energy units, the entropy increment $dS = \delta Q/T$ would be, evidently, dimensionless.

Thus, to transform to thermodynamic entropy, it is necessary to multiply (9) by the Boltzmann constant

$$S = k \nu \ln \frac{V^{\circ}(2\pi \ell m k T)^{3/2} \ell}{h^3 N} . \tag{10}$$

This point of view leads, quite naturally, to the Gibbs theorem for the entropy of mixtures and accounts for Gibbs paradox (Terentjeva, Kobozev, 1976). In addition, the entropy (9) conforms to the sign of the Boltzmann H-function.

These results, as well as the simplicity and directness of the approach gives confidence in the proposed method.

Returning to the problem formulated by Jaynes, one can state that the link between thermodynamic entropy and the entropy of information theory is the concept of chaotic motion entropy of an individual particle as a measure of the uncertainty of the state of this particle.

Optimal Thermodynamic Processes Under Chemical Reactions and Diffusion

L. I. Rozonoer

One of the central problems of thermodynamics is, as well known, the problem of limiting energetic possibilities of thermodynamic processes (the problem of maximum work which can be 'extracted' from a non-equilibrium system under certain conditions; the problem of maximum efficiency of thermodynamic cycles, etc.). The main result obtained within classical thermodynamics is that optimal processes must be reversible, i. e. proceed without entropy increase. This result implies, however, that the time of an optimal process is 'infinitely large'. Heat and matter transfer proceeds in such a proces with 'infinitesimal' differences of temperature and chemical potential, and therefore, an 'infinitely large' time is required to transfer finite amounts of heat and substance. The problem of seeking for optimal thermodynamic processes of a given finite duration naturally arises. This problem oversteps the limits of equilibrium thermodynamics since now the optimum must be sought for among the processes which are known to be irreversible. The problem can be formulated in the framework of non-equilibrium thermodynamics by assuming a non-equilibrium system to consist of equilibrium subsystems (which are, generally speaking, not in equilibrium with one another), and the laws of heat and matter transfer and chemical reactions to be given. From the mathematical point of view the problem of optimal thermodynamic processes is typical of the optimal control theory, which was mentioned by Rozonoer (1973) (for the optimal control theory see e. g. Pontryagin et al., 1969; Alekseev et al., 1979).

In recent years a number of papers have appeared where optimal irreversible heat transfer processes in the absence of matter transfer and chemical

Thermodynamics and Regulation of Biological and Processes

reactions are considered (see the cycle of papers by Rozonoer, Tsirlin (1983) and the bibliography presented there).

The present paper is aimed at formulating the problem of optimal thermodynamic processes in the presence of chemical reactions and diffusion, which, for clear reasons, can be of considerable interest for bioenergetics. The latter particularly refers to energetics of 'thermodynamic pumps'. The thermodynamic pump is understood here as a system which maintains thermodynamic non-equilibrium and thus provides energy or matter pumping from a reservoir with lower temperature or electrochemical potential into that with higher values. The role played by thermodynamic non-equilibrium in living organism functioning was particularly emphasised by Bauer (1935). The role of non-equilibrium thermodynamic processes in biology is now generally accepted, but concrete mechanisms of maintaining 'stable non-equilibrium' in living organisms have been studied very little. From this point of view a theoretical investigation of thermodynamic pumps (specifically, clarification of their limiting possibilities) is necessary for an understanding of real processes that provide biological 'stable non-equilibrium'.

Fundamental thermodynamic relations

A non-equilibrium system will be considered as a set of multicomponent subsystems, among which heat and mass exchange is possible; within each subsystem chemical reactions can proceed. First we describe a separate subsystem. We characterise its thermodynamic state at each time moment t by a set of extensive quantities: entropy $S(t)$, internal energy $E(t)$, and the number of moles $N_1(t)$ ..., $N_m(t)$ of the components among which chemical reactions can proceed, and also by a set of mechanical parameters, such as volume $V(t)$. Subject to the nature of the system, the mechanical parameters may also include other quantities, but hereafter we consider, for simplicity, a single mechanical parameter, the volume. If a subsystem is not electrically neutral, the extensive macroquantities characterising this subsystem should also include electrical charge Q. All the parts of the subsystem at each time moment are considered to be in equilibrium (i. e. pressure, temperature, and electrochemical potential of each component inside the subsystem are equal), although the subsystem is not, generally speaking, at chemical equilibrium. This means that the velocity of establishing thermal, mechanical, and electrodiffusion equilibrium within a subsystem greatly exceeds the velocity with which chemical equilibrium is established in this system.

At each moment the subsystem entropy can be expressed as a function of other extensive quantities and volume

$$S = S(E; N_1, \ldots, N_m; Q; V). \tag{1}$$

From (1) the intensive macroquantities characterising the subsystem (temperature T, pressure P, chemical potentials μ_i, i = 1, ..., m, and electric potential) can be expressed through the values of the extensive quantities E, N_i, Q and V with the help of the relations

$$\frac{1}{T} = \frac{\partial S}{\partial E}, \quad \frac{p}{T} = \frac{\partial S}{\partial V}, \quad \frac{\mu_i}{T} = -\frac{\partial S}{\partial N_i}, \quad \frac{\varphi}{T} = -\frac{\partial S}{\partial Q}. \tag{2}$$

The electric charge of a subsystem can be expressed through the number of moles on the components

$$Q = \sum_{i=1}^{m} e_i N_i, \tag{3}$$

where e_i is the charge of one mole of the i-th component (if the i-th component is neutral, $e_i = 0$). Note that if charges are transported by free electrons, the latter must be regarded in such a description as one of the components ('a purely electric substance'), and the ionisation process as a chemical reaction. Substituting (3) into (1), one can express entropy through E, N_i, V

$$S = S(E; N_1, \ldots, N_m; \sum_{i=1}^{m} e_i N_i; V) \equiv \tilde{S}(E; N_1, \ldots, N_m; V). \tag{4}$$

Using the function $\tilde{S}(E; N_1, \ldots, N_m; V)$, we obtain instead of (2)

$$\frac{1}{T} = \frac{\partial \tilde{S}}{\partial E}, \quad \frac{p}{T} = \frac{\partial \tilde{S}}{\partial V}, \quad \frac{\tilde{\mu}_i}{T} = -\frac{\partial \tilde{S}}{\partial N_i} \tag{5}$$

where $\tilde{\mu}_i = e_i \varphi + \mu_i$ is the electrochemical potential of the i-th component. If N_i, V and S (but not E) are taken as independent thermodynamic variables, one can obtain from (4) and (5)

$$E = \tilde{E}(S; N_1, \ldots, N_m; V), \quad T = \frac{\partial \tilde{E}}{\partial S}, \quad p = -\frac{\partial \tilde{E}}{\partial V}, \quad \tilde{\mu}_i = \frac{\partial \tilde{E}}{\partial N_i}. \tag{6}$$

To describe the dynamics of a subsystem, one should set up a time-dependent variation of the extensive quantities E and N_i. Expressing time derivatives by a superscript, we write on the basis of the energy and matter conservation law

$$\dot{E} = q - P, \quad \dot{N}_i = v_i + W_i \qquad i = 1, 2, \ldots, m \tag{7}$$

where P is mechanical power, q and v_i are fluxes of energy and i-th substance, respectively, which flow into the subsystem from without, and W_i is the velocity of the increase of the i-th substance due to chemical reactions. If V is the only mechanical parameter, then

$$P = p\dot{V}, \quad A = \int_{t_1}^{t_2} p\dot{V}\,dt, \tag{8}$$

where A is the work done by the subsystem for the time $[t_1, t_2]$.
We should emphasise that the energy flux, q, includes both the heat flux obtained by the subsystem at the expense of heat transfer and the irreversible energy flux associated with the substance (at the expense of diffusion of the substance into the subsystem).

Differentiating (4) with respect to time and taking into account (5) to (8), we derive the following expression for the entropy derivative

$$\dot{S} = \frac{1}{T} q - \sum_{i=1}^{m} \frac{\tilde{\mu}_i}{T} (v_i + W_i). \tag{9}$$

By virtue of electric charge conservation in the process of chemical reactions, one can write the relation

$$\sum_{i=1}^{m} e_i W_i \equiv 0, \tag{10}$$

whence $\sum \tilde{\mu}_i W_i = \sum \mu_i W_i$, and accordingly (9) can be rewritten in the form

$$\dot{S} = \frac{1}{T} q - \sum_{i=1}^{m} \frac{\tilde{\mu}_i}{T} v_i - \sum_{i=1}^{m} \frac{\mu_i}{T} W_i . \tag{11}$$

If energy and matter fluxes, velocities of chemical reactions, and mechanical parameters were given directly as time functions $q = q(t)$, $v_i = v_i(t)$ $W_i = W_i(t)$, $V = V(t)$, then, under the initial conditions $E(0)$, $N_1(0)$ the above relations would completely determine the time dependence of all the thermodynamic quantities. However, the quantities q, v_i, W_i are usually not known directly as time functions: W_i is determined by the values of thermodynamic parameters of the subsystem, and q, v_i depend, in addition, on the boundary conditions. In accordance with the general phenomenological scheme of the thermodynamics of irreversible processes (Rozonoer, 1982; Orlov, Rozonoer, 1981), we consider that W_i are determined by the values of the intensive quantities (5)

$$W_i = W_i(p; T; \tilde{\mu}_1, \dots, \tilde{\mu}_m) . \tag{12}$$

To set up the quantities q, v_i, which determine heat and matter transfer, one should characterise the conditions of contact of a given subsystem with other subsystems. If a subsystem is in thermal and electrodiffusion contact with another subsystem which has temperature T', pressure p', and electrochemical potentials $\tilde{\mu}'_i$, then

$$\begin{aligned} q &= q(T, T'; p, p'; \tilde{\mu}_1, \tilde{\mu}'_1; \dots; \tilde{\mu}_m, \tilde{\mu}'_m) \\ v_i &= v_i(T, T'; p, p'; \tilde{\mu}_1, \tilde{\mu}'_1; \dots; \tilde{\mu}_m, \tilde{\mu}'_m) . \end{aligned} \tag{13}$$

(12) and (13) set up phenomenological laws of chemical reactions, heat and matter transfer. They must meet the requirements of the Second Law of Thermodynamics. From (9), by virtue of the requirement $\dot{S} > 0$, at $q = v_i = 0$, we have

$$- \sum_{i=1}^{m} \frac{\tilde{\mu}_i}{T} W_i(p; T; \tilde{\mu}_1, \dots, \tilde{\mu}_m) \geq 0 . \tag{14}$$

Since the rates of chemical reactions are independent of the contacts of a given subsystem, the inequality (14) must hold irrespective of the values of q, $_i$. A similar inequality for the laws of heat and matter transfer is obtained from the requirement of non-decrease of the total entropy of two contacting subsystems. From the conditions $\dot{S} + \dot{S}' \geq 0$ we derive the relation

$$\left(\frac{1}{T}-\frac{1}{T'}\right)q(T,T';p,p';\tilde{\mu}_1,\tilde{\mu}_1';\dots;\tilde{\mu}_m,\tilde{\mu}_m') - \\ -\sum_{i=1}^{m}\left(\frac{\tilde{\mu}_i}{T}-\frac{\tilde{\mu}_i'}{T'}\right)\nu_i(T,T';p,p';\tilde{\mu}_1,\tilde{\mu}_1';\dots;\tilde{\mu}_m,\tilde{\mu}_m') \geq 0. \tag{15}$$

In addition, because there is equilibrium between adjacent systems energy and matter fluxes are absent and, therefore,

$$q(p,p;T,T;\tilde{\mu}_1,\tilde{\mu}_1;\dots;\tilde{\mu}_m,\tilde{\mu}_m) \equiv 0 \\ \nu_i(p,p;T,T;\tilde{\mu}_1,\tilde{\mu}_1;\dots;\tilde{\mu}_m,\tilde{\mu}_m) \equiv 0 . \tag{16}$$

Our further conclusion is based on the fact that relations of the type (12), (13) are known and satisfy the conditions (14), (15), (16).

Near equilibrium relations (13) are determined by the known linear relations between 'fluxes' and 'forces'

$$q = \alpha\left(\frac{1}{T}-\frac{1}{T'}\right) + \sum \beta_i\left(\frac{\tilde{\mu}_i'}{T'}-\frac{\tilde{\mu}_i}{T}\right) \\ \nu_i = \beta_i\left(\frac{1}{T}-\frac{1}{T'}\right) + \gamma_i\left(\frac{\tilde{\mu}_i'}{T'}-\frac{\tilde{\mu}_i}{T}\right) \tag{17}$$

which satisfy the symmetry condition and the positive requirement of the matrix of kinetic coefficients. In the general case relations (1) to (16) can also be used to describe systems 'far' from equilibrium in accordance with the general phenomenological scheme of thermodynamics of irreversible processes (Rozonoer, 1982; Orlov, Rozonoer, 1981).

Thermodynamic process as a controlled system

We should recollect that a controlled system is described by a system of differential equations in the form

$$\dot{x}_s = f_s(x_1,\dots,x_n;u_1,\dots u_r) \qquad s = 1,2,\dots,n \quad (*)$$

where $x_1, \ldots, x_n$ are phase coordinates determining the state of the system, and $u_1, \ldots, u_r$ are control actions (Pontryagin et al., 1969). By choosing control actions as time functions, one can guarantee certain properties of the phase trajectory $x_1(t), \ldots, x_n(t)$. The control actions being chosen, the phase trajectory is unambiguously determined by equations (*) for the given, initial, conditions $x_s(0)$. Under optimal conditions the control actions are so chosen that they provide a maximum or a minimum of some functional. Examples of such functions which characterise the 'quality' of the control process are expressions of the type $\int_{t_0}^{t} F(x_1, \ldots, x_n; u_1, \ldots u_2)dt$, $F(x_1(\bar{t}), \ldots, x_n(\bar{t}))$ etc., where $[t_o, \bar{t}]$ is a time interval for which control actions are chosen. The problem of the minimum or maximum of the function is formulated under certain limitations up on the control actions and also under certain additional conditions imposed on phase coordinates at the beginning and at the end of the process (at $t = t_o$ and $t = \bar{t}$).

We shall show that a system consisting of thermodynamic subsystems as characterised in the previous section can be regarded as a controlled system.

Let the system as a whole be adiabatically isolated, so that heat and matter transfer can be realised only among subsystems. The superscript k stands for quantities which refer to the k-th subsystem - S^k, E^k, N_i^k, etc., $k = 1, \ldots, n$, where n is the overall number of subsystems. According to (12), the velocity of the change in the amount of the i-th component in the k-th subsystem is determined by the following relation

$$W_i^k = W_i^k(p^k; T^k; \tilde{\mu}_1^k, \ldots, \tilde{\mu}_m^k). \qquad (18)$$

If the k-th and l-th subsystems are in contact, the fluxes of energy q^{kl} and matter v_i^{kl} from the l-th to the k-th subsystem are determined, according to (13), by relations of the form

$$q^{kl} = q^{kl}(p^k, p^l; T^k, T^l; \tilde{\mu}_1^k, \tilde{\mu}_1^l; \ldots; \tilde{\mu}_m^k, \tilde{\mu}_m^l)$$
$$v_i^{kl} = v_i^{kl}(p^k, p^l; T^k, T^l; \tilde{\mu}_1^k, \tilde{\mu}_1^l, \ldots; \tilde{\mu}_m^k, \tilde{\mu}_m^l). \qquad (19)$$

If there is no contact between the k-th and the l-th subsystems, then $q^{kl} = v_i^{kl} = 0$. By virtue of the energy and matter conservation laws,

$$q^{kl} = -q^{lk}, \quad v_i^{kl} = -v_i^{lk} \tag{20}$$

Taking into account th additivity of fluxes, for the resultant fluxes of energy q^k and matter v_i^k into the k-th subsystem one can write

$$q^k = \sum_{l=1}^{n} q^{kl}, \quad v_i^k = \sum_{l=1}^{n} v_i^{kl}. \tag{21}$$

For simplicity we sum over all the indices, but assume here and below that $q^{kk} = v_i^{kk} = 0$.

For each of the subsystems the following equations can be written (c.f. (7) to (9))

$$\dot{E}^k = q^k - p^k \dot{V}^k \tag{22}$$

$$\dot{S}^k = \frac{1}{T^k} q^k - \sum_{i=1}^{m} \frac{\tilde{\mu}_i^k}{T^k} (v_i^k + W_i^k) \tag{23}$$

$$\dot{N}_i^k = v_i^k + W_i^k \tag{24}$$

where W_i^k, q^k, v_i^k ($k = 1, \ldots, n$; $i = 1, \ldots, m$) is determined through the intensive parameters p^k, T^k, $\tilde{\mu}_i^k$ of the subsystems by the relations (18) and (19) to (21). The quantities p^k, $\tilde{\mu}_i^k$, T^k, can be expressed through the extensive parameters N_i^k, E^k (or S^k) and the mechanical parameters V^k with the help of thermodynamic equations of the form of (5) or (6). Therefore, the set of relations (22), (24) or (23), (24) with $V^k(t)$ taken as time functions is a complete system of differential equations for the unknown functions $N_i^k(t)$, $E^k(t)$ (or $S^k(t)$). This simply means that the thermodynamic system under consideration is a controlled system, where the role of phase coordinates is played by extensive quantities, and the role of control actions by mechanical parameters. Establishment and interruption of intersystem contacts can also be considered as control actions (e. g., the work of heat engines is based on this principles). To take into account such a possibility, we introduce new control actions, namely, time functions u^{kl} (t), which can take only two values: 1 in the presence and 0 in the absence of contact between the k-th and l-th subsystems ($u^{kl} = u^{lk}$). Then in all the above relations one should replace q^{kl} and v_i^{kl} by $u^{kl}q^{kl}$ and $u^{kl}v_i^{kl}$, respectively. We can also consider a more general case when established and interrupted contacts are partial - for example, when energy, but not matter can be transferred, or when only certain components can be transported ('semitight barriers') etc. Partial contacts are also easily taken into account by introducing additional control actions. For simplicity we restrict ourselves, however, to introducing the control actions u^{kl}. As a result, taking the quantities S^k and N_i^k for phase coordinates, we obtain (see (21), (23), (24)):

$$\dot{S}^k = \frac{1}{T^k}\sum_{l=1}^{m} u^{kl}q^{kl} - \sum_{i=1}^{m}\frac{\tilde{\mu}_i^k}{T^k}\Big[\sum_{l=1}^{n} u^{kl}v_i^{kl} + W_i^k\Big]$$
$$\dot{N}_i^k = \sum_{l=1}^{n} u^{kl}v_i^{kl} + W_i^k \qquad (25)$$

Here W_i^K and q^{kl}, v_i^{kl} are expressed by means of (18) and (19) through intensive thermodynamic quantities, and the latter in their turn can be expressed through the phase coordinates S^k, N_i^k and the control actions V^k from the relations (see (6))

$$T^k = \frac{\partial \tilde{E}^k}{\partial S^k}, \quad p^k = -\frac{\partial \tilde{E}^k}{\partial V^k}, \quad \tilde{\mu}_i^k = \frac{\partial \tilde{E}^k}{\partial N_i^k} \qquad (26)$$

Thus, a controlled thermodynamic system is described by the set of relations (25), (18), (19), (26).

By virtue of (25) the total entropy $S^{\Sigma} = \sum_{k=1}^{n} S^k$ of the adiabatically isolated system under consideration does not decrease. Indeed, taking into account (20), we have

$$\dot{S}^{\Sigma} = \frac{1}{2}\sum_{k,l=1}^{n}\Big[\Big(\frac{1}{T^k}-\frac{1}{T^l}\Big)q^{kl} - \sum_{i=1}^{m}\Big(\frac{\tilde{\mu}_i^k}{T^k}-\frac{\tilde{\mu}_i^l}{T^l}\Big)v_i^{kl}\Big]u^{kl} + \sum_k\sum_i\Big(-\frac{\tilde{\mu}_i^k}{T^k}W_i^k\Big) \geq 0, \qquad (27)$$

since chemical reactions, heat and matter transfer satisfy inequalities of the form (14) and (15).

In those cases where the system is not adiabatically isolated, but is in contact with the 'reservoir' (the external medium) which has a given temperature T^{res}, pressure p^{res}, and electrochemical potentials $\tilde{\mu}_i^{res}$ of the components, it is convenient to include the reservoir to the system as a subsystem, and to consider the entire system, again, as adiabatically isolated. To this end, the reservoir entropy should be given in the form

$$S^{res} = \frac{1}{T^{res}}(E^{res} + pV^{res}) - \sum \frac{\tilde{\mu}_i^{res}}{T^{res}} N_i^{res}, \qquad (28)$$

where T^{res}, p^{res}, $\tilde{\mu}_i^{res}$ are given constants. To use this procedure for consideration of, for example, an isothermal and isobaric process, it is sufficient to put $T^k = T^{res}$ and $p^k = p^{res}$ (physically this is equivalent to the assumption that thermal and mechanical equilibrium in an entire system is achieved much faster than diffusion and chemical equilibrium).

Different statements of problems of optimal control over a given thermodynamic system differ in the functionals which are to be optimised. As such a functional one can consider: the work done by the system over a fixed time; the amount of heat or the amount of a certain component in a given subsystem at the final moment; 'efficiencies' determined in different ways, etc..

Solution of optimisation problems for thermodynamic systems requires, in the general case, the whole arsenal of methods of optimal control theory (including calculation methods). In a number of simple, but important, cases such problems can, however, be completely solved purely analytically. Possible simplifications are based on the following specificity of equations of controlled thermodynamic systems (Rozonoer, Tsirlin, 1983). The right-hand sides of (25) depend on the extensive quantities (phase coordinates S^k, N_i^k) only through the values of the intensive quantities p^k, T^k, $\tilde{\mu}_i^k$ (see (18), (19)). If these quantities had not been connected by the relations (26), the intensive quantities could be taken directly for control actions, and the right-hand sides of (25) would not depend on the phase coordinates. The optimisation problems in which the right-hand sides of such equations depend only on control actions and do not depend on phase coordinates are the so-called Lyapunov type problems and are solved comparatively simply. Although in the general case the constraints imposed by relations (26) cannot be ignored; but if there are 'enough' mechanical parameters, and they can change arbitrarily, relations (26) impose no limitations upon the intensive quantities. For example, this is just the case with purely thermodynamic processes (in the absence of chemical reactions and diffusion) or with diffusion-mechanical processes where only one component diffuses ($m = 1$). In such cases it is not the volumes, but the temperatures or chemical potentials of the subsystems that can be taken as control actions (along with the variables u^{kl}), and the corresponding optimisation problems are of the Lyapunov type. But even when, for exact solution of a problem, it is necessary to take into account constraints (26), neglect of relations (26) still allows a useful estimate of the optimal value of the functional (extension of the variety of admissible controls may only increase the maximum and decrease the minimum value of the optimising functional). Therefore, the limiting possibilities of optimal control over a thermodynamic system can be estimated only on the basis of relations (25), (18), (19). In particular, in this way it is easy to obtain the known classical results of thermodynamics for maximum work and maximum efficiency of thermodynamic cycles - one need not even know the concrete form of the relations (18) and (19), but it is sufficient to use the non-decrease of the total system entropy, S (inequality (27)). If one takes into account the concrete laws of chemical reactions (18), of heat and matter transfer (19), these classical results can be essentially specified by an exact solution of the optimisation problem for the system (25), (18), (19) (disregarding (26)). Such specified estimates transform into classical estimates whenever the duration of control tends to infinity since, in this case, optimal processes approach reversible processes (differences in temperatures and in chemical potentials of subsystems and chemical affinity in reversible processes are 'infinitesimal'). We should emphasise the following interesting specificity of such estimates: they do not depend on matter-state equations, but are determined only by the laws of heat and matter transfer, and, chemical reactions since state equations are contained only in relations (26) which are ignored. This circumstance is, in particular, the basis of the 'universality' (independence of the equations of state) of classical thermodynamic estimates.

Leaving aside the technique for solution of Lyapunov type problems, we note only that they can be written in the form

$$\int_{t_0}^{\bar{t}} F(u_1,\ldots,u_r)\,dt \to \min \text{ (or max)}$$
$$\int_{t_0}^{\bar{t}} f_i(u_1,\ldots,u_r)\,dt = a_i \qquad i=1,2,\ldots,n \tag{**}$$

in which phase variables are absent (a_i are given constants). Note the following characteristic feature of optimal control in such problems: optimal control actions take a finite number of values and 'jump over' from one value to another during the time of control. The value of the functional is not affected by the order in which such values replace one another; essential are only the fractions of the time intervals within which the control actions take this or that value. Hence, optimal control in Lyapunov type problems is, as a rule, not unique (except for the case when a control action take only one value, i. e. is a constant).

Optimal thermodynamic cycles and thermodynamic pumps

The general scheme for statement of the problem of optimal thermodynamic cycles can be described as follows. Let there exist two or more reservoirs with given values of intensive parameters (temperature, pressure, electrochemical potential). Consider, in addition, the 'actuating medium' - a subsystem which can be brought into contact with the reservoirs accompanied by changes in the mechanical parameter values. A thermodynamic process is called a cycle if the state of the actuating medium at the beginning and at the end of the process is the same. Cyclic behaviour does not refer, of course, to reservoirs, in an irreversible process cyclical behaviour of an aggregate system (the actuating medium + reservoirs) cannot be realised at all since the resultant entropy increases. A cycle can be characterised by the values of different physical quantities, functionals of the process - such as: the work done by the actuating medium during the cycle; the amount of heat or some substance obtained ore released by a certain reservoir during the cycle; the amount of a substance produced in the actuating medium due to chemical reactions during the cycle. The values of these quantities depend on the process, i. e. on how the contacts of the actuating medium with the reservoir are established and interrupted, and how the mechanical parameters change. The problem consists in finding such a cycle among those with a fixed duration t, in which one of the functionals acquires a maximum, or a minimum, value under the condition that the values of some other functionals are fixed.

The problems of optimal cycles are special cases of optimisation problems characterised in the previous section. To describe reservoirs, one should use relations of the type (28). If the thermodynamic quantities referring to the actuating medium are denoted by corresponding letters without superscripts (E, N_i, S, T, etc.), the cyclical conditions can be written in the form

$$E(0)=E(\bar{t}), \quad S(0)=S(\bar{t}), \quad N_i(0)=N_i(\bar{t}) \tag{29}$$

From this it follows (see (7) to (9)) that

$$\int_0^{\bar{t}} q\,dt = A, \quad \int_0^{\bar{t}} (v_i + W_i)\,dt = 0, \tag{30}$$
$$\int_0^{\bar{t}} \frac{1}{T}\Big[q - \sum_{i=1}^{m} \tilde{\mu}_i (v_i + W_i)\Big]\,dt = 0$$

where q(t) and $v_i(t)$ are fluxes of energy and the i-th substance from the reservoir into the actuating medium with which it is in contact at a given moment t; A is the work done by the actuating medium during the cycle. In addition, from the fact of non-decrease of entropy in the whole system and, from the cyclical condition of actuating medium entropy there follows the relations

$$\Delta S^{res} = \Delta S^{\Sigma} \geq 0 \tag{31}$$

where ΔS^{res} is the resultant increase in entropy of the reservoirs during the cycle.

When stating the promblems of optimal cycles, it is convenient to use the mean quantities over the cycle. Denoting time averaging by a superscript bar then

$$\bar{z} = \frac{1}{\bar{t}} \int_0^{\bar{t}} z(t)\,dt,$$

and one can rewrite the cyclical condition (30) in the form

$$\bar{q} = \bar{P}, \quad \bar{v}_i + \bar{W}_i = 0, \qquad (32)$$

$$\overline{\left(\frac{q}{T}\right) - \sum_{i=1}^{m} \frac{\tilde{\mu}_i}{T}(v_i + W_i)} = 0,$$

where $\bar{P} = A/\bar{t} = 1/\bar{t}\int_0^{\bar{t}} p\dot{V}dt$.

is the mean power. Not only the work, but all other quantities characterising the cycle (the amount of heat or matter lost or gained by the reservoirs, etc.) are conveniently taken as per unit cycle time. It turns out that if an optimisation problem is of the Lyapunov type, the optimal values of time averages do not depend on the absolute cycle duration t. If the cycle approaches a reversible cycle then all such averages tend to zero.

Let us now consider some examples.

1. Thermomechanical cycle

The problem of an optimal thermodynamic cycle was first formulated by Carnot. Let there exist two reservoirs - a 'hot' one, H, with a high temperature, T^H, and a 'cold' one, L, with a low temperature, T^L, with which the actuating medium can be in contact. The ratio of the work done, A, by the actuating medium during the cycle to the heat, Q^H, gained per cycle from the reservoir H: $\eta = A/Q^H$ is called the cycle efficiency. Carnot formulated the problem of the optimal thermomechanical cycle which provides a maximum, η, and showed that the optimal cycle must approach a reversible cycle. The efficiency of a reversible Carnot cycle is equal to

$$\eta_{rev} = 1 - T^L/T^H \qquad (33)$$

and does not depend on the equation of state of the actuating medium. The structure of a Carnot cycle is 'two isothermals and two adiabatics', where the isothermals correspond to contact of the actuating medium with H and L with an 'infinitesimal' difference between the actuating medium and reservoir temperatures.

Representing as time averages, the efficiency is

$$\eta = \bar{P} / \bar{q}^{H}(t) \tag{34}$$

where $q^{H}(t)$ is the heat flux from H to the actuating medium at time t (if at the time t there is no contact between the actuating medium and H, then $q^{H}(t) = 0$). In a reversible Carnot cycle the averages $\bar{P}$ and $\bar{q}^{H}$ are 'infinitesimal'. The problem of optimal efficiency can be formulated under the condition that the average cycle power, P, has a given, non-zero value. In this case, the maximum efficiency (34) corresponds to the minimum q^{H}. Taking into account the cyclic condition (32) (in the absence of matter transfer and chemical reactions $\underline{v}_i = W_i = 0$), we come to the optimisation problem

$$\overline{q^{H}} \to min; \quad \bar{q} = \bar{P}, \quad \overline{(q/T)} = 0. \tag{35}$$

Here, the heat flux, q, is determined by the heat transfer law, $q = q\ (T, T^{res})$, where $T^{res}(t)$ is the temperature of the reservoir with which the actuating medium is in contact at time t. For control actions one can take directly T(t) and $T^{res}(t)$, where the latter variable can have only two values, T^{H} and T^{L}. Therefore, (35) is a Lyapunov type problem (see (**) at the end of the previous section). The optimal control and the optimal efficiency value in (35) depend, of course, on concrete heat transfer laws. It turns out, however (Rozonoer, Tsirlin, 1983), that for a wide class of heat transfer laws the structure of the optimal cycle remains the same as that of the Carnot cycle - 'two isothermals and two adiabatics' - with the only essentially difference being that in the isothermal regions the temperatures of the actuating medium and of the reservoirs differ not by an 'infinitesimal', but by some finite, quantity (determined from the heat transfer laws and dependent on a given mean power $\bar{P}$). The fractions of the cycle when the actuating medium is in contact with H and L can also be found; adiabatic processes must proceed 'instantaneously' (in a jump). Evidently, an optimal efficiency, $\eta^{*}(\bar{P})$, is always smaller than the value of η_{rev} (33) (since the optimal cycle is irreversible), but $\eta^{*}(\bar{P}) \to \eta_{rev}$ as $\bar{P} \to 0$ (since an optimal cycle approaches a reversible Carnot cycle as $\bar{P} \to 0$). Using (31) one can easily show (see below for an analogous derivation for a diffusion-mechanical cycle) that the efficiency (34) can be given by

$$\eta = \eta_{rev} / (1 + T^L \bar{\sigma} / \bar{P}) \,, \tag{36}$$

where $\bar{\sigma} = \Delta S^{\Sigma} / \bar{t} > 0$ is the mean entropy increase velocity. From this it follows that to the maximum efficiency there corresponds a minimum, $\bar{\sigma}$.

As concerns the thermo-mechanical cycle, one can also formulate another optimisation problem, the one of maximum, mean power. With account taken of (32), this problem can be written in the form

$$\bar{q} \to max \,, \quad (\overline{q/T}) = 0 \,. \tag{37}$$

The optimal cycle structure turns out to be the same as that in the problem of maximum efficiency (Rozonoer, Tsirlin, 1983) (but, of course, with other values for the cycle parameters).

2. Diffusion-mechanical cycle

Let two reservoirs, H and L, have certain values of electro-chemical potential, $\tilde{\mu}^H$, and, $\tilde{\mu}^L$, of some substance; $\tilde{\mu}^H > \tilde{\mu}^L$. On contact between the actuating medium and the reservoirs this substance can be transferred (diffused). The dependence of substance fluxes on the electrochemical potentials of the actuating medium and reservoirs (the law of mass transfer) is assumed to be known. The temperature, T, of the working medium and the reservoirs is assumed to be equal and constant.

Quite similar to the thermo mechanical cycle, where the work done is due to the difference in the reservoir temperatures, in the diffusion-mechanical cycle the work done can be due to the difference in the reservoir electro-chemical potentials. The cycle efficiency can be defined as the ratio of the work, A, to the amount of substance, ΔN^H, transferred to the actuating medium from H (and then given to L) during the cycle, or, as the ratio of the corresponding mean values:

$$\eta = A / \Delta N^H = \bar{P} / \bar{\nu}^H(t) \tag{38}$$

where $\bar{\nu}^H(t)$ is a substance flux from H to the actuating medium

($\bar{v}^H(t) = 0$ if at time t the contact between H and the actuating medium is absent). Let us formulate the problem of maximum efficiency at a fixed mean power $\bar{P}$. Taking into account that the maximum defined by (38) is reached at the minimum in $\bar{v}^H$, and, using the cyclical conditions (32) at W = 0 (m = 1) and a constant T, we arrive at the problem

$$\bar{v}^H \to \min; \quad \overline{\tilde{\mu} v} = \bar{P}\ ; \quad \bar{v} = 0. \tag{39}$$

Problem (39) is analogous to problem (35) in the optimal thermomechanical cycle. As control actions one can take the quantities $\tilde{\mu}(t)$ and $\tilde{\mu}^{res}(t)$ (through which the flux v is expressed), where $\tilde{\mu}^{res}(t)$ takes only two values, $\tilde{\mu}^H$ and $\tilde{\mu}^L$. The optimal cycle can also be shown to consist of four 'regions': two of them correspond to contacts of the actuating medium with H and L, the volume increasing and correspondingly decreasing so that the electrochemical potentials of the actuating medium should be constant; the other two 'regions' are jumps in the volume and the electrochemical potential of the actuating medium. At $\bar{P} \to 0$ the cycle approaches reversibility, and the optimal efficiency approaches its limiting value,

$$\eta_{rev} = \tilde{\mu}^H - \tilde{\mu}^L \tag{40}$$

irrespective of the mass transfer law. Expression (40) can easily be derived from (31). Indeed, the resultant reservoir entropy variation per cycle is

$$\Delta S^{res} = -\frac{Q}{T} + \frac{\tilde{\mu}^H}{T}\Delta N^H - \frac{\tilde{\mu}^L}{T}\Delta N^L,$$

where Q is the energy transferred from the reservoirs to the actuating medium. From this, and accounting for the cyclical condition, Q = A, and (31), we have

$$-\frac{A}{T} + \frac{\Delta N^H}{T}(\tilde{\mu}^H - \tilde{\mu}^L) = \Delta S^{\Sigma} > 0,$$

and for $\eta = A/\Delta N^H$

$$\eta = (\tilde{\mu}^H - \tilde{\mu}^L)/(1 + T\Delta S^{\Sigma}/A).$$

This relation can be rewritten in the form

$$\eta = (\tilde{\mu}^H - \tilde{\mu}^L)/(1 + T\bar{\sigma}/\bar{P}) \tag{41}$$

where $\bar{\sigma} = \Delta S^{\Sigma}/\bar{t}$ is the mean entropy increase rate per cycle. From (41) we arrive at (40).

For the diffusion-mechanical cycle one can also state and solve the problem of maximum power. Using the cyclical conditions, one can easily reduce it to the form

$$\overline{\tilde{\mu} v} \rightarrow max , \quad \bar{v} = 0 . \tag{42}$$

The structure of the optimal cycle remains the same in this problem ('four regions') as in the problem of maximum efficiency. Let us present the result of the solution of the problem (42) for the case when the mass transfer law is linear

$$v = \gamma(\tilde{\mu}^{res} - \tilde{\mu}) ,$$

where γ is a constant. The maximum mean power is now equal to

$$\bar{P}_{max} = \frac{\gamma}{16} (\tilde{\mu}^H - \mu^L)^2$$

and the cycle efficiency is half the limiting value (40). Differences in the electrochemical potentials of the actuating medium and the reservoirs whilst in contact are equal to 1/4 $(\tilde{\mu}^H - \tilde{\mu}^L)$, and the time of contact with H and L to 1/2.

3. Thermomechanical (thermal) pump

As distinguished from a 'direct' thermomechanical cycle (example 1), where work is done at the expense of heat transfer from a hot reservoir H to a cold one, L, in a thermomechanical pump (an 'inverse' cycle) work is expended for heat transfer from L to H. To this end, a contact with L on the actuating medium must have a temperature $T < T^L$, and contact with H, a temperature $T > T^H$. The thermomechanical pump efficiency is defined as a ratio of the heat transferred per cycle to the reservoir H, to the work done. Transforming to mean values, one can write

$$\eta = |\overline{q^H(t)}| / |\overline{P}| , \tag{43}$$

where $q^H(t)$ is, as before, a heat flux from H to the actuating medium, but now $q^H(t) \leq 0$ since heat is transferred, in fact, from the actuating medium to H, in addition, $\overline{P} < 0$ (work is done). On the basis of (31) it can easily be shown that

$$\eta = \frac{T^H}{T^H - T^L} \left(1 - \frac{T^L \overline{\sigma}}{|\overline{P}|}\right), \tag{44}$$

and, for the limiting efficiency, in a reversible process, this gives

$$\eta_{rev} = T^H / (T^H - T^L). \tag{45}$$

The problem of an optimally efficient thermodynamic pump is set up by analogy with (35)

$$\overline{q^H} \to min; \ \overline{q} = \overline{P} ; \ (\overline{q/T}) = 0. \tag{46}$$

This problem has been considered by Rozonoer and Tsirlin (1983).

4. Diffusion-mechanical pump

As distinguished from the diffusion-mechanical cycle (example 2), in the diffusion-mechanical pump work is expended for substance transfer from L to H. When the actuating medium is in contact with L, its electrochemical potential $\tilde{\mu}$ should be lower than $\tilde{\mu}^L$, and when in contact with H, it should be higher than $\tilde{\mu}^H$. The pump efficiency can be determined as

$$\eta = |\overline{\nu^H}| / |\overline{P}| , \tag{47}$$

where the substance flux $\nu^H(t)$ from H to the working medium is negative, the same as for the mean power, $\overline{P}$. Using (31), one can easily show that

$$\eta = (1 - T\overline{\sigma} / |\overline{P}|) / (\tilde{\mu}^H - \tilde{\mu}^L) \tag{48}$$

and for a reversible process we obtain the limiting value

$$\eta_{rev} = 1/(\tilde{\mu}^H - \tilde{\mu}^L). \tag{49}$$

Similarly to (39), the problem of maximum efficiency of the diffusion-mechanical pump with a given mean power can be written in the form

$$\overline{v^H} \to min; \quad \overline{\tilde{\mu}v} = \overline{P}; \quad \overline{v} = 0. \tag{50}$$

Its solution can be obtained similarly to the solution of problem (39).

5. Diffusion-chemical pump

Let us consider a system containing four reservoirs and an actuating medium. Two reservoirs, H_A and L_A, are filled with a substance A with high, $\tilde{\mu}_A^H$, and low, $\tilde{\mu}_A^L$, values of the electrochemical potential ($\tilde{\mu}_A^H > \tilde{\mu}_A^L$). Similarly, the reservoirs H_B and L_B are filled with a substance B with values $\tilde{\mu}_B^H$ and $\tilde{\mu}_B^L$ of the electrochemical potential ($\tilde{\mu}_B^H > \tilde{\mu}_B^L$). The actuating medium contains both A and B, and a reversible chemical reaction A $\rightleftharpoons$ B may also proceed in it; let the electrochemical potentials of substances in the actuating medium be designated by $\tilde{\mu}_A$ and $\tilde{\mu}_B$ (without superscripts). When the actuating medium is in contact with H_A and L_A, A, but not B, is transferred; on contact with H_B and L_B, B, but not A, is transferred ('semitight barriers'). Mechanical parameters of the actuating medium are fixed; all the processes proceed at constant temperature. The only control actions are the establishment and interruption of contacts between the actuating medium and the reservoirs.

The system under consideration can also be used as a 'pump', whose work results in a decrease of the amount of A in L_A, and, the same increase of A in H_A, accompanied by a decrease of the amount of B in H_B, and the same increase of B in L_B. Indeed, on contact of the actuating medium with H_B at $\tilde{\mu}_B < \tilde{\mu}_B^H$, B is transferred from H_B into the actuating medium, and according to Le Chatelier's principle the reaction A $\rightleftharpoons$ B shifts to the left. That is why the electrochemical potential increases. If $\tilde{\mu}_A$ exceeds $\tilde{\mu}_A^H$ and, if, simultaneously, the contact between the actuating medium and the reservoirs H_A is established, the latter acquires some amount of A. Similarly, on contact between the actuating medium and L_B at $\tilde{\mu}_B > \tilde{\mu}_B^L$ the reaction shifts to the right and $\tilde{\mu}_A$ decreases. If $\tilde{\mu}_A$ becomes less than $\tilde{\mu}_A^L$ and, if, at the same time, contact with L_A is established, A will be extracted from L_A. Therefore, the cycle is, in principle, possible which provides transfer of substance A from a low value of the electrochemical potential to a high one at the expense of B transfer from a high to a low potential (in fact, such a 'transfer' is, of course, realised only indirectly since the direct mechanism of the process are the transformations B → A and A → B). For this cycle, different optimisation problems can be formulated, e. g. the problem of the maximum, mean velocity of transfer of substance A to H_A.

Let us denote the difference between the velocities of the forward and backward reactions by $\xi(\tilde{\mu}_A, \tilde{\mu}_B)$. The velocities, W_A and W_B, of changes in the amount of substances A and B in the actuating medium under chemical reactions are equal to

$$W_A = -\xi(\tilde{\mu}_A, \tilde{\mu}_B), \quad W_B = \xi(\tilde{\mu}_A, \tilde{\mu}_B) \tag{51}$$

The equations for changes in the amount of substances in the actuating medium can be written in the form (e. g. (7))

$$\begin{aligned} \dot{N}_A &= -\xi(\tilde{\mu}_A, \tilde{\mu}_B) + \nu_A(\tilde{\mu}_A, \mu_A^{res}) \\ \dot{N}_B &= \xi(\tilde{\mu}_A, \tilde{\mu}_B) + \nu_B(\tilde{\mu}_B, \mu_B^{res}) \end{aligned} \tag{52}$$

where ν_A, ν_B are the fluxes of substances A and B to the actuating medium, $\tilde{\mu}_A^{res}$ (t) and $\tilde{\mu}_B^{res}$(t) are the electrochemical potentials of the reservoirs with which the actuating medium is in contact at time t ($\nu_A = 0$ and $\nu_B = 0$ if the actuating medium is not in contact with the reservoirs L_A, H_A and, correspondingly, L_B, H_B). The control actions in (52) are the functions $\tilde{\mu}_A^{res}$ (t) and $\tilde{\mu}_B^{res}$ (t), which take two values each, $\tilde{\mu}_A^L$, $\tilde{\mu}_A^H$ and $\tilde{\mu}_B^L$, $\tilde{\mu}_B^H$, respectively. (52) must be supplied with expressions for $\tilde{\mu}_A$ and $\tilde{\mu}_B$ expressed through N_A and N_B and, therefore, the right-hand sides in (52) depend, in fact, not only on the control actions, but also on phase coordinates. Thus, the optimisation problems for the controlled system (52) are not

Lyapunov type problems. However, in some particular cases they can be solved easily.

Let us consider system (52) under the assumptions that the velocity of substance transfer from the reservoirs to the actuating medium greatly exceeds the chemical reaction rate, and that the cycle duration is sufficiently large. In this case the electrochemical potentials of substances in the actuating medium become, for a negligibly small time, practically equal to the electrochemical potentials of the reservoirs with which the actuating medium is in contact, and $N_A(t)$, $N_B(t)$ acquire at this time essentially constant values (determined by the values of the electrochemical potentials). Therefore, during the main cycle time the relations

$$\nu_A = \xi(\tilde{\mu}_A^{res}, \tilde{\mu}_B^{res}), \quad \nu_B = -\xi(\mu_A^{res}, \mu_B^{res}) \tag{53}$$

are approximately fulfilled. The problem of the maximum mean velocity of transfer of the substance A to H_A takes here the form

$$\bar{\nu}_A^H \to \min; \quad \bar{\xi} = 0, \tag{54}$$

where $\nu_A^H(t)$ is the flux of A to the actuating medium on contact with H_A at time t ($\nu_A^H \leq 0$ and ν_A^H (t) = 0 in the absence of contact with H_A at the moment t). Since ν_A^H, like ξ, is expressed through μ_A^{res}, μ_B^{res}, (54) is a Lyapunov type problem which is easily solved. Note that work performance by the pump is only possible under the conditions

$$\xi(\tilde{\mu}_A^H, \tilde{\mu}_B^H) < 0, \quad \xi(\tilde{\mu}_A^L, \tilde{\mu}_B^L) > 0.$$

Systems, in which more complicated chemical reactions are used for substance transfer from L_A to H_A, can be considered in an analogous way.

Examples 1 to 5 illustrate the possibility of exact formulation and solution for various problems of optimal thermodynamic cycles.

Control and Regulation in Biological Systems

Order and Disorder in Biological Control Systems

R. Rosen

Two basic characteristics of biological systems are the complex developmental processes which give rise to them, and the equally complex physiological processes which maintain them. These have been the subject of intensive study over the years because of their dramatic nature and their extreme practical importance. Not nearly so much attention, however, has been given to processes like senescence and mortality, which are in their way equally dramatic, equally important, and in a certain sense even more puzzling. Why should a physiological system, which is a collection of interacting homeostats designed precisely to buffer the organism against externally imposed stresses, not be capable of indefinite persistence? Why should organisms have rather well-defined life-spans, and why should these life-spans vary by factors of two, three, or more, between species otherwise taxonomically similar and obviously conforming to the same general *bauplan*?

Perhaps one reason that degenerative processes have been neglected in comparison with generative ones is that we intuitively expect all systems to fail eventually. Our own mechanical artifacts, with which we have had the greatest experience, do break down regularly; even with the most careful maintenance, machines and electronic devices malfunction and ultimately fail. We expect that the general trend of things will always go from 'order' to 'disorder'. Biological phenomena, like development and physiology, are so striking precisely because they appear to violate this trend; that is why we must strive so hard to explain them. Senescence and mortality, on the other hand, appear to be fully consistent with the trend, and thus apparently (but only apparently) require no special explanations.

Thermodynamics and Regulation of Biological and Processes

If development and physiology are respectively the processes through which order is generated and maintained, then we may regard senescence as in some sense a disordering. We are going to take the viewpoint that, in general, 'disordering' in a system is manifested by the deviation of system behaviour from some nominal standard of behaviour. Furthermore, this standard against which system behaviour is measured is, in the most general sense, a model of the system itself. This model or standard tells us how the system ought to behave; i. e. it generates expectations or predictions about the system. When actual system behaviour coincides with that of the standard, this coincidence represents order in the system. When system behaviour departs from that of the standard, we shall say that the system is disordering with respect to the standard; the degree of disorder is measured by the extent of the departure from standard behaviour.

The reader may here object that in science, if a discrepancy arises between the actual behaviour of a system and that predicted on the basis of some model, then it is the model which is to be rejected. That is, the traditional scientific method takes actual system behaviour as the standard, and judges the model with respect to this standard. However, in biology, we freely invert this tradition. Thus, we speak of 'errors' in the genetic code (e. g., mistranscriptions, mistranslations) whenever there is a discrepancy between, say, an actual generated polypeptide sequence and that expected on the basis of the code; we speak of neuronal 'error' when a neuron fires even though its excitation is below threshold, or fails to fire when its excitation is above threshold. In all such cases, the 'error', which is the discrepancy between actual behaviour and nominal behaviour, is imputed to the system and not to the standard or model. In all these cases, and many others which might be cited, we thus regard the system as disordering with respect to the standard.

Let us look at these notions in a limited but familiar context. A typical model of a biological process might take the form

$$dx/dt = F(x(t), \alpha(t)) \qquad (1)$$

where x(t) is some appropriate state vector, and $\alpha(t)$ is a vector of controls or inputs or forcings. When integrated, these equations of motion specify trajectories of the form

$$x(t) = \phi(x(0), \alpha(t)). \tag{2}$$

Thus, knowing an initial state, we can predict the system response to any particular control sequence $\alpha(t)$.

Now a control sequence $\alpha(t)$ generally arises because our given system is interacting with other systems; i. e. is interacting with its environment. The existence of such interactions means that our system of interest is open, in the sense of thermodynamics. However, these equations of motion also specify precisely how open our system is assumed to be; i. e. it expresses absolute limitations on the character of system-environment interactions. That is, a model like (1) can tell us what happens if we make our system thermodynamically more close (i. e. if we abolish certain system-environment interactions characterised by (1)), but it cannot in principle tell us what happens if we make our system more open.

Now any model (1) of an organic process is necessarily an abstraction, in the sense that features present in the system itself are neglected or omitted in the model. This is especially true with respect to potential interactive capabilities. In this sense, our model of such a system must necessarily be more (thermodynamically) closed than is the system itself. A basic corollary of this fact is the following: under very general conditions, the behaviour of such a system will necessarily ultimately begin to deviate from that predicted on the basis of the model (1), and the magnitude of this deviation will tend to grow in time. In other words, if we take the model (1) as our standard of order, the behaviour of the actual system will appear to become progressively disordered in time.

These considerations have some important implications, which we shall now briefly sketch:

1. What we have called 'disordering' (i. e. the deviation of the behaviour of a system from that expected on the basis of a given standard or model, which arises from the fact that the system is more open to interaction than is the standard) subsumes such concepts as error, system failure, and emergence. In all of these we find an unpredictable and apparently causeless departure from expected behaviour, arising precisely because the interactions responsible for such behaviour in the actual system have been abstracted out of our model, and hence are invisible in it.

2. Such 'disordering' arises from the very nature of abstraction, and hence cannot be removed by simply passing to a more comprehensive model. Any model, however comprehensive, allows disordering with respect to it as long as it is an abstraction. Disordering of this kind is hence an essential fact of life. In this sense, it is reminiscent of the Gödelian limitations of axiom systems in arithmetics; in fact, the two kinds of limitations are closely related, but we cannot enlarge on this matter here.

3. The employment of predictive models for the purpose of controlling a system (of which medical therapeutics is a specific instance) will generally result in a variety of unpredictable effects on system behaviour ('side-effects') arising from the implementation of a control strategy. The appearance of such 'side-effects' is in fact a familiar feature of attempts to control complex system behaviours. As we have seen above, their appearance is a consequence of using an abstraction as the standard of system behaviour, and cannot be eliminated by simply passing to a more comprehensive standard.

4. The disorderings we have described pertain to any behaviour which is model-based. This is true not only of man-made models, but also of the intrinsic models which biological systems generate and utilise to modify their own behaviour. Indeed, much of the adaptive physiological behaviour which has heretofore been considered exclusively in cybernetic terms (i. e. on feedback regulation through the generation of an 'error signal') is best regarded as feedforward regulation based on such intrinsic predictive models. Such important physiological systems as the endocrine system and the immune system are almost entirely governed by such feedforward,

model-based regulators. Accordingly, they are subject to exactly the same kinds of considerations as have been developed above. Ultimately, the systems they control will disorder relative to the control models, resulting in maladaptive responses and ultimate system failure. We have elsewhere (Rosen, 1978 a) proposed that such disordering is responsible for what we recognise as senescence in complex systems; we refer the reader to that work for fuller details and examples. We may note here, however, that one of the main features of system failure arising from such disorderings is in a sense that the failure is global and not local; i. e. the system as a whole can fail, even though each part of the system, considered in isolation, does not fail in any detectable way.

From these brief considerations, we can see that the 'disordering' which occurs in biological systems can be vastly different from that characteristic of physical systems, which are related to increase of entropy via the Second Law of Thermodynamics. The disorderings we have considered above arise from the fact that a real system is more open to interaction than any abstract standard or model can be, and is thus more closely akin to the ideas of 'wear-and-tear' which have often been invoked to account for systems failures in biology and elsewhere. In physics, on the other hand, disordering arises from closing and isolating a system. Since thermodynamic ideas have been so influential in establishing the trend of systems from 'order' to 'disorder' which we mentioned at the outset, we will conclude our discussion by relating our considerations to those developed for thermodynamic systems.

In thermodynamics, the standard systems are those which are closed and isolated (i. e. those which are exchanging neither matter nor energy with their environements). Such systems can be shown to relax towards a unique equilibrium state (thermodynamic equilibrium). In simple systems, like fluids, this state of thermodynamic equilibrium is one of spatial and temporal uniformity possessing no gradients of any kind. This state of thermodynamic equilibrium is chosen as the standard of disorder; hence any non-equilibrium state of such a system is necessarily ordered with respect to it. Moreover, the dynamical process of relaxation towards equilibrium in such a system, which must begin inititally with an ordered state and asymptotically approach equilibrium, is one of disordering.

A certain thermodynamic state function the entropy, always increases as such a closed, isolated system relaxes towards equilibrium and assumes its unique maximum on the equilibrium state itself. In fact, entropy is a Lyapunov function for the dynamics; indeed, the assertion that entropy is a Lyapunov function, together with the necessary linearity of this dynamics near equilibrium, are essentially sufficient to completely determine the near-equilibrium dynamics; this is how the rate equations of 'irreversible thermodynamics' are derived (e. g. de Groot and Mazur, 1963; Prigogine, 1967).

For closed, isolated systems, then, we have three equivalent propositions

1. Such systems autonomously relax towards a state of thermodynamic equilibrium;
2. The state of thermodynamic equilibrium is the standard of disorder for states; hence every non-equilibrium state is ordered with respect to it. Further, the relaxation towards equilibrium is necessarily a disordering of the system itself;
3. A certain specific state function (entropy) is a Lyapunov function for dynamics in such systems.

It is this last proposition which gives entropy its special role in thermodynamics, and which allows us to identify 'disorder' in thermodynamic states with the value of the entropy on those states.

If we now take such a closed, isolated system and open it up to environmental interaction. several things happen. First of all, the new forcings imposed on the system by virtue of these interactions will obviously change the dynamics of the system. The entropy function, which was a Lyapunov function for the autonomous dynamics when the system was closed and isolated, will in general have no special relation to the new dynamics, and loses thereby all its orginal theoretical significance.

Second, because the dynamics has been change, new stable attractors appear, and the system will now autonomously move towards the one in whose basin of attraction it happens to lie initially. From the thermodynamic standpoint, of course, such movement away from thermodynamic equilibrium represents an ordering of the system with respect to the thermodynamic standard of disorder. For this reason, there have been numerous attempts to regard biological morphogenesis, and other kinds of pattern generation, in this light, simply on the grounds that biological systems are ordered in some absolute sense.

On the other hand, from the standpoint of the new dynamics in the open system, the autonomous movement of the system towards an attractor is exactly analogous to the relaxation of a closed isolated system towards equilibrium (which is the only available attractor for this dynamics). If we take the thermodynamic parlance seriously, the proposition (2) above says that the conditions of maximal disorder for an arbitrary dynamical system are determined by its attractor; these are what the system autonomously tend towards, and what energy expenditure is required to keep them away from. That is, if some open system is chosen as standard, then that system necessarily disorders with respect to itself as it approaches its attractors. Moreover, with respect to this new choice of standard, any change in the character of system-environment interactions (whether their effect is to open the system further, or to close the system) will cause a deviation from this standard, and hence by definition will order the system.

Thus, from our standpoint, the thermodynamic characterisation of order and disorder is inappropriate and unnatural in several ways. First, it chooses the dynamics of closed isolated systems as standard, It then chooses the attractors of such dynamics (thermodynamic equilibria), which are the states to which such dynamics autonomously tend, as standards of disorder. Finally, it measures disorder, according to these standards, by the values of entropy, which happens to be a Lyapunov function for the standard dynamics, hence identifying entropy with disorder.

On the other hand, we have argued that there is nothing special about closed, isolated systems; indeed, from the standpoint of biology, they are most unnatural systems to choose for standards. We have seen that any dynamics can be chosen as standard, and once we have left the thermodynamic realm, it is much more natural to regard this standard as establishing the norm of order and not disorder. From our viewpoint thermodynamics represents a very special case of our treatment, if we interchange the terms 'order' and 'disorder'. Indeed, the thermodynamic usage of these terms has no particular justification, except that of historical tradition.

As we have seen, the concept of entropy loses all significance once we depart from the choice of closed, isolated systems as standards. We can retain the term for open systems, but only at the cost of inventing an increasingly unwieldy formalism of 'entropy fluxes', which are attempts to create new Lyapunov functions for the open system dynamics. If the system is open enough, this whole approach fails; the problems associated with open systems are dynamical problems and not thermodynamic ones.

Instead of entropy, the treatment sketched above gives as a measure of disorder the discrepancy between the actual behaviour of the system and that of the standard or model which we have chosen to characterise order (or any numerical function of this discrepancy). Indeed, thermodynamic entropy is a very special case of this, in which the standard is the equilibrium of the system when closed and isolated, and with the terms 'order' and 'disorder' interchanged.

The discussion we have provided above is but one small part of the larger problem of establishing the relation betwen a real system (physical, biological, or social) and the models we can make of that system (see Rosen, 1978 b). This is not only the basic problem of theoretical science; it is also central in understanding the behaviour of systems which contain models, either of themselves or of their environments, and utilise these models for purposes of control. The existence of such internal models of self and/or environment is one of the essential differences between biological systems and non-biological ones, but their presence has been essentially neglected in our reductionistic haste.

Simple Feedback System Used to Model Biomolecular Reactions in Drug-Receptor Populations

A. Seese, P. P. Mager

Introduction

Recent developments in the study of biological phenomena as processes at the molecular level have concerned the action of the so called receptor unities. These specific molecules play a crucial role as structural entities, in certain biological objects, which are responsible for the initiation of various activities. Generally, the interaction between a bioactive agent and its receptors can be regarded as a change of chemical and physical order leading to the stimulation of a certain effect. Recently, some progress has been achieved in the identification and isolation of various receptor types. But often, the concrete molecular characterisation of the receptors and the exact chemically reactive forms are still unknown because the receptors are very difficult to probe. Nevertheless, all the published material clearly proves the importance of specific receptors and their ability to induce activity. On that basis, some regulatory properties and control effects in biological systems can be understood. To clarify the fundamental processes leading to receptor excitation and connected phenomena, a theoretical background for the analysis of the experimental data is required. The development of the several models is still progressing and the aim of such treatment is the presentation of a suitable model from which satisfactory interpretation of observed phenomena is possible taking into account a variety of physical and chemical conditions. For instance, the magnitude of the receptor output signal is an interesting quantity which has to be calculated via its dependence on some parameters and from such a theory, it must also be possible to formulate some practical consequences which permit experimental confirmation. It is clear that the proposed models cannot describe the whole complexity of a

Thermodynamics and Regulation of Biological and Processes

biological system. They only analyse a specific problem. Transition from the empirical assumptions to a well-founded theory takes place in a stepwise fashion.

In this paper, interest is directed to pharmacon-receptor interaction; an important field of research for the development of active agents as well as for the understanding of elementary processes in order to clarify the nature of drug action. The task consists of the study of the influcence of an exogenous substance on receptor recognition sites. Physically, this corresponds to an open system situation in a many body problem. The direct drug receptor interaction in the target tissue is assumed to be a chemical reaction governed by kinetic features. Thus, one has to deal with a system operating far from equilibrium. The most important feature of this pharmacodynamic interaction is that all events occurring on a molecular level lead to a macroscopic effect measured as a response from the total system. Usually, this effect is highly dependent on a unique parameter. Often, a slight change of a certain control variable yields an extreme global reaction. Such a phenomenon is characterised by the appearance of a cooperativity between the population unities expressed by non-linear terms in the interaction structure. The possible formation of new kinds of ordered phases is an essential, intrinsic property of a biological regulating system such as a receptor arrangement. For this reason, the pharmacon-receptor interaction may be classified within the region of the dynamic critical phenomena. The description of this field is carried out with the help of several theories described by the term 'synergetics' (Haken, 1980). In this way, it is hoped that one may interpret some drug-receptor interaction related phenomena on the basis of non-equilibrium considerations.

The next section is devoted to the presentation of models. The model ansatzes and the model parameters are discussed with respect to their physical and chemical realisation. For simplicity, some restrictions are assumed within the parameter set. The local mass balance equations for the reacting species are written down. The special feedback effect is introduced from a plausible physical picture. The steady state values are determined characterising the different non-equilibrium phases. Then, linear stability analysis is carried out to find the critical value of the control variable and some parametric restrictions yielding an instability point. The consideration is limited to a continuous transition between homogeneous concentration ordered phases. Next, in the following section, using the reductive pertubation method, the transition, valid for the description of the behaviour of the relevant order parameter variable in the neighbourhood of the instability, from the deterministic kinetic equations to the stochastic Langevin type equation is performed. The steady state solution of the corresponding Fokker-Planck equation is a probability distribution which may be extracted and, from this, some interesting results relating to certain state variables can be calculated. Some consideration is given to a possible dynamic interpretation of the system and finally, in the last section, the results obtained are interpreted with respect to their practical realisation in the drug-receptor interaction process. As an example, the acetylcholine receptor will serve to demonstrate a possible successful application of the proposed models. Some suggestions are made to stimulate measurements which may support the predicted results. Finally a critical review of the models is given and some conclusions regarding a more realistic extension are presented.

Models, steady states and stability

In this section, some simple models are proposed for a possible treatment of the pharmacon-receptor interaction. The drug is characterised as a certain chemical compound with a well-known chemical and physical ordering structure. The drug input is an externally controlled variable. The receptor system consists of special molecules arranged in a certain manner in the biological object. The receptors have to accept the message from the drug molecules, resulting in the stimulation of a signal which excites a biological activity. In the first sense, the drug-receptor interaction can be regarded as a chemical reaction process and a sequence of bimolecular steps is assumed as the simplest, and most often used, reaction scheme for the illustration of the pharmacodynamic phase. The drug and the receptor associate to form an active intermediate complex which releases the effect which will then be degraded into some other product and the pure receptor form made available for the next reaction. The absolute stoichiometries of the interaction process are unknown because of the complicated *in vivo* complexity. As the starting point of the investigations, the following sequence of reactions is proposed

$$D_1 + R \underset{k_{-1}}{\overset{k_1}{\rightleftharpoons}} [D_1 R] \xrightarrow{k_2} R + P_1 \tag{1}$$

where R is identified with the receptor sites. D_1 denotes the drug molecules. D_1R stands for the active complex and P_1 means any product. The k_i's represent the rates. D_1 is an externally controlled variable and may be kept in excess at d_1° through constant supply. Also, the initial concentration of the receptors available for the reaction is limited at the fixed value, r_o. Thus, the following restrictions can be formulated

$$r = (r_o - x) \qquad d_1 = d_1^\circ . \tag{2}$$

x, r denote the concentrations of the complex $[D_1R]$, the interesting variable, and the receptors, respectively. Introducing an additional bimolecular reaction scheme corresponding to a competitive action,

$$D_2 + R \underset{g_{-1}}{\overset{g_1}{\rightleftharpoons}} [D_2 R] \xrightarrow{g_2} R + P_2 \tag{3}$$

where now, the restrictions take the form

$$r = (r_o - x - y) \qquad d_1 = d_1^\circ , \; d_2 = d_2^\circ \tag{4}$$

and where y is the concentration of the complex $[D_2R]$. Again, the concentration of the competitive substance is maintained constant so that $d_2 = d_2^\circ$.

The evolution of the local concentrations $X_i(r, t)$ of the reacting species are given by the common balance equations

$$\frac{\partial X_i}{\partial t} = D_i (\nabla X_i)^2 + F_i(\lambda_\alpha, X_j) \tag{5}$$

where $D_i(\nabla X_i)^2$ is the diffusion contribution; approximated from Fick's law. The term F_i contains the chemical rates for the forward and backward flow rates. The λ_α's are the parameters, i. e. as for the external concentrations as well as for the intrinsic rate constants. Considering the restrictions (4), the reaction diffusion equations for the combined systems (1) and (3) now become

$$\frac{\partial x}{\partial t} = k_1 d_1^\circ r_0 - x(k_{-1} + k_2 + k_1 d_1^\circ) + D_x \frac{\partial^2 x}{\partial r^2} - y k_1 d_1^\circ \quad (6)$$

$$\frac{\partial y}{\partial t} = q_1 d_2^\circ r_0 - y(q_{-1} + q_2 + q_1 d_2^\circ) + D_y \frac{\partial^2 y}{\partial r^2} - x q_1 d_2^\circ . \quad (7)$$

These equations contain linear dependiencies only. Such a model does not exhibit the possibility for the appearance of a phase observe transition. To observe critical behaviour, non-linearities are required. There are several mechanisms which can produce instability. The well-known example is an autocatalytic reaction containing a trimolecular step. But such a reaction has a very small probability, especially in receptor kinetics. A feedback can also be realised by the influence of physical factors on certain model parameters. This seems more probable than to fulfill some stoichiometric conditions. In the models here proposed, a feedback will be introduced by the concentration dependence of a certain rate. The magnitude of the forward rate k_1 may depend on the instantaneous concentration of the active complex. The system acts kinetically upon itself so that the reaction is accelerated. This positive feedback could be the consequence of a change of the polar properties within the media during the complex forming process. In this way, the reaction energy is modified. Formally, a Taylor expansion of the rate k_1 related to a small complex concentration, x, has to be carried out to describe such an effect

$$k_1 \rightarrow k_1 = k_{10} + k_{11} x + k_{12} x^2 + O(x) \quad (8)$$

In this analysis only terms up to the quadratic contribution were considered but replacing k_1 by the non-linear relationship (8), the evolution equations become

$$\begin{aligned} \frac{\partial x}{\partial t} = {} & k_{10} d_1^\circ r_0 - x(k_{-1} + k_2 + k_{11} d_1^\circ y - k_{11} d_1^\circ r_0 + \\ & + k_{10} d_1^\circ) + x^2 (k_{12} d_1^\circ r_0 - k_{12} d_1^\circ y - k_{11} d_1^\circ) - \\ & - x^3 k_{12} d_1^\circ - k_{10} d_1^\circ y + D_x \frac{\partial^2 x}{\partial r^2} \end{aligned} \quad (9)$$

$$\frac{\partial y}{\partial t} = g_1 d_2^\circ r_0 - y(g_{-1} + g_2 + g_1 d_2^\circ) - g_1 d_2^\circ x + D_y \frac{\partial^2 y}{\partial r^2} . \quad (10)$$

The cubic term in (9) will play an important role for the system behaviour near the instability.

The steady states of the chemical reactions (1) and (3) are determined from the real zeros of (9) and (10). By a suitable choice of the parameter values, the possible multiple roots coincide in the instability point where the system may convert into a new phase. Distinguishing the different cases, the following homogeneous steady states for the concentration of the active complex are obtained

A (linear bimolecular reaction model): $X_{ss}^A = \dfrac{k_1 d_1^\circ r_0}{(k_{-1} + k_2 + k_1 d_1^\circ)}$

B (linear coupled models) $X_{ss}^B = \dfrac{k_1 d_1^\circ r_0 (1 - G)}{(k_{-1} + k_2 + k_1 d_1^\circ (1 - G))}$

with $G = g_1 d_2^\circ / (g_{-1} + g_2 + g_1 d_2^\circ)$

C (linear feedback extended bimolecular reaction model): $X_{ss}^C = \frac{1}{3}(r_0 - k_{11}/k_{12})$ (11)

D (linear coupled and feedback extended model) $X_{ss}^D = \frac{1}{3}(r_0 - k_{11}/k_{12})$

$$Y_{ss}^D = \frac{2}{3} G (r_0 - k_{11}/3 k_{12}) .$$

For the cases A and B, the reaction equations only admit a single steady state solution. In models C and D, the triple root of the conditional equations which correspond to the requirements for the existence of the instability point are determined. Therefore, some parameter restrictions must be employed viz.

$$(r_0 - k_{11}/k_{12})^3 = 27\, k_{10}\, r_0 / k_{12} \tag{12}$$

$$\left(r_0 - \frac{k_{11}}{k_{12}}\right) = \left\{ 3 \left[\frac{k_{-1} + k_2}{k_{12} d_1^0 (1-G)} + \frac{k_{10}}{k_{12}} - \frac{k_{11}}{k_{12}} r_0 \right] \right\}^{1/2} \tag{13}$$

where G, equal to zero, represents model C.

Linear analysis is used to test the stability of the steady states extracted. This method investigates the behaviour of the steady states for infinitesimally small disturbances exp $(\omega t + i(qr))$, for this the rate equations are linearised around their reference states, X_i^{SS} and so the deviations $\bar{x}_i = X_i - X_i^{SS}$ obey the equations

$$\frac{\partial \bar{x}_i}{\partial t} = \left(\frac{\partial F_i}{\partial x}\right)_{X^{ss}} \bar{x}_i - q^2 D_i \bar{x}_i = \hat{K} \bar{x}_i . \tag{14}$$

The eigenvalue of the matrix $\hat{K}$ are estimated from the corresponding dispersion relation. An instability occurs if the conditions of negative, nonzero real roots for all ω's is violated. A homogeneous transition takes place for one eigenvalue $\omega = 0$ at $q = 0$ and $\lambda_{d'} = \lambda_{d'}^{c}$. Returning to the models, the application of the concept outlined above yields the following eigenvalues

$$A: \quad \omega^A = -(k_{-1} + k_2 + k_1 d_1^0)$$

$$B: \quad \omega^B_{1/2} = -\frac{\alpha}{2} \pm \sqrt{(\alpha/2)^2 - \beta}$$

with $\alpha = k_{-1} + k_2 + k_1 d_1^0 + g_{-1} + g_2 + g_1 d_2^0$

$\beta = k_1 d_1^0 g_1 d_2^0 - (k_{-1} + k_2 + k_1 d_1^0)/(g_{-1} + g_2 + g_1 d_2^0)$

$$C: \quad \omega^C = -\frac{(k_{10} d_1^0 + k_{-1} + k_2)}{k_{12} d_1^0} + \frac{1}{3} r_0^2 - \frac{1}{3} r_0 k_{11}/k_{12} + \frac{1}{3}(k_{11}/k_{12})^2$$

$$D: \quad \omega^D = -\frac{1}{2}(\alpha + \beta) \pm \sqrt{\left(\frac{\alpha+\beta}{2}\right)^2 + \gamma\delta + \alpha\beta} \tag{15}$$

with $\alpha = -(k_{-1} + k_2 + k_{10} d_1^0 - 2 k_{12} d_1^0 X_{ss}^D (r_0 - Y_{ss}^D - \frac{3}{2} X_{ss}^D))$

$\beta = -(g_{-1} + g_2 + g_1 d_2^0)$

$\gamma = -k_{10} d_1^0 - k_{12} d_1^0 (X_{ss}^D)^2 \qquad \delta = -g_1 d_2^0$.

For model A, ω^A always remains negative and real because the rates are positive quantities. Also, a real ω_1^B cannot vanish. The necessary condition $\beta = 0$ implies a critical concentration, d_1^{oc}, which is negatively valued and hence practically unrealisable. Thus, in models A and B, a phase transition cannot appear a fact clear from the linear structure of the equations. The system moves along a stable thermal pathway for all parameter values. Applying the condition $\omega = 0$ to cases C and D, a bifurcation becomes possible for the following critical concentrations

$$
\begin{aligned}
C:&\quad d_1^{oc} = (k_{-1}+k_2)k_{12}/\left(\tfrac{1}{3} r_0^2 k_{12}^2 - k_{12}k_{10} + \tfrac{1}{3} k_{11}^2 - \tfrac{1}{3} k_{11}k_{12} r_0\right) \\
D:&\quad d_1^{oc} = (k_{-1}+k_2)/8k_{10}(1-G).
\end{aligned} \tag{16}
$$

In model D, k_{11} will be taken as zero for simplicity. As seen from (16), the d_1^{oc} values are realisable so that a phase transition can happen. The remarkable difference between cases C and D, setting $k_{11} = 0$, is due to the term $(1 - G)$. The presence of $(1 - G)^{-1}$ has an increasing effect on d_1^o in comparison to the value for the uncoupled model. The influence of the second reaction can be interpreted as a constraint which shifts the critical point towards higher values. This shows the inhibitory action of substance D_2 on the receptor excitation. This result will be of importance in any practical proof of the theory. Next, the equations of motion for the deviation variables $\bar{x}_i = X_i - X_i^{SS}$ are written down, with the critical concentrations, in a suitable form

$$
C:\quad \frac{\partial \bar{x}}{\partial t} = -\bar{x} f(k_{ij}, r_0)(d_1^{oc} - d_1^o) - \bar{x}^3 k_{12} d_1^o \tag{17}
$$

with $f = \left(k_{10}k_{12} - \tfrac{1}{3} k_{12}^2 r_0^2 + \tfrac{1}{3} k_{11}^2 - k_{11}k_{12} r_0\right)/k_{12}$

$$
\begin{aligned}
D:\quad \frac{\partial \bar{x}}{\partial t} =& -\bar{x} g(k_{ij}, r_0, d_2^o)(d_1^{oc} - d_1^o) - \bar{x}\bar{y}\left(\tfrac{2}{3} k_{12} d_1^o r_0\right) - \\
& - \bar{x}^2\left(\tfrac{2}{3} k_{12} d_1^o r_0 G\right) - \bar{x}^2\bar{y}(k_{12} d_1^o) - \bar{x}^3(k_{12} d_1^o) - \\
& - \bar{y}\left(k_{12} d_1^o r_0^2 \tfrac{1}{9} + k_{10} d_1^o\right)
\end{aligned} \tag{18}
$$

$$
\frac{\partial \bar{y}}{\partial t} = -\bar{y}(g_{-1} + g_2 + g_1 d_2^o) - \bar{x}(g_1 d_2^o)
$$

with $g = 8k_{10}(1-G)\{1 + 4k_{10}G/((k_{-1}+k_2)/d_1^o - 8k_{10}(1-G))\}$.

Values $d_1^o > d_1^{oc}$ yield a negative term in (17) and (18) behind $\bar{x}$. In the opposite case $d_1^o < d_1^{oc}$, the linear coefficient has a positive sign. Thus, the factor behind $\bar{x}$ changes its sign at the transition point a remarkable property for a phase bifurcation. The critical concentration d_1^{oc} marks the threshold at which the system behaviour is changed in an explosive manner. The evolution equations obtained above are the starting point for a further treatment, especially, with respect to the study of behaviour near the instability.

3. Statics and dynamics near the instability

The aim of this chapter is to study the behaviour of the proposed models near the bifurcation point. In the vicinity of the instability, the statics and dynamics of the system are governed by the critical modes. These excitations represent the relevant part of the modes determining the macroscopic phenomena within the transition. The approach to the bifurcation point is characterised by some anomalies in the properties induced by long range fluctuations. Because of the high correlation of the fluctuations, the system may react very sensitively. For the description of the transition regime, the effect of these fluctuations must be incorporated into the model. For this reason, a stochastic treatment is required. This consists of the reduction of the evolutionary kinetic equation set to a simple, generally single, equation of motion for the relevant critical mode amplitude or so-called order parameter quantity. Such a derivation formalism is given by the reductive pertubation method based on an expansion of the variables in terms of the deviation ($\lambda_{\alpha'} - \lambda_{\alpha'}^c$). Following Nitzan (1977), the final equation of motion for the extracted order parameter $W(R, \tau)$ is given as

$$\frac{\partial W}{\partial \tau} = \varepsilon^2 (a+b) W + \varepsilon b W^2 + (b+c) W^3 \qquad (19)$$

with $\varepsilon^2 = (\lambda_{\alpha'} - \lambda_{\alpha'}^c)$ for the calculation to order ε^3. The coefficients of (19) can be expressed from those of the initial kinetic equations with the help of the n'th order derivatives of the function, F, and the eigenvectors of the matrix $\hat{K}$. (19) describes the variation of the variable W near the transition point on long time and length scales ($\partial/\partial t \to \varepsilon^2 \partial/\partial \tau$: $\nabla_r \to \varepsilon \nabla_R$).

In the next step, W will be regarded as a random variable. This means consideration of fluctuations and their effect on the bifurcation behaviour. For the fluctuation force, a Gaussian noise term is added. In this way, from (19) the generalised Langevin type equation of the problem may be obtained

$$\frac{\partial W}{\partial \tau} = \varepsilon^2(a+b)W + \varepsilon b W^2 + (b+c)W^3 + \eta(R,\tau) = \hat{F} + \eta . \quad (20)$$

For the probability function $P(W(R, \tau))$, the equivalent Fokker-Planck equation holds

$$\frac{\partial P}{\partial \tau} = -\frac{\partial}{\partial W}\{F(W)P\} + \Gamma \frac{\partial^2}{\partial W^2}\{P\} \quad (21)$$

where Γ is connected with the Gaussian distribution by

$$\langle \eta(R,\tau)\,\eta(R',\tau)\rangle = \Gamma\,\delta(R-R')\,\delta(\tau-\tau') \quad (22)$$

and will be assumed to be constant. The steady state solution from (21) is extracted as

$$P(W(R)) = \frac{1}{N}\exp\{-\frac{1}{N}\int dR\,\hat{\Phi}(W(R))\} = \frac{1}{N}\exp(-\Phi) \quad (23)$$

$$\text{with } \hat{F}(W) = -\partial\hat{\Phi}(W)/\partial W .$$

The normalisation factor N is determined by $\int P(W)\,dR = 1$. (23), now, allows the calculation of some interesting results.

For models C and D, the formalism outlined above yields the following results. In model C, one has to deal with a single state variable. A reduction of the number of the observables is not required. The relevant order parameter is the concentration of the active complex, i. e. the concentration deviation from the steady state value. In stochastic terminology, the random quantity, X, obeys the equation

$$\frac{\partial \bar{x}}{\partial t} = -\bar{x}\,f(k_{ij}, r_0)(d_1^{oc} - d_1^{o}) - x^3 k_{12} d_1^{o} + \eta(\,,\,) = \hat{F} + \eta . \quad (24)$$

The potential ϕ in the steady state probability distribution function is obtained from the volume integral over the integral $(- \int F(X)d\bar{x})$ as

$$\phi = \frac{1}{\Gamma}\int dr\{(d_1^{oc} - d_1^o) f(k_{ij}, r_o)\frac{\bar{x}^2}{2} + k_{12} d_1^o \frac{\bar{x}^4}{4}\}. \qquad (25)$$

To calculate the mean value for the stochastic variable $\bar{x}$, one has to minimise the Ginzburg-Landau type functional (25) after having averaged over some volume. This requires the substitution of $x(\vec{r})$ by the $\vec{r}$-independent average justified in the homogeneous steady state transition case considered. Thus, the minimisation procedure takes the form

$$\frac{\partial\phi}{\partial\langle\bar{x}\rangle} = 0 = (d_1^{oc} - d_1^o) f(k_{ij}, r_o)\frac{\langle\bar{x}\rangle}{\Gamma} + k_{12} d_1^o \frac{\langle\bar{x}\rangle^3}{\Gamma} \qquad (26)$$

For the region above the instability, (26) yields

$$\langle\bar{x}\rangle = \{(d_1^{oc} - d_1^o) f(k_{ij}, r_o)/k_{12} d_1^o\}^{1/2} \qquad (27)$$

For $d_1^o < d_1^{oc}$, the mean value of x vanishes. The order parameter characteristic is illustrated in Fig. 16.

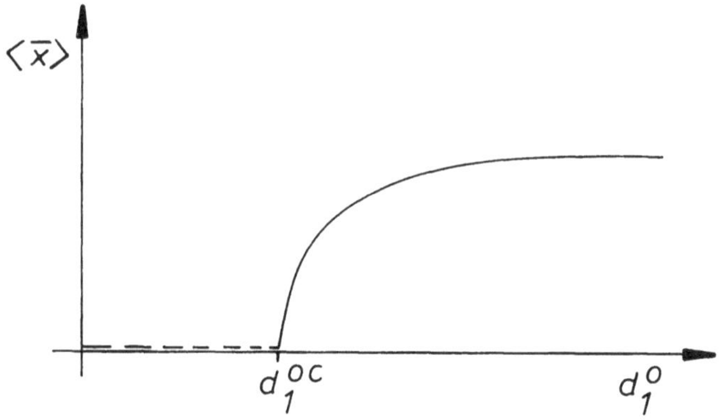

Fig. 16

Considering only the Gaussian part of ϕ , the mean value of the variance $\bar{x}^2$ can easily be calculated as

$$\langle\bar{x}^2\rangle = \frac{1}{N}\int d\bar{x}\, \exp(-\phi)\bar{x}^2 = 1/f(k_{ij}, r_o)(d_1^o - d_1^{oc}). \qquad (28)$$

The result (28) shows the divergence of the static correlation function responsible for the sudden onset of some ordering in the system. Notice, that the found power laws deduced represent a mean field result. The exponents will be changed if the non-linearities of the distribution function are considered. This is carried out within for example the so-called renormalisation group procedure (Wilson & Kogut, 1974). Thus the conventional or classical exponents are modified by some corrections in a first order theory so that

$$\beta = 1/2 - \bar{\varepsilon}/6 ; \quad \gamma = 1 + \bar{\varepsilon}/6 \tag{29}$$

where $\bar{\varepsilon}$ = 4-d and d represents the dimension of the system.

In model D, the order parameter must be extracted via the reductive pertubation method. One obtains a linear combination of the variables $\bar{x}$ and $\bar{y}$ for the calculation of the relevant mode W

$$W(R,\tau) = \frac{1}{\varepsilon}\left(\bar{x} + \frac{\alpha}{\delta}\bar{y}\right) / \sqrt{(1 + \alpha^2/\gamma\delta)} \tag{30}$$

with α, γ, δ from (15) for case D.

Following the calculation scheme of Nitzan (1977), the coefficients of the equation of motion (19) are

$$\begin{aligned} a &= \bar{\alpha}\left(-\frac{4}{3} k_{12} d_1^0 v_0 \left(\frac{\alpha}{\gamma} + G\right)\right) \\ b &= \frac{-\bar{\beta}}{(\gamma\delta - \alpha\beta)} \left(\frac{2}{3} k_{12} d_1^0 v_0\right)^2 2\beta \left(\frac{\alpha}{\gamma} + G\right)^2 \\ c &= -\bar{\beta}\, 6 k_{12} d_1^0 (1 + \alpha/\gamma) \qquad \bar{\alpha}, \bar{\beta} = \mathrm{const.} \end{aligned} \tag{31}$$

From the condition $\partial\bar{\phi}/\partial\bar{W} = 0$, one gets the behaviour of the order parameter by

$$\bar{W}_{1/2} = \frac{-\varepsilon b}{2(b+c)} \left\{ 1 \pm \sqrt{1 - \frac{4(a+b)(b+c)}{b^2}} \right\} . \tag{32}$$

Next, consideration is given to the possible treatment of the dynamics of the system. With the use of ϕ, the Langevin equation may be rewritten as

$$\frac{\partial W}{\partial t} = - \frac{\partial \phi}{\partial W} + \eta(\vec{r}, t) . \tag{33}$$

Such an equation describes the relaxation process of the order parameter to its equilibrium value after a small pertubation. Considering the quadratic term of ϕ only, the relaxation frequency for the model C is obtained as

$$\tilde{\omega} = f(k_{ij}, r_o)(d_1^o - d_1^{oc}) . \tag{34}$$

(34) shows the critical slowing down behaviour typical of transition phenomena. The dynamic correlation function is extracted by

$$G(\omega) = \Gamma / (\omega^2 - \tilde{\omega}^2) . \tag{35}$$

The important benefit in this treatment of the time-dependence is the reduction of the evolutionary kinetic set of equations to only one equation.

Practical applicability, discussion

In this section, the results of the last chapter will be tested with respect to their practical relevance for the explanation of some concrete effects in the drug-receptor interaction. The assumed conditions and some derived statements are examined through measured quantities and detected phenomena. The experimental data of the well-known acetylcholine receptor (ACh-R) are used to demonstrate a possible application of the models proposed. From biochemical analysis, the ACh-receptors are identified as membrane proteins. By freeze-etching techniques, it is found that the ACh-R of the postsynaptic membrane are sharply localised at a high density of approximately 6000 - 7000 μm^{-2}. This fact provides the justification of the term 'receptor population' with a given concentration r_o as used for the models. The function of the ACh-receptors consists in the recognition of the neurotransmitter molecules and the subsequent formation of an open ion channel resulting in a membrane permeability change. Electrophysiological methods are a need in the determination of static receptor properties, and the following results are well established (Dreyer, 1982). The dose-response curves for the ACh-action yield a sigmoidal relationship between the endplate response and the agonist concentration. This fact supports the statement that cooperativity must exist within the receptor excitation mechanism. The binding of agonist molecules to the receptors increases (Colquhoun, 1973) the affinity of subsequent molecules for other recognition sites. Such a picture corresponds to the chosen concentration dependence of the rate k_1 in the presented models. Thus, the inclusion of non-linear terms is also a practical requirement. The absolute value for the complex dissociation constant $K_1 = (k_{-1}/k_1)$ can be determined from the endplate currents curves and is different for various agonists. Drug-receptor complex formation is governed by a very fast reaction velocity, in most cases, in order to produce the effect in a short time. For rate k_2, some estimates are available from the study of cholinesterase kinetics. The stoichiometries of the reaction are an interesting problem in the model equations used. For instance, from experiments with radionuclide α-bungarotoxin associating with ACh-R, two binding sites are located on one receptor. But, it cannot clearly stated how many molecules are required to activate one receptor. Therefore, a bimolecular reaction seems also probable. To extend the model in that direction in (1) D_1 must be replaced by nD_1.

The problems discussed above are associated with the conditions assumed in the model. However, the interesting point is the detection of possible phase transition behaviour. The fact that a brief increase of the ACh concentration in the synaptic cleft activates about 2000 ion channels within 3000 μs may indicate a non-equilibrium phenomenon in the system. This aspect points to the appearance of a very large coherence length in the receptor population obtained as one result in the model treated. To produce such an explosive reaction or, so called, all-or-none transition, only a minor change of the ACh concentration is necessary. Such an effect falls into the field of critical phenomena. The ACh concentration corresponds to the externally controlled variable D_1 in the model representation.

Another proof for the occurrence of the homogeneous phase transition sought could be obtained by some concentration measurements to find the relevant mode amplitude. Notice that this will be difficult because the order parameter is often an unreal quantity given by a linear combination of concentrations; as extracted in model D. Besides, such an experimental study can only be carried out in artificial receptor systems. For these in vitro conditions, the parameters are regulatory quantities necessary for systematic research. A significant fact which reflects a possible transition is a threshold behaviour for a certain response to a control variable. The onset of the slope in the response curves must be studied. The existence of a threshold becomes especially important in view of a competitive action caused by a substance which can block the receptor sites. The shift of the critical concentrations d_1^{oc} found could serve as confirmation of the validity of the theory. For instance, from (26) one gets for, models C and D,

$$(d_1^{oc})^D/(d_1^{oc})^C = 1 + d_2^o/\{g_4/g_1 + g_2/g_1\} \approx 1 + d_2^o/K_2 \qquad (36)$$

corresponding to the Schild equation derived by taken into account co-operativity. The detection of non-classical, critical behaviour, however, hardly seems possible because the so called critical region is too small for the fast kinetics and slow diffusion (Dewel et al., 1980) which the receptor interaction exhibits.

Another domain of research are time-dependent phenomena. For the ACh-receptor mechanism, for example, membrane potential noise investigations and relaxation methods are used to determine channel dynamics. Certain time constants for channel characteristics and elementary current are known. From these results, conclusions can be drawn as to the direct agonist-receptor interaction. Thus, the mean life time of a channel is related to an active complex state. Further, it is demonstrated that the membrane current is directly proportional to the number of open channels. This implies a linear dependence of the effect on the receptor output signal. In this way, from time dependent phenomena, some clues can be sought to describe non-equilibrium transitions in receptor systems.

Summarising, the following conclusions can be drawn. The main point of the study was to consider the receptors as a system driven in the vicinity of an instability far away from equilibrium. This corresponds to an extension of the usual equilibrium points of view for the description of receptor interaction as applied, for instance, by Ariens et al. (1979). This additionally, allows the inclusion of such phenomena as cooperativity and fluctuation influence, certainly present in complex biological systems. The proposed models contain the possibility of an explanation of certain effects by a suitable modification of the parameter set used. For example, the externally controlled variable may become time dependent. In such a case, chaotic behaviour of the system may result (Tomita, Daido, 1980). This shows that there is great variability within possible model extensions representing a flexible reservoir for the theoretical treatment of some measured effects in the pharmacodynamics.

Towards a Stochastic Theory of Photoreception

J. Schnakenberg, W. Keiper

Introduction

The photoreceptor cell transforms light signals into electric signals, the so-called receptor potential, which depend on the intensity and the duration of the light signal. The information contained in the receptor potential is passed on to nerve cells in the form of action potentials which are the information units of the nervous system.

One of the striking effects of a photoreceptor is its high degree of amplification. A photoreceptor is able to detect a single photon which causes a number of ions up to 10^5, to move across the cell membrane and thereby to generate its electrical response - the so-called 'bump'. The fact that one really observes single photon events is deduced from the phenomenon that ordinary receptor potentials at high light intensities are dissolved into single bumps as the intensity is continuously decreased and that, at very low intensities, the number of bumps is proportional to the number of photons. Further evidence comes from the fact that the times at which bumps occur conform to a Poisson distribution. A photoreceptor cell could thus be called the biological version of a photomultiplyer. Apart from amplification it shows the capability of adaptation, i. e. a shift of its input-output characteristics towards a lower amplification after a certain period of illumination at high intensities, and also latency behaviour, i. e. a time delay between the beginning of the light stimulus and that of its response. The biological significance of the latency behaviour is unclear. For further biological details of photoreception the reader is referred to a review article by Stieve (1982).

Thermodynamics and Regulation of Biological and Processes

In this contribution we shall focus our interest on the phenomenon of amplification in photoreceptors. To this end it seems appropriate to study bumps as single photon responses rather than receptor potentials at high intensities sinde (a) the bumps are small signal responses and are thus expected to show a less complicated behaviour than full receptor potentials and (b) their experimental observation does not only produce information on their mean properties, but includes the possibility of also measuring fluctuations of their parameters. The corresponding distributions and correlations will, hopefulle, supply us with additional, and independent, information on the amplification mechanisms involved. The experiments that we refer to were performed mainly by Stieve and co-workers in the Institute of Neurobiology (KFA Jülich). The special experimental object is the ventral photoreceptor of *Limulus polyphemus*. Our conclusions are thus restricted at least to invertebrate photoreceptors, since it is known that vertebrate photoreceptors show qualitatively different behaviour in many respects.

In the following section we shall give a short survey of the principal statistical properties of bumps, i. e. distributions and correlations of their parameters, which we have evaluated from Stieves experiments. Later sections will present a particularly simple model of a chemical amplifier and some of the methods applied to it to obtain predictions for the expected bump behaviour. The comparison with our experimental evaluation in section 5 will lead us to necessary extensions of the proposed model. It will turn out that a complete evaluation of a bump model involves the application of the full store of techniques of the stochastic theory of non-equilibrium systems and poses some, as yet unsolved problems.

Measurement and statistical properties of bump parameters

In order to illustrate the measurement of the electrical responses of the photoreceptor cell to light stimuli, we represent the cell by an equivalent electric circuit as shown in Figure 17.

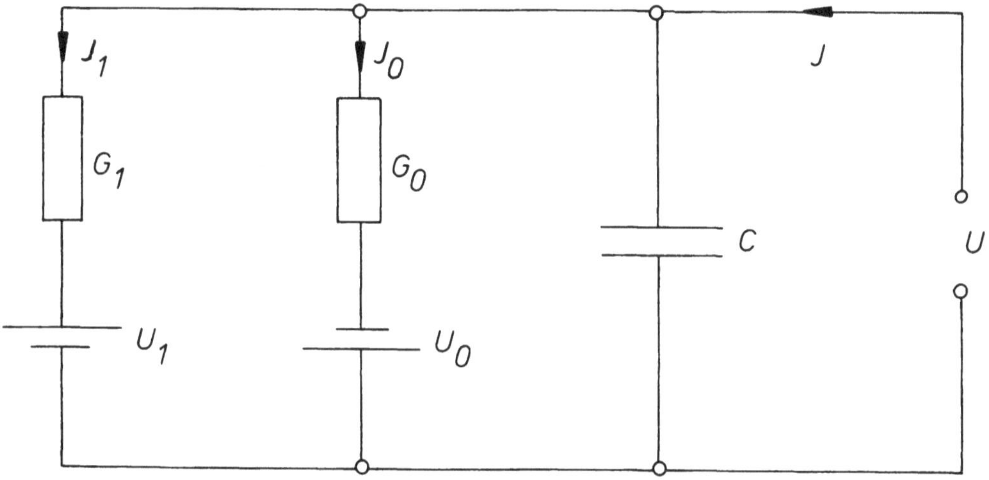

Fig. 17 Equivalent electric circuit for the photoreceptor.

U is the overall membrane potential, i. e. the potential difference between the inside and outside of the photoreceptor, G_o is the dark conductivity of the cell membrane (mainly Na^+) with a reverse potential of the order $U_o \approx -35$ mV and a resistance $R_o = 1/G_o \approx 5$ MΩ and G_1 denotes the light activated conductivity (mainly K^+) with a reverse potential $U_1 \approx +15$ mV.

The light current, J_1, is a function of the stimulating light intensity I(t) and of $U-U_1$. Since the cell membrane has a thickness of ≈ 100 Å the applied electric fields are of the order of 100 mV/ 100 Å = 10^5 V/cm. C is the overall capacitance of the cell membrane with a value C ≈ 4 nF.

The electric responses of the photoreceptor can be measured either as voltage signals U(t) of the free cell, or as current signals $J_1(t) = J(t)-J_o$ if the membrane potential is clamped to a fixed value, $U = U_c$, by inserting a second electrode into the cell which is connected to a suitable electric calibration circuit. Whereas the voltage signal involves a further integration due to the capacitance, C, of the equivalent circuit in Figure 17. The current signal gives direct information on the conductivity changes of the cell membrane during illumination. In the following we shall therefore refer only to voltage clamp measurements of the current.

The responses to the absorption of a single photon, the so-called 'bumps', are observed as noisy deviations of the current from its dark value J_o, which, after appropriate smoothing have a shape such as shown in Figure 18.

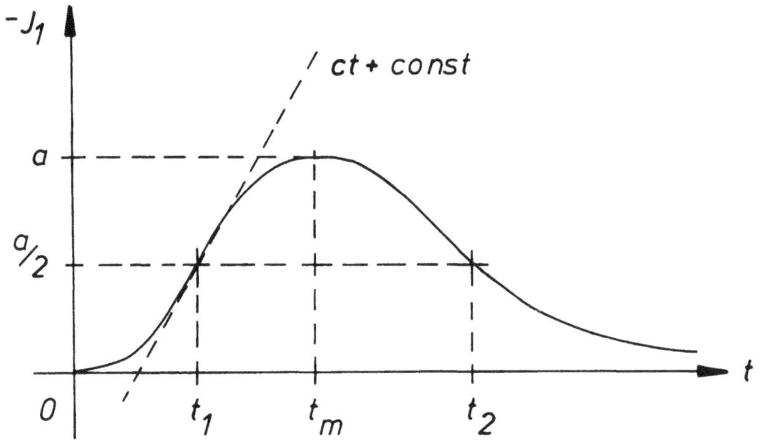

Fig. 18 Typical smoothed shape of a current bump.

The bump parameters which have turned out to be useful for a statistical treatment are the amplitude, a, the rise time, t_m, the times, t_1 and t_2, at which the bump has half the value of a in its rising and decaying phases, respectively, and the slope, c, during an almost linear part of the rising phase. In addition to these parameters, which can be extracted from a continuous record of the current, and, which fluctuate for the different bumps, we introduce a shape parameter, q, defined as $q = t_1/t_m$. Typical distributions of a, t_m, c and q are shown in Figure 19.

A correlation analysis of the experimental data shows that t_m, c and t_2 are almost uncorrelated. This result can be interpreted by assuming that only t_m is related to the excitation process caused by the absorption of the photon whereas c describes a property of the membrane region where the photon is absorbed. The decay phase is approximately exponential,

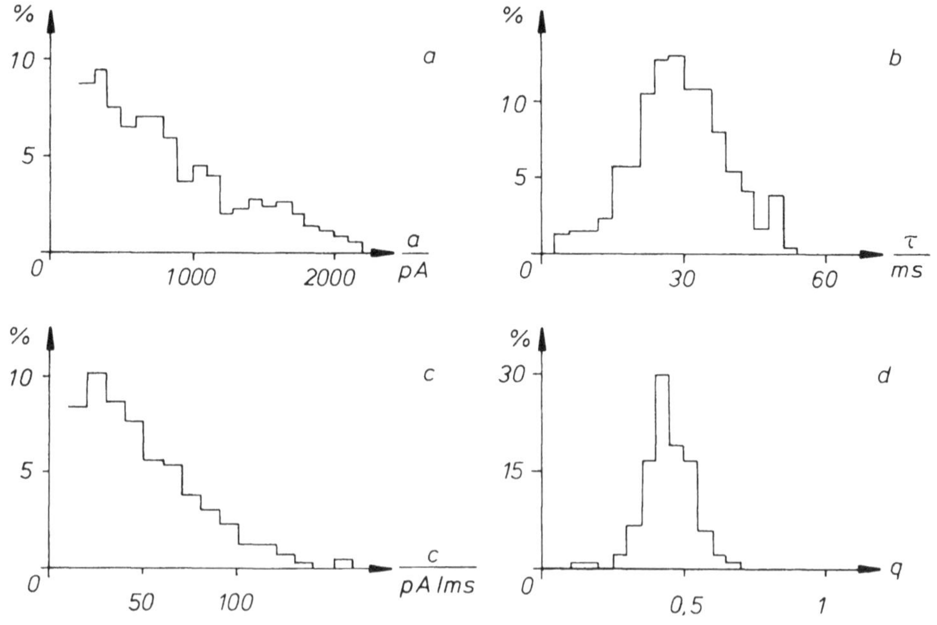

Fig. 19 Distribution of *Limulus* bump paramters, as determined from 500 current bumps.
a - amplitude a
b - rising time t_m
c - slope of the linear rising phase, m
d - q
Very small bumps and slopes are not included.

$\sim e^{-\lambda t}$, where λ is a fluctuating rate constant for the closing of the light channels. The amplitude a, however, will then be correlated with c and t_m.

The experimental results mentioned above lead us to scale the rising phase of the individual bumps b(t) by writing

$$b(t) = a\varphi(t/t_m; \alpha_1, \alpha_2, \ldots) \tag{1}$$

where $\alpha_1, \alpha_2, \ldots$ are form parameters and for $\varphi(s; \alpha_1, \alpha_2, \ldots)$ we assume, as a first approximation in order to keep the number of parameters as low as possible

$$\varphi(s;\alpha_1,\alpha_2)=\begin{cases} a_1 s^2 & 0\leq s\leq \alpha_1 \\ \gamma s-a_2 & \alpha_1\leq s\leq \alpha_2 \\ 1-a_3(1-s)^2 & \alpha_2\leq s\leq 1 . \end{cases} \tag{2}$$

The constants a_1, a_2, a_3 are determined by postulating that $\varphi(s;\alpha_1,\alpha_2)$ and its derivative with respect to s, are continuous functions of s. Noting that $c = a\ \gamma\ /t_m$ and $\varphi(q;\alpha_1,\alpha_2) = 1/2$ we thus eventually obtain

$$a=\frac{1}{2}c\,t_m(1-\alpha_1+\alpha_2) \qquad q=\frac{1}{4}(1+\alpha_1+\alpha_2) \tag{3a, b}$$

Figure 19 shows that the q-distribution is rather narrow with a mean $\langle q\rangle \approx 0.45$. The experimental evaluation of α_1 yields a mean value of $\langle\alpha_1\rangle \approx 0.25$ and consequently from (3b) $\langle\alpha_2\rangle \approx 0.55$. In order to describe a very rough outline for the subsequent construction of a model we neglect the fluctuations in α_1 and α_2 and replace them by their mean values which turns (3a) into

$$a \approx 0.65\,c\,t_m . \tag{4}$$

Summarising the considerations of this section we conclude that a model for bump generation ought to satisfy at least the following conditions:

a) a continuous rising phase of the bump with a q-value ≈ 0.45 and a linear regime with fluctuating slopes c and a c-distribution as shown in Figur 19c,

b) a fluctuating rise time t_m with a distribution as shown in Figure 19b,

c) individual bump amplitudes $a \sim ct_m$ with a distribution as shown in Figure 19a,

d) an exponential decay of the bumps.

The c-distribution as stated in a) can be realised by assuming that the local membrane properties fluctuate according to Figure 19c. The remaining conditions will turn out to be rather restrictive for the models chosen.

Enzymatic amplifiers and their statistical treatment

A particularly simple model for a chemical amplifier is a catalytic reaction of the type

$$Z + X \longrightarrow X + Z \qquad\qquad Z \longrightarrow \overline{Z} \qquad (5)$$

In (5) it is proposed that molecules $\bar{Z}$ are activated into Z by the catalytic action of an enzyme X. Simultaneously, Z-molecules spontaneously decay into their inactivated state, $\bar{Z}$. For a bump model we shall interpret $\bar{Z}$ and Z as closed and open states of a transport protein for light channels across the photoreceptor cell membrane such that the observed light current J_1 is proportional to the number of open channels, $J_1 \sim Z$. For $X = 0$ all channels will be closed, $Z = 0$. Let us assume now, that the absorption of a photon causes the appearance of an enzyme X for a limited life-time, τ . During the period τ the number of Z will increase and afterwards decay to $Z = 0$ again and accordingly also $J_1 \sim Z$.

The model (5) does not include the phenomena of latency and adaptation as described in the first section. We assume that latency is caused by a chain of non-amplifying reactions, perhaps a series of conformational changes of proteins which occur between the absorption of the photon by the rhodopsin molecule and the appearance of the enzyme X. With this assumption our model differs from the earlier models of Fuortes and Hodgkin (1964) and of Borsellino and Fuortes (1968) which assume a chain of simultaneously delaying and amplifying enzymatic reactions. Recently Goldring and Lisman (1982) have shown that the simulation of bumps by models of the latter type does not lead to realistic bump shapes. Our assumption of separated latency and amplification has been reasoned and worked out in detail by Schnakenberg and Tiedge (1979), Tiedge (1981). Keiper (1981) has calculated distributions of latency times on this basis and has found a satisfactory agreement with measurements by Wong (1977). In order also to include adaptation one needs to introduce an appropriate feed back which perhaps limits the life time of X after large amounts of J J_1- currents (see Schnakenberg, Tiedge, 1979; Tiedge, 1981).

A rigorous statistical treatment of the model (5) starts by formulating transition probabilities per time for $\Delta Z = \pm 1$. Making the usual assumption that the transition probabilities are proportional to the number of molecules involved in the corresponding step we have

$$W(\Delta Z=+1)=kX\bar{Z} \qquad W(\Delta Z=-1)=\lambda Z \tag{6}$$

If $\bar{Z} \gg Z$, we can replace $k\bar{Z}$ by A ($k\bar{Z} \approx A$) where A is a constant. X is treated as an external parameter with X = 1 for $0 \leq t \leq \tau$ and X = 0 otherwise. An immediate consequence of (5) and (6) is that the rise time, t_m, coincides with the life time, τ . If X spontaneously decay with a rate constant, μ, after having been created at t = 0 we would expect an exponential distribution law for τ and hence for t_m, i. e.

$$w(t_m) = \mu e^{-\mu t_m} \tag{7}$$

in contradiction to the observed distribution as shown in Figure 19b. A possible way out of this contradiction is to assume that the enzyme X exists in three active conformations which are linked by conformational changes, $X_1 \rightarrow X_2 \rightarrow X_3 \rightarrow$ inactive, such that the effective life time would have a distribution of the type

$$w(t_m) = \frac{1}{2}\mu^3 t_m^2 e^{-\mu t_m} \tag{8}$$

which would be able to fit the experimental data of Figure 19b.

With the transition probabilities given in (6) we can formulate a master equation which can be solved analytically for the conditional probabilities $p(Z, t|Z', t')$ of having Z open channels at time t if their number was Z' at t' (Bailey, 1964; van Kampen, 1981), but this result would tell us very little about the probabilities of the paths of the bumps which have to be interpreted as realisations of the stochastic process in (6). Thus we actually would have to solve a random walk problem as indicated in Figure 20.

Because of the non-homogeneous transition probabilities an analytical solution is possible only in the linear case. Even if it were known, we then would have to reduce the large amount of information characterising a

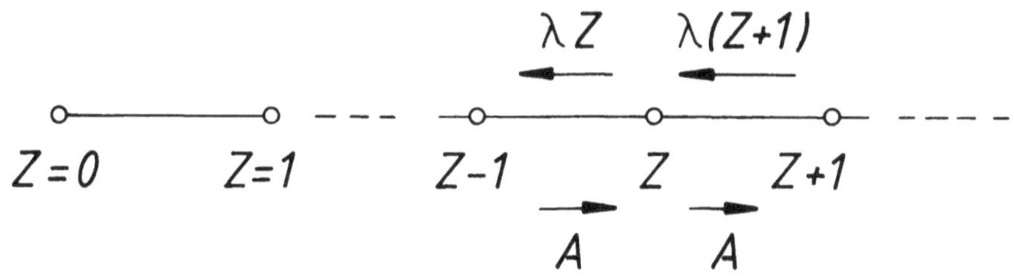

Fig. 20 Random walk problem of the rise phase of a bump.

single bump to the few parameters which are observed in the experiments. This would pose another severe analytical problem. At this stage, the only possibility is a computer simulation as described for systems like (6) by Gillespie (1976, 1978). Such simulations usually consume extremely large amounts of computer time. With respect to the experimentally significant parameters, their results are expected to differ only little from a semi-stochastic method to be presented in the following section.

Semi-stochastic treatment of enzymatic amplifiers

The number of light channels involved in a single bump has been estimated by Wong (1978) as of the order of 10^3. Interpreting the variable Z in our model (5) as the number of open channels this means that we expect the fluctuations of Z inherent in the Z-dynamics of (5) to be at least an order of magnitude smaller than those caused by the fluctuations of the life time τ of the enzyme X, or of the local membrane properties. On the basis of this consideration we neglect the Z-fluctuations due to the stochastic character of the channel opening processes and describe the time evolution of Z by the differential equation for its mean value which follows from (5) and (6).

$$\frac{dz}{dt} = \chi(t) - \lambda z, \quad \chi(t) = \begin{cases} 1 & 0 \leq t \leq \tau \\ 0 & \text{else} \end{cases} \tag{9}$$

where z = Z/A. This kind of approximation has been proposed for the evaluation of bump models by Eckert and Kramer (1978). It yields a solution

$$z(t) = \begin{cases} \frac{1}{\lambda}(1 - e^{-\lambda t}) & 0 \leq t \leq \tau \\ \frac{1}{\lambda}(e^{\lambda\tau} - 1)\, e^{-\lambda t} & t \geq \tau \end{cases} \tag{10}$$

which depends parametrically on the actual life time τ . Keeping in mind that $J_1 \sim Z = A \cdot z$ for the light current J_1, we see that the bump described by (10) begins directly with a constant slope, z = t, at small t without an initial t^2-phase and has a discointinuity at its maximum value for $t = \tau$. Although it is thus a rather poor simulation of real bumps let us establish its relationship to the experimental parameters c = slope of the linear phase, a = amplitude and $q = t_1/t_m$ as defined in the second section:

$$c \sim A, \qquad a \sim \frac{A}{\lambda}(1 - e^{-\lambda\tau}) \tag{11}$$

such that

$$a = \frac{c}{\lambda}(1 - e^{-\lambda\tau}) \tag{12}$$

and

$$q = -\frac{1}{\lambda\tau} \ln \frac{1 + e^{-\lambda\tau}}{2} \tag{13}$$

where (13) follows from the definition of q by $z(q \cdot \tau) = z(\tau)/2$ for $0 < q < 1$. The fluctuations of the slope c are thus ascribed to local fluctuations of $A = k\bar{Z}$, i. e. of the surplus of closed channels. For $\lambda\tau \gg 1$ we have from (12), $a = ct_m$, instead of (4) for $t_m = \tau$ but in this case the shape of the bump (10) is unrealistic, since it changes discontinuously from a linear phase to exponential decay.

Finally let us evaluate the distributions of amplitude a and the q-value. If the life time τ and the slope c are statistically independent as discussed in the second section we obtain, for the conditional distribution, $w_a(a|c)$, of finding an amplitude a at given slope c

$$w_a(a|c) = \int_0^\infty d\tau\, w(\tau)\, \delta\left(a - \frac{c}{\lambda}(1 - e^{-\lambda\tau})\right) \qquad (14)$$

where $w(\tau)$ is the distribution of the life time. The evaluation of (14) for an exponential distribution (7) gives

$$w_a(a|c) = \frac{\mu}{c}(1 - x)^{\nu - 1} \qquad (15)$$

with $x = \lambda a/c$ and $\nu = \mu/\lambda$ and for the three stage models (8) for the enzyme X,

$$w_a(a|c) = \frac{\mu}{2c}\left[\nu \ln(1 - x)\right]^2 (1 - x)^{\nu - 1}. \qquad (16)$$

(15) is a monotonically decreasing function whereas (16) suppresses small amplitudes and has a maximum at $x = 1\text{-}\exp(-2/(\nu - 1)$. Since the experimental registration of small bump amplitudes is rather ambiguous both results (15) and (16) could be used to fit the experiments quite well.

The q-distribution

$$w_q(q) = \int_0^\infty d\tau\, w(\tau)\, \delta\left(q + \frac{1}{\lambda\tau} \ln \frac{1 + e^{-\lambda\tau}}{2}\right) \qquad (17)$$

cannot be calculated analytically. Its numerical evaluation for both types of the τ -distribution is shown in Figure 21 for $\nu = \mu/\lambda = 10$.

The q-distribution for the exponential τ -distribution (7) has a sharp peak at $q = 0.5$ whereas that for the Poisson-type τ -distribution (8) has a maximum near $q = 0.47$ and resembles the experimental distribution of Figure 19 d slightly better.

The three-stage model (8) can be improved by proper choice of individual decay constants and different amplification rates for the three enzymes E_1, E_2, E_3. Both the initial and the linear bump phase can be simulated. This model yields the correct distributions of τ and a. The q-distribution, however, is somewhat off the mark.

Extended model for an enzymatic amplifier

In the preceding section we saw that our simple model for an enzymatic amplifier produces many difficulties when compared with the experimental results. These difficulties can be resolved considerably by an extension of the model, namely by the introduction of an additional intermediate enzymatic step in the following way:

$$\begin{aligned} \bar{Y} + X^* &\to X^* + Y, \qquad & Y &\to \bar{Y} \\ \bar{Z} + Y &\to Y + Z & Z &\to \bar{Z} \end{aligned} \tag{18}$$

(18) is a two-fold 'enzymatic cascade'. A ten-fold cascade of this type has been assumed in the model of Fuortes and Hodgkin (1964) mentioned above in order to include latency. As before, $\bar{Z}$ and Z are interpreted as closed and open channels, respectively, and $\bar{Z}$ is assumed to be present at large excess, $\bar{Z} \gg Z$. For the intermediate enzyme Y, however, we assume that $Y + \bar{Y} = C$ is finite such that Y saturates over the time course of a bump. Neglecting the fluctuations of Y and Z due to the stochastic processes of Y-activation and channel opening and replacing the time evolution of Y and Z by their mean values as in (9) we write

$$\left.\begin{aligned} \frac{dy}{dt} &= \alpha X^*(t)\left(\frac{\alpha+\beta}{\alpha} - y\right) - \beta \cdot y \\ \frac{dz}{dt} &= \kappa y - \lambda z \end{aligned}\right\} \tag{19}$$

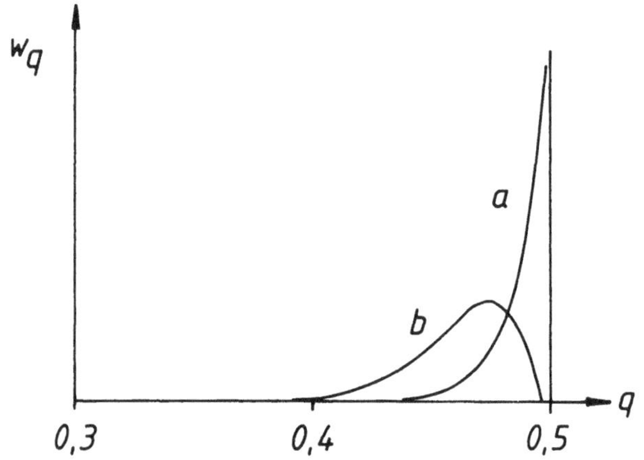

Fig. 21 q-distributions calculated from the model (5) for an exponential τ-distribution (7), curve (a), and for a Poisson-type τ-distribution (8), curve (b).

where $y = Y/Y_m$, $Y_m = \alpha C/(\alpha + \beta)$ is the saturation value of the number of activated Y and z is the density of open channels as before. The local fluctuations of the total number of channels are now expressed by fluctuations of κ such that the light current of the bump J_1 is proportional to z, $J_1 \sim z$. For X(t) we set $X(t) = 1$ for $0 \leq t \leq \tau$ otherwise $X(t) = 0$ as in (9).

At $t = 0$, y will initially be linearly related to t, $y \sim t$, and hence $z \sim t^2$ which represents the initial rising phase of the bump. At a time $t' \approx 1/(\alpha + \beta)$, y will reach its saturation value, $y = 1$, and hence z will increase linearly, $z \sim t$, and this represents the linear part of the rise phase with a slope dependent on κ. At $t = \tau$, y will start to decrease exponentially, but z will continue to increase for a certain amount of time until it reaches its maximum and then start to decrease exponentially, too. This coupled time evolution of X, y and z is shown in Figure 22.

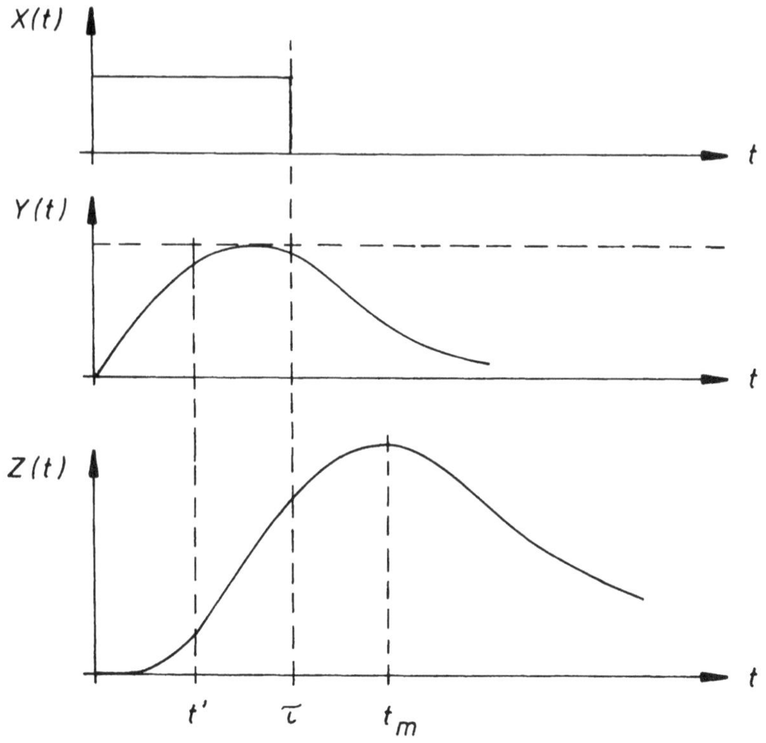

Fig. 22 Time evolution of X(t), y(t), z(t) in the extended model (18), (19).

The intermediate enzyme thus causes inertial behaviour, due to which, the rise time, t_m, of the bump is no longer identical with the life time, τ, of X. As a consequence, our model yields a t_m-distribution with a maximum at $t_m > 0$ even for an exponential distribution of τ. This means that we no longer need a three-stage hypothesis for enzyme X as argued in (8).

We have evaluated the semi-stochastic version (19) of the extended model by generating X(t)-processes with an exponential distribution of the life time, τ, and calculating the corresponding bumps, z(t), and its parameters a, t_m, c and q on a computer. It appears that we can fit the calculated distributions of the parameters to the experimental parameters by an appropriate choice of the model parameters α, β, κ, λ and μ (μ = rate of the spontaneous decay of X, see (7)). This remark should not give rise to the suspicion that an appropriate choice of the model parameters could bring about agreement with any experimental results since from the experiments we cannot only extract the distributions shown in Figure 19, but also those of further bump parameters and, even, correlations between them. This work will be continued in the future and is expected to lead to more precise definition of the mechanisms involved.

Acknowledgements

The authors are deeply indebted to H. Stieve and his coworkers M. Bruns, J. Klomfaß for very fruitful cooperation and suggestions and to R. Neikes who has performed a considerable amount of the analysis of the experimental data. They would also gratefully like to mention the continuous support of this work by the Deutsche Forschungsgemeinschaft (SFB 160).

Temperature Homeostasis and Principles of its Regulation in the Living Organism

K. P. Ivanov

The temperature of peripheral tissue in man, even in the comfort (thermoneutral) zone, is 2 °C to 4 °C lower than in central body regions. Temperatures of inner organs and, even of some regions of the same organ (brain, liver) differ by several tenths of a degree. These differences are smoothed in small animals, but they are greater in animals with a body mass above that of man. In addition, in man, as well as in animals the body temperature shows significant variation during circadian rhythm; in the course of intensive muscle activity; with abrupt changes of environmental temperature. That is why determination of a notional 'homeostatic temperature' is associated with certain difficuties.

The Commission on Thermobiology and Physiology of the International Union of Physiologists has attempted to find a solution to this problem. In the glossary of thermoregulation published by the Commission in 1972 *homeothermia* is characterised by an inner temperature amplitude not exceeding 2 °C. It is unlikely that such regulations will help to define a real boundary between homeo- and poikilothermal animals and correctly formulate the notion of temperature homeostasis. However, the latter is closely related to homeothermia. An attempt will be made to consider this relationship.

The opinion is often advanced that the single, and most important, feature of homeothermia is the ability of animals to maintain a constant inner body temperature. This, however, is not quite correct. Homeothermia and temperature homeostasis of homeothermal organisms are characterised not only by relative temperature stability, but also by the temperature itself. There are no animals with constant body temperatures of 15 °C, 25 °C or 50 °C,

Thermodynamics and Regulation of Biological and Processes

although a somewhat lowered body temperature would be useful for the animals inhabiting polar regions, and the higher temperature for the animals of hot climatic zones. But this is not the case. According to Morrison and Riser (1952) the long line of homeothermal animals from the mouse to the whale, is characterised by an average body temperature of 37.8 °C ± 0.4 °C. Statistical deviation from the mean value is small. This shows, not so much the differences between animals, but a surprising temperature similarity between organisms whose mass differs by factors of tens, thousands and even hundred thousands. Consequently homeothermia is characterised not only by stability of body temperature, but by the absolute temperature, established by nature for homeothermic organisms. However, this additional parameter does not exhaust the characteristics of homeothermia and homeostasis.

Homeothermia and heat production

According to Benedict's data (1932) for rabbits of 2.5 kg weight, body temperature 37 °C, and at complete rest, heat production equals 2.21 W/kg. In the rattlesnake of the same weight and with the same body temprature and also at complete rest heat production is only 0.35 W/kg. The pattern of these differences in poikilo- and homeothermal organisms is maintained even if their body temperature is decreased by up to 10 °C to 20 °C (Adolph, 1951). Although the true reasons for such striking differences are still unknown, it can be said that a high level of energetic expenditure and heat production is also the most important feature of homeothermia. This statement is sometimes rejected because under certain conditions energy expenditure in poikilothermal animals is similar to that in homeothermal species. Thus, for example, oxygen consumption per weight unit of the summer frog and crayfish, at a sufficiently high environmental temperature, is almost equal to that of man. It should be remembered, however, that metabolic intensity in homeothermal organisms of similar size is not the same. To provide temperature homeostasis and homeothermia the homeothermal animal with a body mass equal to that of the frog or crayfish has the intensity of energetic expenditure per mass unit several times higher than that in man. We shall return to this problem later.

The question arises as to the causes of such high energy expenditure in homeothermal organisms. It is well known that all energetic transformations in the organism are, if no work is performed by the organism, finally completed by heat production. In man, of average weight and age, who performs no physical work, energy expenditure (heat production) amounts to circa 2000 kcal per day (100 W). These rather high energy expenditures are sufficient to heat three buckets of water fom 0 °C to 100 °C, or, to set in motion a 20 ton tram car. But the energy expenditures in small animals at relative rest are 2 to 4-fold greater than in man; in very small animals, such as the mouse, for example, they are 6 to 8-fold greater. Energy expenditures in the pigmy shrew weighing 3.4 g exceed those of man at rest by 25 to 30-fold. To provide such high metabolism and heat production levels the amount of food consumed per day by the shrew is 1.5 to 2 times larger than its body mass (Tupikova, 1949). As is known the energy is spent to produce work. It can, therefore, be concluded that, even at rest, life is hard work.

The question arises as to what work is produced by an organism at rest. Lehninger (1965) believes that there are three types of biological work in a living organism:

- contraction of smooth and striated muscle
- ion transport against concentration gradients
- chemical work during the synthesis of various organic and inorganic compounds.

Muscle contractions at rest involve heart activity, maintenance of vessel and bronchus tonus, gastro-intestinal peristalsis and tonus. Ion transport against a concentration gradient is constantly carried out in excitable structures of the nervous system and muscles. This work is obligatory for maintenance of the excitability of corresponding tissues and providing their functions. Chemical work of synthesis proceeds permanently due to the continuous renewal of body structures. For example, about 100 g of protein is renewed each hour in man's body. Other organic and inorganic compounds are constantly renewed. It should be noted that all these types of work have a very low efficiency. So for example, the efficiency of chemical work for ATP synthesis is 40 to 50 % (Krebs, Kornberg, 1957). This means that in the course of synthesis 50 to 60 % of free energy spent on such synthesis is immediately lost in the form of heat dispersed in the tissues. In terms of the thermodynamics of cellular processes these losses are natural and inevitable (Green, Goldberger, 1967). Using literature data we have tried to discover the efficiency for different kinds of biological work in organisms (Ivanov, 1972). The efficiency of mechanical work of the heart proved to be 10 to 20 % and less. The efficiency of chemical work spent on synthesis is approximately 10 to 13 %. The efficiency of the work of the ionic pump in the cell seems to be about 10 %. It is clear that the comparatively low efficiencies of these work processes in different tissues and organs is one of the major causes of the comparatively great total energetic expenditure and of the high level of heat production of homeothermal organisms at rest.

Heat production and liberation and mean body temperature

The energetic losses of homeothermal animals are, comparatively, very large. One must agree with Burton and Edholm (1955) that the energetic cost of homoiothermia is rather high. Evidently, this is due to an extremely intensive energetic metabolism and that, in the course of evolution a constant (genetic) adaptation of animals to low and super-low environmental temperatures never took place at the expense of raised energetic expenditure and increased heat production. Animals of related species with approximately similar body mass living in cold or hot climatic zones do not differ significantly in basal metabolic heat production. The polar fox is a comparatively small animal; it has no special shelters (holes), and therefore maintenance of temperature homeostasis is rather complicated for this animal. Nevertheless at environmental temperature of -20 °C the polar fox preserves it's normal body temperature without increasing heat production. Beautiful fur is its only protection from cold! It provides maintenance of valuable energy without spending it on heat production.

According to Khalabukhov (1946), saving energy is the basis of any adaptation to the environment. It is noteworthy that for urgent thermoregulation or for short-term-adaptation the homeothermal organism appears to forego the rules of conservation and increases energetic expenditure for heat production in order to stay alive and to maintain temperature homeostasis. The well-known effect of cold muscle shivering is such a specific reaction. Since no external work is performed thereby and the mechanical energy of contraction is spent on reciprocal friction in muscle fibre, total muscle contraction energy is transformed into heat. The reaction of microshivering, or thermoregulation tonus, is weak, irregular contraction of individual muscle fibres without any external visible sign of muscle contractile activity. A detailed study of the quantitative aspects of the heat efficiency of shiver and thermoregulation tonus has been presented earlier (Ivanov, 1965, 1972). Regulation of energy expenditure and heat production at the expense of specific forms of muscle activity is a feature of homeothermal organisms alone. The power involved in these reactions amounts to 300 to 600 % of that of basal metabolism.

However, we believe that the regulation of the efficiency of muscle contractile activity is the most striking phenomenon in the field of regulation of energy expenditure and heat production. First, indirect evidence in favour of such phenomena following adaptation to cold was obtained by us as early as the beginning of the 60's (Ivanov, 1965). Afterwards we investigated, in detail, changes in muscle efficiency (muscles in situ, isolated muscles and heart) after adaptation to cold of intact animals; these changes were analysed quantitatively and the proposed mechanisms were examined (Ivanov, 1980). This decreased efficiency permits a 1.5 to 2 fold increase in heat production during muscle contraction following adaptation to cold or to the effect of some hormones. It seems to us that this fact is most significant and not only for temperature homeostasis. It shows that even the most important thermodynamic constants of life, such as energetic expenditure and heat production of muscle contraction, can be the object of a rather extensive physiological control. An interesting mechanism of heat production increase in the organism is the rise in metabolism of the, so-called, brown fat. According to the latest evidence obtained in our laboratory this function of brown fat is closely associated with the regulation of muscle efficiency. An increased release of fatty acids from brown fat into the blood is one of the causes of a decrease in muscle contraction efficiency and increase in heat production in the entire organism.

The living organism is not a heat machine. Therefore heat liberated as a result of metabolism is not used for the performance of biological work and must be immediately conducted out of the organism. As an example we have found that in experiments on the animal brain cortex, and by calculation, that neuron bodies consume about 360 µl O_2 per gram per min, or, they liberate correspondingly 2.1 mW/g (Ivanov, Kalinina, 1972; Ivanov, 1972; Ivanov, Kislyakov, 1979). It is easy to calculate that, theoretically, in the absence of heat removal the neuron body will be heated at a rate of 2.16 °C per min, and within 3 to 4 min it will cease functioning. Therefore it is essential to emphasise such a simple and obvious fact that, at the established temperature, and at the cellular or organism level heat removal always equals heat production. In the process the level of the body temperature depends on the value of heat insulation (heat resistance), which determines the value of the temperature gradient (drop) between the body and the environment. If a room is well heat-insulated, it can be maintained at a comfortable temperature by a rather weak heat sources. If it is not properly insulated, and some heat escapes through windows, the comfort temperature is achieved by increasing the power of the heat source. In both cases, at the desired temperature, the amount of heat produced e equals that of the heat removed. The body temperature of homeothermal animals, i. e. 37.8 ± 0.4 °C in a wide range of animals, from mouse to whale, depends on a suitable ratio between heat production, power per body mass unit and the value of total heat insulation of the body. The latter, in turn, depends on the surface area per unit body mass. This rule underlies the so-called law of the body surface postulated by Rubner; but many unsolved problems remain.

It is believed that if the only result of a decrease of animal mass were an increase in relative body surface, then heat production power should increase as weight to the power 2/3. However, as Schmidt-Nielsen (1972) pointed out, heat production in fact, varies to the power 3/4. This means that heat exchange between the organism and it's environment is determined not only by the geometric ratio: surface area/body mass. The heat transfer coefficient of outer integuments (i. e. the amount of heat passing through the unit surface per unit time subject to a temperature gradient of 1 °C) changes in conformity with physiological regulation of blood circulation. Heat exchange is realised with varying intensity along respiratory pathways etc.. Knowing real values of heat production in the thermoneutral zone, the body surface and heat removal coefficients for a number of animals of different body mass one can calculate their theoretical body temperatures. The difference between the theoretically calculated temperature and their actual temperature could show, what quantitative role, in homeothermia, the body mass/body surface ratio, plays, and what is the role of the integument's properties and what are the mechanisms of physiological control of heat insulation and production in the organism. Unfortunately this important problem has not yet been resolved. There are difficulties in the determination of the coefficient of heat production of the outer integuments in different homeothermal organisms. Exact determination of the body surface also presents certain difficulties. Recently new methods were developed in our laboratory for the determination of the coefficient of heat transfer by integuments in various animals, and satisfactory measurements have been made. These data and extensive computation has made possible an approach to this problem. An important fact was discovered during the process, which is directly related to the theory of homeothermia, the fact that, although blood equalises the temperature regions in the body, small animals nevertheless preserve certain temperature differences between parts of their body. These differences are quite large in large organisms. To resolve the problem under study it would be necessary in these cases to calculate some average body temperature. However, it is almost impossible to do so. Therefore, firstly one must calculate heat energy remaining in the body accounting for all parameters and conditions given above. Then, from this quantity and the body's heat capacity the average temperature can be found. This complex reasoning leads us to infer that an average temperature is the best way, to characterise temperature homeostasis and that this also underlies regulation of temperature homeostasis.

Average body temperature and localisation of thermosensitive structures

Let us imagine a large laboratory thermostat as a bath filled with water which has a heater and contact thermometer adjusted to a temperature of 37 °C. Let us assume that at a given time the water temperature at all points is about 37 °C, and the heaters are switched off. If the surrounding temperature is low enough (20 °C, 10 °C or lower), the water temperature will gradually decrease. Let our thermostat have no device for stirring the water. In this case, at first, the outer water layers will get colder, then the layers located nearer to the centre follow, and finally the temperature in the middle of the bath, where the contact thermometer is placed, will decrease. When a certain threshold is reached, the thermometer will switch on the heaters. It is absolutely clear that the contact thermometer will be switched on after a significant delay, because, before decreasing in the centre, the water temperature falls significantly at the periphery. Thus temperature changes in the centre of the bath will reflect less well the real average temperature of the total mass of water, the greater the volume of water in the bath, and apparently, the lower the surrounding temperature.

This behaviour can be analysed in mathematical terms. However, it is clear, even without calculation, that there are at least two ways possible to achieve high accuracy in maintaining the average water temperature at some constant level in the thermostat. The first: constant stirring of the water; the second: adjustment of an additional contact thermometer (or thermometers) in the periphery of the bath and uniting them in a single measuring system. Combination of these two ways seems to be most effective. There are grounds to believe that placing additional thermometers in the water layers between the centre and periphery of such a thermoregulation system will not increase the accuracy of maintenance of average temperature in the bath.

The living organism is characterised by the presence of thermo sensitive structures in the central nervous system (CNS; hypothalamus, midbrain, spinal cord) and on the periphery (skin, subcutaneous vesicles). The question of specific thermoreceptors in deep tissues and inner organs beyond the CNS has not yet been solved. Such data as exists are summarised in the well-known monographs by Minut-Sorokhtina (1972), Bligh (1973) and others. They contain interesting data, but give no grounds to believe that there are thermoreceptors in the inner organs whose sensitivity approximates that of the specially thermosensitive elements of the CNS or skin. Therefore one can say that the organism analyses body temperature by means of a powerful system of thermosensors in the CNS and over a large temperature field, the skin. It should be noted that the hypothalamus with its extremely sensitive thermosensors is located in the central part of the cranium, and the spinal cord is near the body axis in man and mammals. Thus, inner thermosensitive elements are located in the central parts of the body. The blood circulating in the body, as we have stressed already, largely equalises the body temperature fields. This construction is reminiscent of the thermostat described above. The latter is designed for maximally accurate regulation of the average temperature of the thermoregulated water mass whereas localisation of thermosensitive structures in the organism suggests that it is the average body temperature that is measured and regulated in the organism. Certainly, such a conclusion cannot be drawn only on the basis of the localisation of thermosensors. What do these physiological data imply?

Localisation of thermosensitive structures and physiological regulation of temperature homeostasis

The existence of thermoreceptors in skin integuments was known long ago. Specific thermosensitive elements were discovered in the hypothalamus at the beginning of this century. The question as to what is the main role in thermoregulation of such peripheral or central structures, has been discussed for a long time. Experimental resolution of this problem has been achieved quite recently. It has been shown, that the organism continuously controls its heat state by integrating (summing), at the thermoregulation centre, the temperature signals from both central and peripheral thermosensors. The temperature sensitivity of the hypothalamus was shown to depend on skin temperature, and the effect of thermal stimulation of the skin on thermoregulation depends on the temperature of the hypothalamus and other parts of the CNS where thermosensors are located. These relationships were demonstrated in animals and man (Chatonnet, Cabanac, 1965; Hellstrom, Hammel, 1967; Dymnikova, Ivanov, 1969; Ivanov, Slepchuk, 1971; Benzinger, 1969; Bligh, 1973, and others). Almost simultaneously, in the frontal and rear lobes of the hypothalamus of animals were found nerve cells, which, having no temperature sensitivity of their own possessed the ability to sum temperature signals from the skin and central thermosensors (Hensel, 1973). Continuing these investigations into neural mechanisms of thermoregulation we recently discovered in a chronic experiment, on animals neurons in the thermoregulation centre which perceived temperature signals from all thermosensitive body structures (Ivanov et al., 1981). The fact of continuous, close interaction between temperature signals from the central and peripheral thermosensitive structures at the thermoregulation centre in the hypothalamus and an extremely high concentration of temperature information in some nerve cells of this centre make us believe that the main object of regulation of temperature homeostasis is the average body temperature or some derivative of this value. At moderate temperature disturbances, outside the thermoneutral (comfort) zone, in man shifts of temperature in different regions of the skin are 2 °C to 4 °C or a little more. The inner temperatures (hypothalamus/or auditory meatus for man, rectum) are changed by an order less, i. e. by 0.1 °C to 0.4 °C. We could not record the mean body temperature; one can suppose, however, that in man it is changed by 0.5 °C to 2 °C or somewhat more. It seems to be the temperature shift which triggers the switching on of thermoregulatory reactions which oppose freezing or heating of the body (abrupt narrowing of vesicles, shivering, vesicle dilation, sweat insulation). In terms of such sensitivity the hypothesis outlined above explains well the phenomenology of thermoregulation. Observations under different temperature conditions permit an estimation of the relative contribution to the regulation of temperature homeostasis made by the skin and central thermosensors. Further mathematical investigation along these lines may provide the key for understanding the operations performed by the summing (integrating) neurons in the thermoregulation centre; calculating the shifts in the mean body temperature or of a similar parameter.

The main difficulties in understanding physiological mechanisms in the regulation of temperature homeostasis for a system with high sensitivity is under study. We have managed to show, recently, that, in the thermoneutral (comfort) zone introduction of small amounts of heat (about 150 - 300 J/kg) into the animal or man induces a pronounced reaction of skin vesicle dilation and evacuation of excessive heat. Such an increase in heat regulation changes the mean body temperature of the organism by 0.05 °C to 0.08 °C. It has been observed that the reaction begins and proceeds independently of the change in the hypothalamus temperature for animals and in the auditory canal temperature for man. Skin temperature can rise by only 0.1 °C to 0.3 °C, but irregularly. The question as to the mechanisms of such a sensitive reaction arises. Allowing that the organisms removes 70 % of the heat introduced into the environment (determined by a special calorimeter), the question also arises as to the mechanisms of the strikingly precise work of this autoregulation system. There are no answers to these questions as yet. Nevertheless we succeeded in showing recently, contrary to commonly held opinion, that thermoreceptors in the skin are located in several layers (Ivanov et al., 1982). This revives an old hypothesis that the organism estimates the intensity of heat current through the skin by measuring the temperature gradient between the deep and surface layers of the skin. In such a case the skin of a homeothermal organism could be compared with the cover of a gradient calorimeter, where the thermosensitive gradients are located in two layers. The sensitivity of this device is extremely high. If the organism really estimates the intensity of heat flow, the mean body temperature can also be measured in the same way. Changes in the body temperature will be in proportion to the changes in heat flow. The checking and development of this hypothesis will be the subject of future investigations.

Extremal Principles

An Extremal Principle for Biomass Diversity in Ecology

D. Lurié, J. Wagensberg

Introduction

In both statistical mechanics and in information theory the constrained maximisation of an entropy function of the form $\Sigma\ p_i \log p_i$ has proved to be an effective method for characterising the stationary state of a complex system. The derivation of the canonical ensemble in equilibrium statistical mechanics involves the maximisation of entropy subject to a constraint on the total energy of the system. In information theory the maximisation of Shannon's entropy function the 'missing information', subject to a constraint on the total word length (Brillouin, 1962) leads to Shannon's noiseless coding theorem which prescribes the probability distribution of word lengths required for optimal matching of an information channel to the message source. In both disciplines one can view the constrained maximisation of an entropy function as a means of characterising the stationary state of a system that has 'adapted' itself to its environment. Mathematically the constrained maximisation represents the conciliation of two conflicting pressures: on the one hand a tendency toward entropy maximisation which, were the system to be isolated, would proceed unchecked and on the other hand an opposing pressure due to the fact that the system does interact with its environment giving rise to a constraint on the maximisation of the entropy function.

One is tempted to ask whether the constrained maximisation of an entropy-like function might not provide a fruitful method of approaching certain biological problems. Two concepts are fundamental for the understanding of living systems: adaptation and self-organisation. It is the latter concept which chiefly distinguishes biological systems from those studied in equilibrium statistical mechanics and information theory. This is in fact the object of the current intensive research in non-equilibrium thermo-

Thermodynamics and Regulation of Biological and Processes

dynamics and non-linear kinetics whose object is the analysis of self-organisation in both living and non-living systems. It is also one of the reasons for the more or less veiled scepticism with which most naturalists, indeed since Darwin, have viewed the usefulness of variational principles in attacking biological problems.

Nevertheless the behaviour of living systems does include an adaptive component, adaptive in the sense of exhibiting a tendency to evolve toward stationary states which are compatible with the imposed external constraints. It is for the description of the adaptive behaviour of complex systems, i. e. systems that consist of a large number of degrees of freedom, that the constrained maximisation of an entropy-like function might be expected to yield useful insights. In accordance with this philosophy we propose here an approach to the analysis of ecological systems which, as we shall see, yields results that are in remarkable agreement with the observed data.

To proceed we shall require both a definition of the appropriate entropy-like function and a formulation of the constrained variational principle.

Ecological diversity

The concept of diversity was first introduced by MacArthur (1955) and by Margalef (1958) and has become an important analytical tool in ecology. It makes use of the Shannon formula for information (Shannon, 1948) to measure the richness or variety of an ecological system. If the number of individuals of species 1, 2, ..., s is N_1, N_2, ..., N_s and $N = \sum_{i=1}^{s} N_i$ then the diversity per individual is defined as

$$D = - \sum_{i=1}^{s} P_i \log_2 P_i \tag{1}$$

with $P_i = N_i/N$. An alternative definition is

$$D = - \frac{1}{N} \log_2 \frac{N!}{N_1! N_2! \ldots N_s!} \; . \tag{2}$$

Through use of Stirling's formula, both forms can be shown to be equivalent for large populations.

For most ecological populations the values of the diversity index computed according to (1), or (2), tend to be concentrated within a fairly well defined range with an upper limit of roughly 5 bits per individual (Margalef, 1972). There is as yet no fully understood reason for this. It simply appears that nature tends to avoid either very high or very low diversity. An analogy can be drawn with the measure of information in linguistic systems; Margalef (1972) points out that:

'The building of diversity in objects and concepts can best be seen in the symbols used for communication. We use 10 numerical symbols and 25 to 30 letters in our alphabets. Numerical symbols are equifrequent, not so letters. The result is that average information per symbol is between 3 and 4 bits, as well in numbers as in letters. Except for our Babylonian mode of computing time, no systems of numeration of a base higher than 20 have passed the screens of cultural selection, and the alphabets with a great number of symbols have many that are used sparsely, on holidays only, so to speak, so that diversity does not reach very high values. It seems that the upper limit of 5 bits is also present here'.

Similar analogies can also be drawn with diversity in social systems. Margalef writes:

'There is a functional and self-regulated ratio between the numbers of people engaged in different professions: so many farmers for every lawyer, so many science administrators for every scientist (this ratio is rising, in developed as in underdeveloped countries), and so on. Such categoric organisation of human society has been used as a model analogous to diversity in ecosystems; jobs are used as a criterion to make subsets and compute diversities. Now if a sample is taken of people in a seaside resort or in a stadium, where jobs are not functional, diversity computed in terms of jobs may be too high - at least I have the impression from some flimsy evidence in Spain and the system is supposed to be impermanent if it depends on such criteria. That is, persistence of human societies is more dependent on jobs than on fiestas. After the holidays, people disperse and return on their communities where feedback is operative and diversity at normal level, i. e. below 5.'

Although the existence of a feedback mechanism may provide an adequate rationale for the existence of a limited band of diversity values, our approach here will be different. We shall explore the idea that the adaptive behaviour of an ecological system is conditioned by the competition between two opposing tendencies: on the one hand the pressure toward greater diversity that stems from the continuous ocurrence of errors in the genetic code, i. e. biological mutations, on the other hand the pressure toward lower diversity resulting from natural selection, i. e. competition between species and interaction with the environment. To develop this idea the most appropriate vehicle is not species diversity. A new concept, '<u>biomass diversity</u>' must be introduced which is in closer correspondence with the physical parameters of the ecosystem.

The concept of biomass diversity

How can one express mathematically the competition between the biological tendency toward diversity maximisation and the opposing pressure due to the interaction with the environment? The appropriate mathematical technique is that of constrained optimisation of a diversity function where the interaction with the environment serves to provide the constraint. Since the most immediate environmental constraint appears to be the finite nature of the energy flow into the system, i. e. solar radiation, nutrients etc., one should look at the total biomass, M, which the ecosystem can support and this suggests that a new diversity index should be defined in which biomass appears explicitly. We define

$$\mu = -\int_0^\infty p(m) \log_2 p(m)\, dm \tag{3}$$

and call this quantity the <u>biomass diversity</u> per individual. In (3) p(m) is the biomass probability density function of ecosystem. We may also choose to express biomass diversity in terms of discrete biomass intervals: for this we divide the range of biomass values available to the ecosystem into r discrete intervals of size Δm and take

$$\mu = -\sum_{i=1}^{r} n_i/N \log_2 (n_i/N) \tag{4}$$

where n_i is the number of individuals in the ecosystem whose biomass falls in the interval i, independently of species. N is the sum $\sum N_i$
For large populations an equivalent form of (4) is

$$\mu = 1/N \log_2 (N!/n_1! n_2! \ldots n_r!) . \tag{5}$$

Either the continuous or the discrete form can be used in developing the constrained maximisation technique. In the discrete case, the definition of μ depends on both the size of the interval, Δm, as well as on the range, $r \Delta m$, of biomass values available to the ecosystem and we shall address ourselves to this question when we consider concrete examples further on. In the continuum limit, (3), the integral has been taken from 0 to ∞ for mathematical convenience but it is clear that a cut off is imposed at some value of m. Indeed this cut off arises naturally from the form of the p(m) function which we shall derive. More specifically, extremely heavy organisms are found to be extremely improbable.

Extremal principle and biomass distributions in ecosystems

Consider the continuum for (3). Our basic working hypothesis is that the biomass probability density function p(m) for a given ecosystem with N individuals is given by maximising the total biomass diversity $N\mu$ subject to the constraint imposed by the total biomass

$$M = N \int p(m) m dm = N \bar{m} \tag{6}$$

which the system can support. In (6) $\bar{m}$ denotes the mean biomass per individual. The maximisation of (3) subject to the constraint will yield a probability density function p(m) for the given ecosystem which can be tested against observed data.

Before proceeding it may worth noting that the maximisation of an entropy-like quantity subject to constraints is a classic technique for deriving a probability density that is consistent with whatever information one possesses regarding the corresponding random variable. In other words, any information which one does have can be expressed as a constraint on the maximisation procedure. It can be shown for example that the maximisation of the entropy function

$$-\int_{-\infty}^{\infty} p(x) \log p(x)\, dx \qquad (7)$$

subject to the condition that one knows the standard deviation of x to be a fixed value, σ , yields the standard Gaussian distribution

$$p(x) = \frac{1}{\sqrt{2\pi}\,\sigma} \exp(-x^2/2\sigma^2). \qquad (8)$$

A similar procedure is used in statistical mechanics: one maximises the entropy subject to the constraint that the mean energy E is given. This leads to the well known canonical distribution function (Landsberg, 1978)

$$P_i = \frac{1}{Z} \exp(-E_i/kT) \qquad (9)$$

where Z is the canonical partition function, $\sum_i \exp - E_i/kT$. An entirely analogous procedure is used in information theory to obtain the optimal matching of channel and code (Brillouin, 1962). In this case one finds that optimal coding is achieved when the symbol probabilities, P_i, satisfy

$$P_i = \exp(-ct_i/k) \qquad (10)$$

where C is the channel capacity, t_i is the duration of symbol i and $k = \log_2 e$. We shall find that the biomass distribution function for our ecological problem has a structure very similar to (9) or (10).

Returning now to the details of the maximisation procedure for the biomass constraint equation (6) can be written

$$\bar{m} = \int_0^{\infty} p(m)\, m\, dm \qquad (11)$$

and that in addition we have the further (normalisation) constraint

$$1 = \int_0^\infty p(m)\,dm \,. \tag{12}$$

Constrained maximisation of (3) makes use of the Lagrange multiplier technique. We write

$$-\delta\int_0^\infty p(m)\ln p(m)\,dm - \alpha\delta\int_0^\infty p(m)\,dm - \\ -\beta\delta\int_0^\infty p(m)\,m\,dm = 0 \tag{13}$$

where the variations are to be taken with respect to the function p(m) in the sense of the calculus of variations, and, where we have substituted natural logarithms for base 2 logarithms for convenience. The difference is a factor $k = \log_2 e$. The Lagrange multipliers α and β are paramaters whose values are determined *a posteriori* by substituting the value found for p(m) from (13) into the constraint equations (12) and (11). Carrying out the variational differentiation we obtain

$$1 + \ln p(m) + \alpha + \beta m = 0 \tag{14}$$

or

$$p(m) = e^{-\alpha-1}\, e^{-\beta m} \,. \tag{15}$$

To determine α and β we first substitute (15) into (12) and apply elementary integration to obtain

$$1 = e^{\alpha-1}/\beta \,. \tag{16}$$

A further substitution of (15) and (16) into (11) yields upon integration

$$\beta = 1/\overline{m} \tag{17}$$

and hence

$$p(m) = \frac{1}{\overline{m}}\, e^{-m/\overline{m}} \,. \tag{18}$$

Equation (18) is our result: the biomass probability density which maximises μ subject to the constraints (11) and (12) is a decreasing exponential function of the biomass (Lurie, Wagensberg, 1983). From a qualitative point of view the result is in accord with our intuitive expectation that heavier organisms are less probable than lighter organisms. We shall examine the quantitative evidence for the exponential law (18) in the next section.

For readers who are unfamiliar with the techniques of variational calculus it may be worthwhile to see how our result can be derived in a more elementary manner starting from the discrete formula (1):

$$\mu = -\sum_{i=1}^{r} n_i/N \log_2 n_i/N \tag{19}$$

and setting $n_i/N = P_i$

$$\mu = -\sum_{i=1}^{r} P_i \log_2 P_i . \tag{20}$$

The constraint on the total biomass M is

$$M = \sum_{i=1}^{r} n_i m_i \tag{21}$$

or

$$M/N = \sum_{i=1}^{r} P_i m_i \tag{22}$$

where m_i is a representative biomass for individuals in the interval i. Maximising (20) subject to the constraint (22) and the normalisation constraint

$$\sum_{i=1}^{r} P_i = 1 \tag{23}$$

implies setting

$$\delta\left(-k\sum_{i=1}^{r} P_i \ln P_i - \alpha\sum_{i=1}^{r} P_i - \beta\sum_{i=1}^{r} m_i P_i\right) = 0$$

where α and β are Lagrange multipliers and where the variations are now to be taken with respect to the r variables, P_i, in the sense of ordinary calculus. This yields:

$$1 + \ln P_i + \alpha + \beta m_i = 0 \tag{24}$$

or

$$P_i = e^{-\alpha-1}\, e^{-\beta m_i} \tag{25}$$

Substituting (25) into (21) yields $e^{\alpha+1} = \sum e^{-\beta m_i}$ and therefore

$$P_i = e^{-\beta m_i} / \sum_{i=1}^{r} e^{-\beta m_i} . \tag{26}$$

We again find that P_i is a decreasing exponential function of the biomass m_i. In the discrete case we cannot without proceeding to the continuum limit express either β or the denominator, Z, i. e.

$$Z = \sum_{i=1}^{r} e^{-\beta m_i} . \tag{27}$$

In terms of the mean biomass $\bar{m}$, (26) can be written in the form

$$m_i = -\frac{1}{\beta} \ln P_i - \frac{1}{\beta} \ln Z \tag{28}$$

which shows m_i as a linear function of $\ln P_i$. To recover (28) we go over to the continuum limit writing the mean biomass $\bar{m} = M/N$ in the form

$$\bar{m} = \sum_{i=1}^{r} m_i e^{-\beta m_i} / \sum_{i=1}^{r} e^{-\beta m_i} \tag{29}$$

multiplying numerator and denominator by Δm and replacing the sums by integrals ($\Delta m \rightarrow dm$). This yields

$$\bar{m} = 1/\beta \tag{30}$$

and similarly

$$Z = \bar{m} / \Delta m \tag{31}$$

so that (28) becomes

$$p(m) = 1/\bar{m}\ e^{-m/\bar{m}} \tag{32}$$

after replacing P_i by p(m) dm.

Observational tests

We have checked the validity of this prediction against various sets of observed data which were made available by the Instituto de Investigaciones Pesqueras of Barcelona on the weight distributions of fish corresponding to five different oceanographic expeditions (ATLOR IV, ATLOR V, ATLOR VI, BENGUELA II, BENGUELA IV) (Mauriquez, Rucabado, 1976; MacPherson, 1981; Grupo de recursos del Instituto de Investigaciones Pesqueras, personal communication). As we have noted earlier, the definition (1) of the biomass diversity depends on both the size of the mass interval Δ m and on the range r . Δ m of biomass values available to the ecosystem. A corresponding dependence on the choice of these two parameters is also evident in (28) since both P_i and Z by their very definition depend on how the value of Δ m is selected. The choice of the range r . Δ m is selectively straightforward, it is suggested by the analysis of the fishery data itself. The ATLOR IV data, for example, consists of 50 haults in a zone of upwelling containing a total of 430,109 fishes. This sample includes at least 20 different species having a minimum representation of at least 1 % of the total population of fish. The lower limit of the biomass range is established by the size of the fishing mesh and the upper limit is determined by the fact that biomass values greater than 0.085 kg are represented by occupation numbers n_i which are less than 0.1 % of the total number of fish, and therefore sufficiently improbable that they may be neglected.

The next step is the selection of the biomass interval Δ m. For this sample, a choice of Δ m = 0.091 kg yields the set of data points exhibited in Fig. 23. Inspection of this figure shows that the points do indeed fall on a straight line as predicted by the model. The best linear regression fit to the data has slope 0.0140 and y-intercept 0.006 with a Pearsonian correlation coefficient of q = 0.998 indicating a near optimum fit. Before discussing how the analysis of the ATLOR IV data would be affected by a different choice of Δ m we note that the parameters of the straight line (its slope and y-intercept) can be computed in the continuum limit as descripted in the previous section. Taking (30) and (31) into account we rewrite the biomass distribution function (32) as

$$m_i = -\bar{m}\ln P_i - \bar{m}\ln(\bar{m}/\Delta m). \tag{33}$$

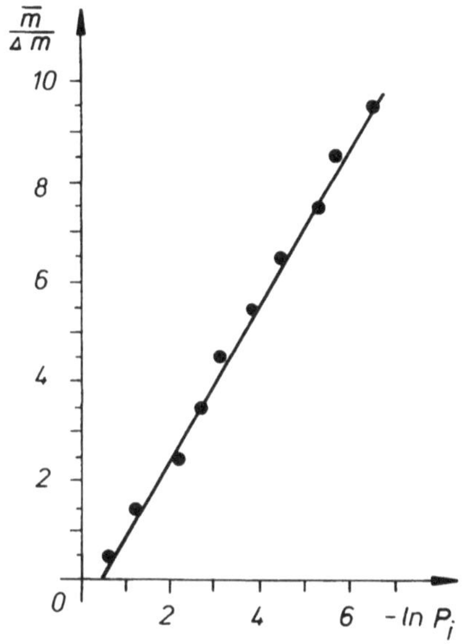

Fig. 23 The biomass distribution of the data of ATLOR VI (N. W. Africa, 1975); comprising 430,190 individuals. The mass interval Δ m is 0.0091 kg and the correlation coefficient 0.998.

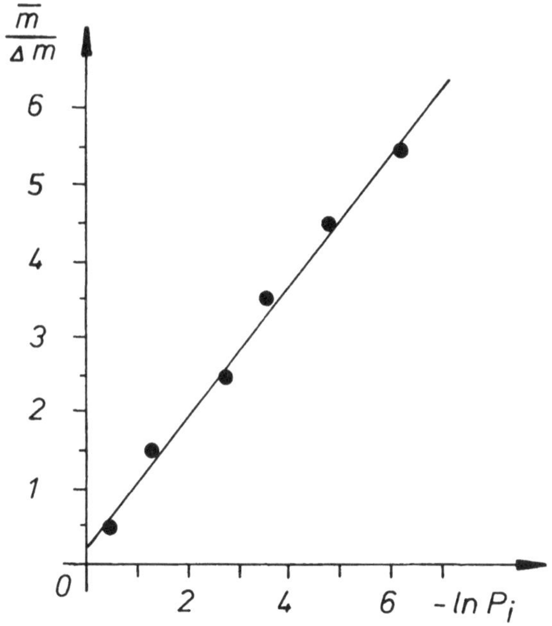

Fig. 24 The biomass distribution of the data of ATLOR IV (North of Namibia, November 1979) comprising 38.355 individuals. The mass interval is 0.032 kg and the correlation coefficient 0.998.

Hence, once $\bar{m}$ is known and Δm is selected any set of data can be tested with (33). For the ATLOR IV data $\bar{m} = M/N$ is found to be 0.0139 kg. The theoretical values for the slope and y-intercept are therefore $\bar{m} = 0.0139$ kg and $-m \ln(\bar{m}/\Delta m) = -0.006$ kg which compare satisfactorily with the observed values cited above. This provides a consistency check on the analysis of the data. Figures 24, 25 and 26 show that the theoretical model (33) provides an equally good fit to the data from the other expeditions. Quite generally, the model performs very well indeed when tested against all available data on fish populations (see detailed information on the figure captions).

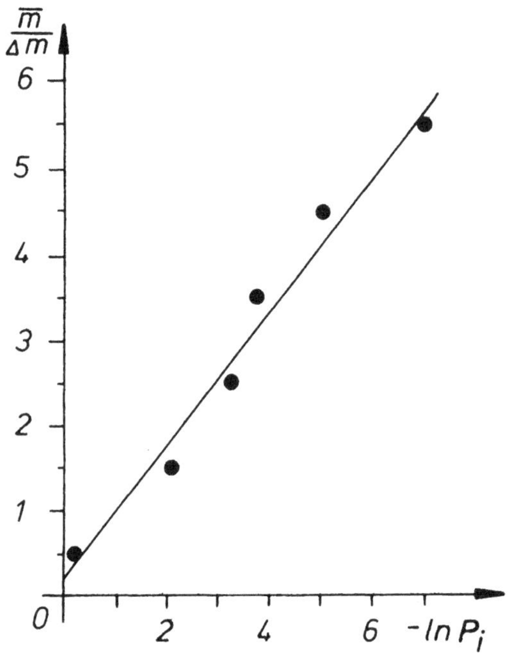

Fig. 25 The biomass distribution of the data of ATLOR V (Cabo Blanco, April 1974) comprising 70,131 individuals. The mass interval is 0.035 kg and the correlation coefficient 0.986.

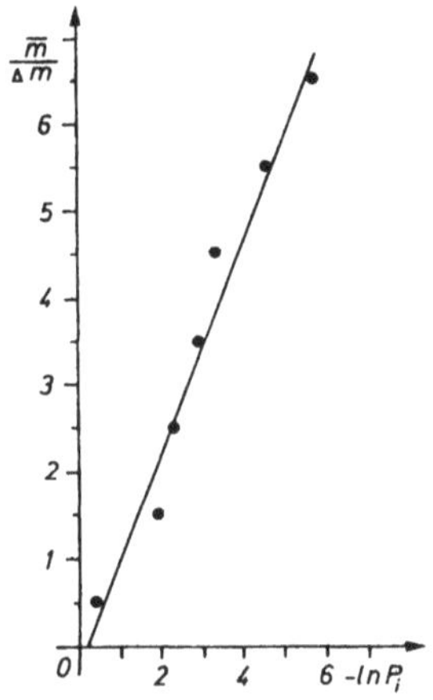

Fig. 26 The biomass distribution of the data BENGUELA II (North of Namibia, September 1980) comprising 482,304 individuals. The mass interval is 0.038 kg and the correlations coefficient 0.989.

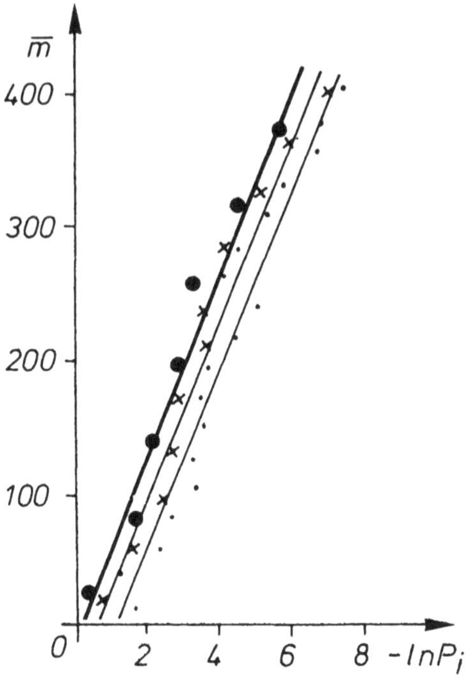

Fig. 27 Consistency check with the Benguela IV data comprising 242,148 individuals. The mass intervals are 0.0225 kg (18 data points), 0.0375 kg (11 data points) and 0.0575 kg (7 data points). The corresponding Pearsonian correlation coefficients are respectively r = 0.971, r = 0.979 and r = 0.986.

A further consistency check can be performed by varying the mass window, Δ m, and observing whether or not the data continue to fall on a straight line having the same slope $\bar{m}$ and a displaced y-intercept in accordance with (33). We have verified that this is indeed the case for the data of BENGUELA IV for mass windows of 0.0225 kg (18 data points), 0.0375 kg (11 data points) and 0.0575 (7 data points), as is shown in Figure 27.

Perspectives

The remarkably good fit of the biomass distribution functions,

$$P_i = e^{-m_i/\bar{m}}(\Delta m)/\bar{m}\,, \tag{34}$$

to the fishery data presented in the previous section provides solid support for the idea that the maximisation of biomass diversity is a promising technique for the analysis of ecosystems. Much more data is, of course, required in order to confirm the validity of this approach for different populations. The data should, in principle, fulfill certain conditions. First, it should cover large population samples and, secondly, populations in which the interaction between different individuals is minimal, as is the case in the fishery examples presented here. It must be recognised that such data is not readily available in the literature. Perhaps the existence of a biomass distribution model will now create the incentive for it; the important point is that the necessary data be organised in terms of numbers of individuals in a given biomass range, independently of genetic classification, with the proviso that one should not mix individuals belonging to the poikilotherm group either with those of the homeotherm group or with unicellular organisms since metabolic efficiency of these three groups is radically different.

We believe that the biomass distribution model presented here is of considerable interest. From the theoretical point of view, our model provides insight into the adaptive component of biological processes referred to in the introduction, and, we feel, it confers a new credibility on the variational principle approach for determining the structure of biological stationary states that are compatible with imposed environmental constraints. Moreover the model suggests new directions for research which could have important practical consequences. In essence the existence of a macroscopic model for the biomass diversity distribution is analogous to the macroscopic thermodynamic approach to many body problems in physics. One is therefore tempted to inquire whether there exists an underlying microscopic model for an ecosystem which would reproduce the results of the present paper on a macroscopic scale. In physics, underlying microscopic models can be developed using either molecular dynamic techniques (Giró et al., 1980) or by Montecarlo simulations (Metropolis et al., 1953). Following this line of reasoning one might expect a Montecarlo simulation technique for biomass production in ecology to be of considerable theoretical, as well as practical, interest. Indeed such a model could be expected to provide naturalists with an accesible tool for the analysis and control of biomass production in both natural and artificial eco-environments, with special emphasis on fishery planning and the design of fish culture plants or hatcheries in general.

How could a Montecarlo simulation technique be set up? Figure 28 exhibits a plausible flow chart, affectionately termed the Barcelonagram (Giró et al., 1982), in which the simultaneous effects of motion, nutrition, reproduction and death in a population distributed over a lattice are handled by a random shock approach. The computer is instructed to generate output data on the structural evolution of the ecosystem after each completion of the loop. We are presently investigating this model for an initial population of roughly 20,000 individuals. The model takes account of known biological facts such as for example the tendency of larger individuals to live longer and reproduce less, and also of the positive correlation between size and distance travelled. Initially the basic food particles, or <u>nutrients</u>, are distributed randomly throughout the grid, and as individuals die off they become part of the nutrient store. Time honoured techniques of molecular dynamics such as periodic boundary conditions and a cellular grid-like structure find a natural application in this approach. The key questions which this Montecarlo simulation technique must answer are

1. Does the system reach a stationary state?
2. If it does, is it stable and how well does it match up with our macroscopic biomass distribution model?

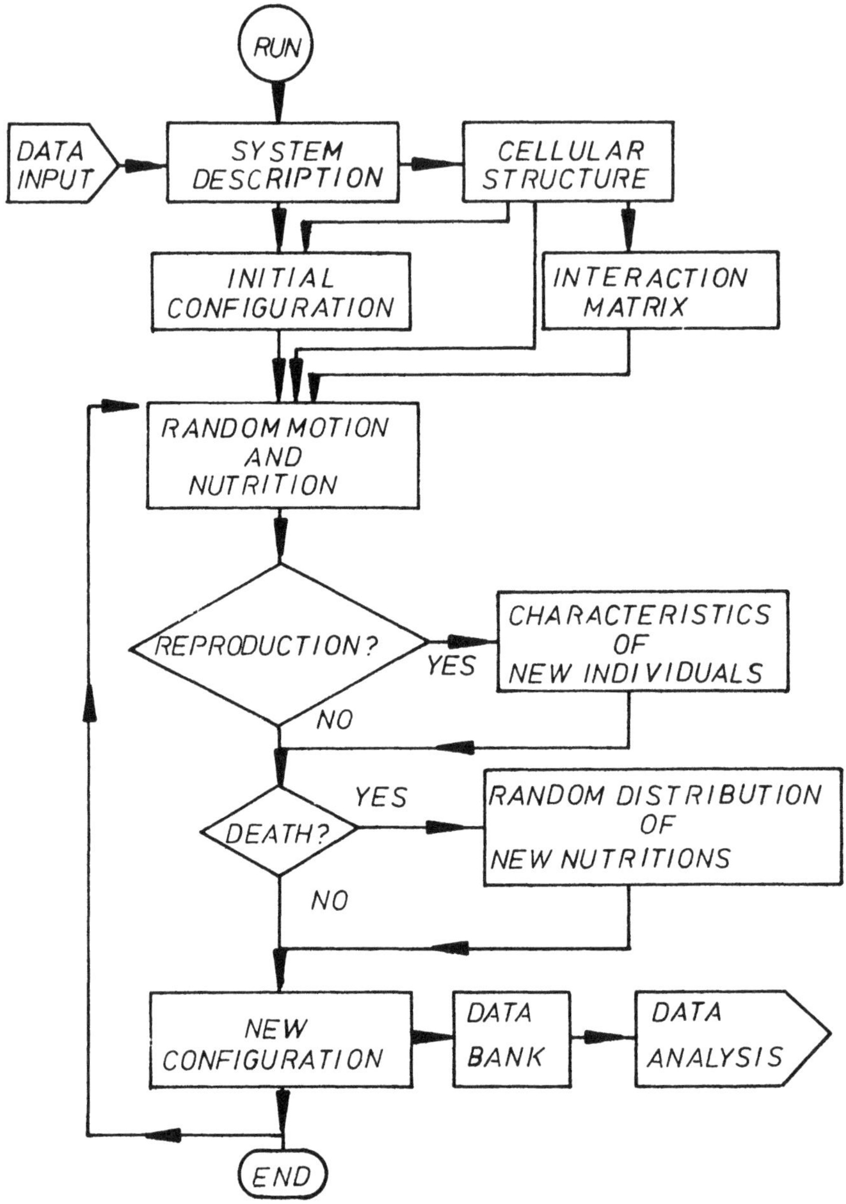

Fig. 28 The Barcelonagram.

3. What conclusions can be drawn about the general evolution law of such ecosystem?
4. Can such a technique provide control and optimisation parameters for use in managing real, natural, or man-made ecosystems?

The intital results of the simulation technique are very encouraging and will be reported elsewhere.

To sum up, we have proposed here three different lines of attack on the problem of describing the adaptive behaviour of a complex ecological system:

a) variational principle technique

b) a numerical simulation approach and

c) an observational method.

Based on the results obtained so far we are confident that the potential of physics-based techniques in biology is far from be exhausted.

Optimal Principle and Temperature Conditions For the Existence of Animals

V. V. Lapkin

The optimalisation principle has not been developed in biology and has only recently been discussed (Novosel'tsev, 1978). Nevertheless the most important principles in the field have been outlined already, and their further development and formalisation are under way. The concepts are underlain by the extremal principle, i. e. by the procedure of finding the minimum and maximum values of some function of a process, or of an integral system, and, more often - of finding a function that is generalised for a number of other functions. The extremum of this function is taken as the optimalisation criterion.

The main difficulties encountered in the application of the optimalisation principle in biology are concerned with selecting the function in it's evolutionary and ontogenetic aspects and also, for transient processes. The absence of an explicitly generalised function describing the directed course of evolution has made some authors use the classical principle of expedient adaptation as a criterion and take the stable functioning of the system for its main quality index (Novosel'tsev, 1978). Taking into account the absence of information-saving systems in populations and biocenoses the adherents of these concepts believe that, even if the optimalisation principle is valid for biology, it is extended only to individual organisms and does not touch upon transient processes (Novosel'tsev, 1978; Insarov, Semenov, 1982).

Another group of investigators assumes the existence of the generalised optimalisation criterion which the developing biosystem strives for (Rosen, 1967; Khanin et al., 1978). The idea of the optimalisation principle in biology most clearly formulated first by Rashevsky (1960) and named by him the principle of optimal construction is based on the following concept:

Thermodynamics and Regulation of Biological and Processes

'Organisms with biological structure optimal in relation to natural selection are also optimal in the sense that they minimise some estimating function determined from the main characteristics of the environment' (Rosen, 1967). The criterion of natural selection means here, obvious evolutionary superiority of some structural-functional parameter established in the system, which brings the latter to a more stable state as compared with other systems (Khanin et al., 1978; Khanin, Dorfman, 1982). Proceeding from a mathematical formulation of the natural selection criterion Khanin and Dorfman (1982) in a simple model of biocenosis (isolated homogeneous population) and taking energetic fulfillment of the population requirements as the basic limiting factor, established such important optimalisation criteria as, the maxima of energetic imbalance, and the relative rate of population number increase. Rabinovich (1982) also arrived at the concept of energetic imbalance analysis of the kinetics of transient processes in living organisms. As an energetic criterion of optimalisation he advanced minimisation of losses of, and maximisation of accumulation of free energy. A similar idea was proposed by Rosen (1967). Using the example of an economic system he considered the profit maximum as optimalisation of the difference between losses related to goods production and their cost. Kogan (1977) observed optimalisation of the program of animal behaviour. He wrote: 'Optimal strategy of a hunting wolf flock is the higher, the larger is their catch and the less energy is spent' (Kogan, 1977). He believes that such a relationship is valid for different organisational levels of the living systems, i. e. cell, organ, individual, population, species and biocenosis, and that it reflects the essence of the natural selection criterion. Such a formulation of the problem unites the search for a generalised optimalisation criterion with the search for a generalised criterion of evolution studied by thermodynamics. The following criteria, those applied to biological problems among them, were advanced at different times:

1. A maximum in the rate of entropy production (Prigogine, Wiame, 1946).
2. The universal evolution criterion of Glansdorff, Prigogine (1971); $d_x\Psi/dt \leq 0$ from which, it follows that entropy production in the system not always decreases, and does not necessarily have the minimum value in the steady state, i. e. $d\Psi/dt \lesseqgtr 0$.

3. A heat production maximum in the system (Zotin, 1974).
4. A minimum in the mixing of useful products of metabolism and wastes (Chernavsky, 1975).
5. A maximum in entropy production (Trinscher, 1967).
6. A maximum in free energy (Rubin, 1976).
7. Power maximalisation in organisms of similar weight (Dontsova, Zotin, 1982).

In the present situation further accumulation of experimental data, at different levels of biosystem organisation, including the organism, is of primary importance in the search for partial extremal values in order to analyse more deeply the properties of the optimal state of these objects, and to give a direct evaluation of the existing theoretical and empirical criteria of evolution.

It is known that the life of animals proceeds over a narrow temperature range (~ 50 °C). It has been also found that, if, animals capable of moving independently were given the right of free selection of temperature conditions, they would be concentrated in a narrow temperature range. For fishes, for example, such temperature variations in the established selection regime is 3 °C to 6 °C (Lapkin et al., 1979).

It has been shown that selection of fixed temperatures by mobile organisms or, the existence of definite temperature zones in which optimal growth and development of immovable organisms may take place, is a characteristic of all classes of living organisms, from unicellular microorganisms to mammals (Ivlev, Leyzerovich, 1960; Symposium, 1979). The basis of these phenomena as well as of steady state establishment in living systems (Parknomenko, 1968) seems to be limitation of the temperature range within which the combined chemical reactions unique to the cells and entire organism proceed.

Therefore we shall now turn to the analysis of these factors in animals under various temperature conditions, including those selected by the animals themselves. For a clearer and simpler understanding of the energetic and structure-function peculiarities of living organisms in relation to the environmental temperature our discussion will concentrate on cold-blooded animals, e. g. fish.

A number of studies are known which have investigated the relationship, for fish, between active (or total) metabolism and basic (or standard) metabolism and the environmental temperature. On the basis of such data Fry (1947) developed an energetic model of the interaction between cold-blooded animals and the environment (Fig. 31 a; curves 4, 5, 6). According to his concept the optimal state of the living system is determined by a maximum difference between active and basic metabolism, i. e. the maximum value of energy spent on producing some work. The value of this energy may be changed through the effect of various external factors, in particular temperature. It has been shown in recent work that in some fish species the temperature dependence of basic metabolism, under threshold loads, is described not by exponential, but by cupola-shaped curves (Brett, 1971; Beitinger, Fitzpatrick, 1979). It is difficult to conclude now whether this phenomenon was due to species specificity of fish, organism size or to the fact that threshold loads were not reached in previous experiments. It is only clear that a turn-over point really exists in the curve of active metabolism, and, according to Brett (1971), it occurs at the temperature selected by the animals (Fig. 29 a). This temperature is fixed for a given animal species and is independent of preceding habitation temperatures (Fry, 1947). Brett (1971) thinks that the temperature corresponding to the turn-over point in the active metabolism curve is optimal. This statement is supported by his experimental data on growth rate, food consumption, blood pressure and other parameters which in red salmon reach a maximum at the selected temperature. As far as warm-blooded animals are concerned, studies of different authors (Khalabukhov, 1950; Ivlev, Leyzerovich, 1960; Prosser, 1973) show that basic metabolism in the threshold zone selected the temperature falls to a minimum; graphically revealed as flattening of the curve (Fig. 29 b). As a rule the temperature range is narrower in small animals and wider in large ones, and it is somewhat lower than the body temperature, and this seems to be due to the heat-conducting properties of the organism. This zone is usually called the thermoneutral zone, and it is related to minimisation of energetic expenditures on thermoregulation at the selected temperature. We have found no studies which have dealt with active metabolism under threshold loads for warm-blooded animals at different temperatures. We believe, however, that the temperature relationship of active metabolism should also be described by a cupola-shaped curve with the turn-over point falling within the thermoneutral zone.

This phenomenon is understandable because in the thermoneutral zone the energetic expenditure on thermoregulation processes is minimised, and additional energy which may be realised under loading is liberated. It should be noted that for the necessary balance between heat exchange and the environment to be maintained, an increase in body temperature due to increased heat production (that related to muscle tension being excluded) the thermoneutral zone of active metabolism ought to be somewhat below that of basic metabolism (Fig. 29 b). Similar phenomena may also be typical for cold-blooded animals of considerable size, if they have no special mechanisms (enzymic complexes) providing for an optimal work regime for the organism at temperatures above those selected in the state of relative rest. Thus in both, warm- and cold-blooded animals maximisation of energy imbalance between active and basic metabolism is observed in the region of the selected threshold temperature.

Having considered the organism's properties at different temperatures as stipulated by its energetic state, we return to it's structure-function properties. It has been shown by a number of researchers that there is a definite temperature region, for fertilised roe, at which yields the maximum number of hatched larvae without morpho-functional defects in the organism (Reznichenko et al., 1962; Kokurewicz, 1970; Gulidov, Popova, 1979). For *Rutilus rutilus* this region is 10 °C to 14 °C. Gulidov and Popova (1979) noted that the highest number of individuals which arrive at the 9th developmental stage, when the organism was most prepared for independent life were hatched at 10 °C to 12 °C, and, that the maximum linear size of larvae was observed when roe were incubated at 12 °C. The temperature range indicated (10 °C to 14 °C) corresponds to the spawning temperatures of rutilus males, and, in relation to the embryo it is considered by us as a passively selected temperature. Thus a maximal size of hatched larvae and a minimum of morpho-functional defects may serve as criteria for the temperature optimum.

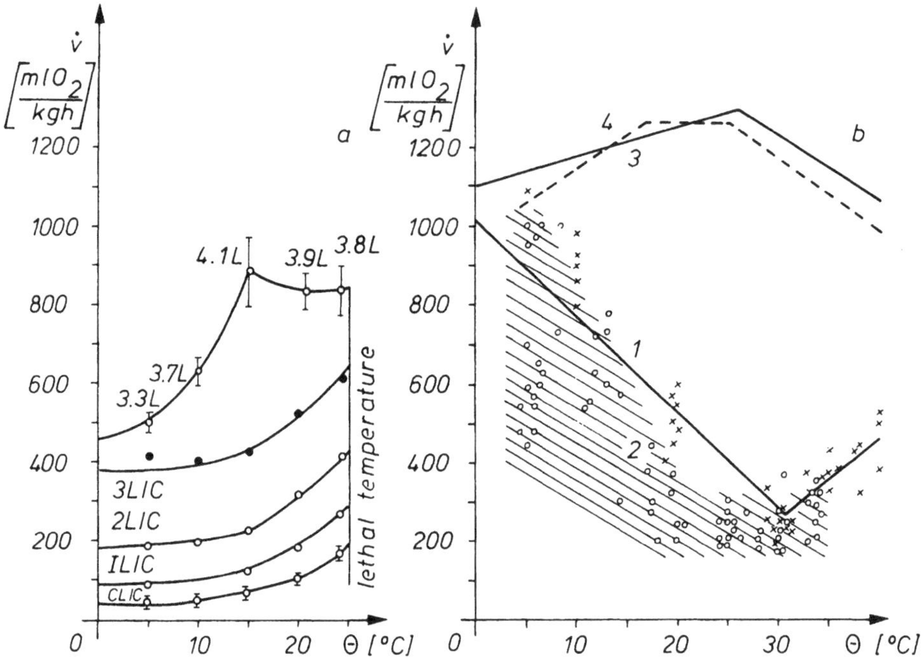

Fig. 29 Dependence of active and basic metabolism of animals on acclimation temperature.

a oxygen consumption by fish (young *Oncorhynchus nerka*) at various temperatures and swimming velocities (expressed in body length per minute) (Brett, 1971)

b oxygen consumption by mammals under various temperature conditions:

1. basic metabolism of *Dasypus novemcinctus mexicanus* (x)
2. basic metabolism of *Tachyglossus acyleatus* (o) (according to Prosser, 1977) hypothetical curve of active metabolism of warm-blooded animals for maximal load
3. armadillo
4. ochidra.

Interesting temperature relationships were obtained in the studies of functional properties of the fish visual system (Pavlov et al., 1967). It has been found that the values of threshold sensitivity in object vision and of perception of the limiting frequency of flashing in *Laucaspius delineatus* and *Cyprinus carpio* are extremal at temperatures of the order of 25 °C (Fig. 30). Such temperatures are in conformity with those selected by the ecologically related carp species (Lapkin et al., 1975).

Attention should be focused on one more fact, the final, selected temperatures are in conformity with the stable, stationary state of the organism, i. e. it is to such constant temperature values (with other constant values for parameters, such as age, season, year, feeding level etc.) which the animal strives from any environmental temperatures (Lapkin et al., 1981). An analysis of the biochemical properties of different subsystems shows that there are definite temperature regions in fish which are characterised by the greatest number of protein molecule isomers. Thus, for *Carrassius auratus* this region is 23 °C (Houstom, Rupert, 1976), and for *Salmo irideus Gibbons* 12 °C (Hochachka, Somero, 1973), and these correspond to the final, selected temperature of the respective adults (Mantelman, 1958; Coutent, 1977).

Let us analyse the experimental results presented, in terms of thermodynamics and cybernetics. Fry (1947) and Brett's (1971) concepts are based on a model of the energetic properties of cold-blooded animals. The model for warm-blooded animals is based on Figure 29 b. It is evident that the temperature dependence curve for an animal's active metabolism under threshold load reflects a maximum in free energy formed in the system; while the basic metabolism curve shows the portion of free energy spent by the organism in maintaining its functions whilst at rest (Fig. 31). The maximal imbalance of free energy in the system under load, and at rest, is reflected in the final, selected or, neutral, temperature. As is known, this energy is spent on motor activity, growth and reproduction of the organism.

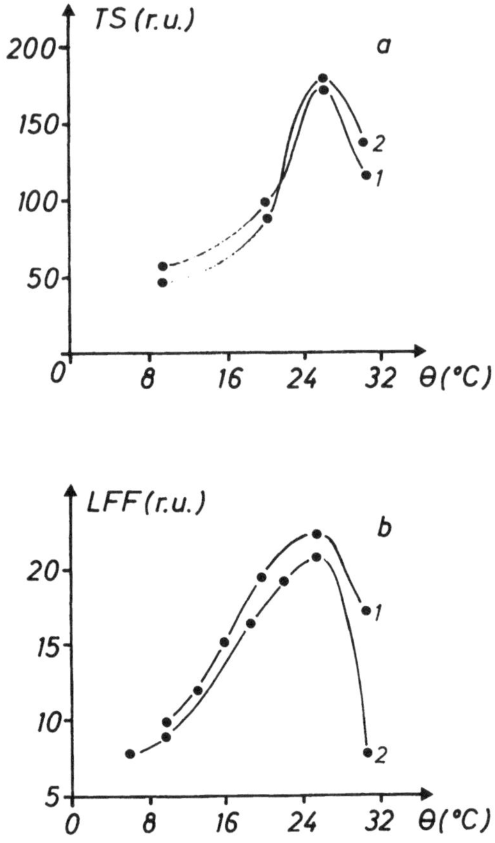

Fig. 30 Temperature relationships of
a threshold sensitivity (TS) of vision
b perception of limiting frequency of flashing (LFF, both in relative units) of fish
1. *Leucaspius delineatus*
2. *Cyprinus carpio* (Pavlov et al., 1967)

It is believed that the intensity of gas exchange in aerobic animals reflects the rate of biochemical reactions (heat production) and of entropy production (Zotin, 1974, 1979 a) and characterises the non-equilibrium of the system. In its turn, non-equilibrium is the source of orderliness (structure-function organisation) for an increase in which, the system entropy decreases (Nicolis, Prigogine, 1977). We think that along with these

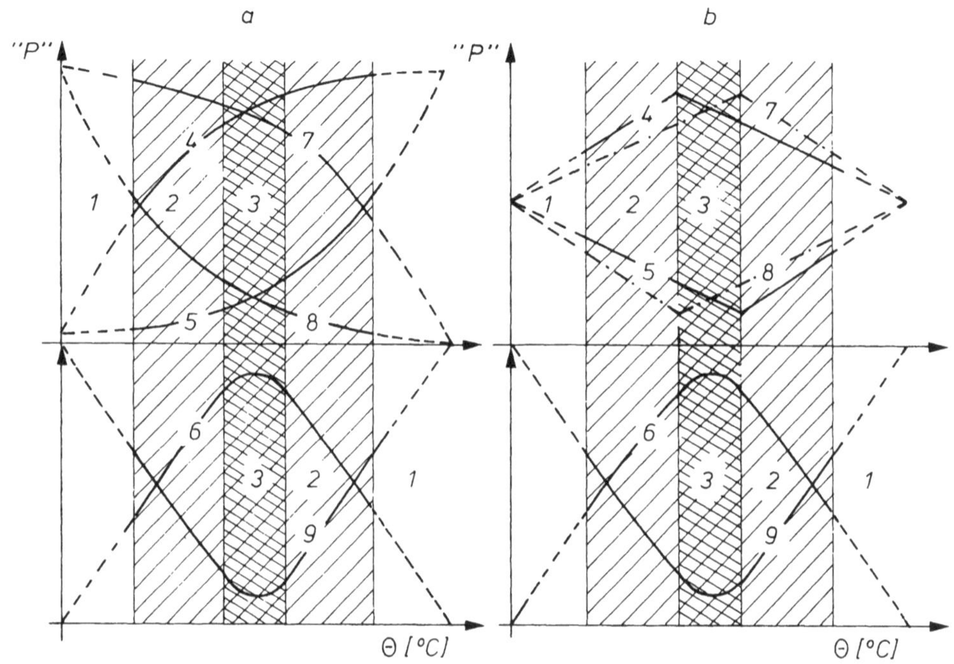

Fig. 31 Scheme of relationship between energetic and functional parameters of poikilothermal (a) and homeothermal (b) animals and environmental temperature Θ

1. resistence zone
2. tolerance zone
3. zone of selected temperatures
4. parameters "P" of the system under limiting load (intensity of oxygen consumption, rate of biochemical reactions (heat production), rate of entropy production, degree of non-equilibrium of the system, free energy, organisation, information quantity)
5. the same parameters as in 4 for the state of rest
6. imbalance of free energy, stability of the system towards external stimuli, number of possible states of the system, fast-response and sensitivity of organisms

7., 8. entropy and efficiency at rest and under load

9. number of morphofunctional defects, time of transient process.

phenomenological parameters the intensity of gas exchange also reflects the extent of system 'open-ness', i. e. the circulation rate of substance and energy and, consequently, it's power. Since, with temperature rise the intensity of active and basic metabolism of fish increases (Fig. 31 a), the rate of entropy production, the degree of the system open-ness, non-equilibrium and structure-function organisation also increase. It's power, developed under load, and 'consumed', i. e. spent by the organism in functioning at rest also rises. It should be noted that the increase of power developed by the animal and of the accompanying parameters depends, not on temperature only, but on training (periodic loading) of the organism, i. e. they exhibit a parametric pattern. Taking into account the fact that entropy and efficiency on the one hand, and non-equilibrium of the system on the other are in inverse proportion, it is suggested that entropy and efficiency in the model under study decrease with temperature rise, and are higher when the organism is at rest (Fig. 31 a). At the same time, entropy as a measure disorder reflects indeterminacy in the system behaviour which decreases with a non-uniform distribution of probabilities of the states acquired by the system. The maxim indeterminacy is reached in the case of a state distribution of equal probability in the system and, is equal to the logarithm of the number of states. The difference between the maximum and the actual indeterminacies is, as is known, referred to as the amount of information. The quantitative measure of organisation, which is also the value of eliminated indeterminacy in the system, is similarly estimated via a probability interpretation. Consequently the temperature relationships of amount of information and structure-function organisation in the system are of a similar pattern (Fig. 31 a).

In technical cybernetics the basic quality indices of the control system are stability, precise work by the system under the established regime and the time of transient processes.

The stability of the system is understood as it's ability to return to some stationary regime from any initial state, where its integrity is preserved. The number of possible states of the system (Ashby, 1956) or the threshold coefficient of strengthening (Zaytsev, 1975) can serve as the stability criterion. An analysis of the model (Fig. 31 a) shows that imbalance in free energy is the very parameter which determines the possible states of the fish under various temperature conditions, i. e. its stability towards disturbances from various sources. Biological systems seem to use a perfect, combined control principle where the controlling effect is, at the same time, a function of the deviation of the control value and the disturbance (adjustment). Any disturbance seems to cause, in the living system, imbalance between the controlled value and the controlling effect. To eliminate this imbalance the organism must spend the additional energy necessary for reconstruction of the system due to switching over to a more intensive working regime. And, since the value of the free energy imbalance is maximised in the finally selected temperature zone, it is evident that the maximum number of possible stationary states is reached only within this temperature range. Maximisation of fish stability at the molecular level at the finally selected temperature seems to be due to an increase in the number of isozymes involved in biochemical reactions. The maximum number of isozymes, apparently, creates more favourable conditions for conjugated reactions which are accompanied by the greatest yield of free energy and hence increases the reliability of the system.

Consider now the realisation in the living system of another index of control - exact establishment of a predetermined regime or minimisation of the mean-square root error of such establishment. In automatic regulation this optimisation criterion is widely used; it consists of selecting such parameters (or working regimes) of the system which eliminate the total error (Zaytsev, 1975). In the case of biological systems this technical optimisation criterion is satisfied by a minimum at the spawning temperature, of morpho-functional defects during fish embryonic development. We think that a minimum of morpho-functional errors in a developing system, at a selected temperature, also results from the maximisation of free energy. It is provision of free energy to the system that permits construction of the organism, strictly according to a predetermined program, to be realised.

Another important index of control in the system is minimisation of the time of its transient processes (Zaytsev, 1975). The limiting frequency of flashing perceived by the fish visual system at fixed temperatures undoubtedly reflects its resolving power. The resolution of an impulse intake is characterised by the time of intake restoration or, by an equivalent notion, the time of device regulation, and these are approximately equal to the time of the transient process (Harvey, 1965; Zaytsev, 1975). It is known that the smaller is the time of intake restoration, the greater is the resolution. As the frequency of flashes perceived by the visual system reaches the maximum within the selected temperature range, the times of the transient processes at these temperatures are at the minimum. In other words, at the finally selected temperature the visual system has maximum fast operation. The maximisation of object vision within this temperature region seems to be due to the fact that the subsystem reaches that maximally possible under the given conditions,i. e. a state of dynamic equilibrium. In engineering analogs of such phenomena are devices based on the compensation principle, e. g. measuring bridges of constant current etc.. The presence of a positive correlation between fast operation of the visual system and it's sensitivity can be explained by the decreased time required to eliminate the disturbance of the dynamic equilibrium, when the output parameter deviates from the established regime at the expense of an increase in sensitivity of the receptor element. We believe that it is not only the visual system of the fish that possesses these properties which are revealed at the finally selected temperature, but also other subsystems, and thus the entire organism (Fig. 31 a). It is assumed that fast operation and sensitivity of individual subsystems bear some fixed relationship with the value of free energy for the whole organism. Therefore their absolute values depend not only on temperature, but also on loading, acquiring their maximal values within the selected temperature range and at a state of relative rest. It was found, for example, that in homeothermal organisms including humans (sportsmen, astronauts) that physical strain impairs vision and hearing, and induces confusion into analytical thought. Similar symptoms are also observed under intellectual and nervous overloads. It seems to us that these phenomena can be explained by disproportionate distribution of free energy between separate subsystems of the organism under overload. The energy is spent mostly by the subsystems most actively involved in the process, and is regulated by blood flow. The reduction in free energy in other subsystems increases the probability of error in functioning and prevents these systems from achieving dynamic equilibrium. This results in a decrease in sensitivity and fast operation of the system.

Thus, for poikilothermal animals, in a strictly functional aspect three main optimisation principles can be distinguished showing the extrema within the selected temperatures:
I. maximisation of stability (or of the number of possible states);
II. minimisation of the time of transient processes;
III. the number of errors in the operation of the system.

In our opinion, such phenomenological relationships between energetic and structure-function aspects are also characteristic of homeothermal animals (Fig. 31 b). The difference is as follows. In the poikilothermal animals one observes, for a rise in temperature, an increase in metabolic intensity and the related rate of entropy production, extent of departure from equilibrium, power etc.. The enzyme sequences of poikilothermal animals are arranged in such a way so as to provide, under various temperature conditions, the appropriate rate required in biochemical reactions for life to proceed. This is performed at the expense of specific enzymes designed for operation at fixed temperatures. In homeothermal animals having a relatively constant body temperature, the enzymic sequence is tuned to a rather narrow temperature range. Maintenance of constant body temperature with changing environmental temperature is concerned with additional energy expenditure and this seems to be provided by definite biochemical processes whose rates are regulated mainly by the concentrations of the enzymes. These very properties of homeothermal enzymic systems enables one to explain the extrema of basal and active metabolic intensities in the neutral temperature region. At temperatures below this neutral region the homeothermal organism may be considered as a thermostat with heating; at temperatures above the neutral region, as a thermostat with cooling, or as a refrigerator.

Thus the analysis of the properties of living organisms at different temperatures suggest a total optimisation criterion and make it possible to formulate the notion of an 'optimum'. The optimum is a steady, stationary state of the system characterised by the maximum imbalance between free energy and maximum, fast action of the system which can function with minimum error. The molecular basis of this optimal state of biosystems appears to be the maximum conjugation of all the biochemical processes in the organisms, and this is possible only at fixed temperatures.

Finally let us consider the application of optimisation theory to ontogenesis and to transient processes. It has been found that during the course of ontogenesis in fish the selected, and lethal, temperatures are minimal during roe fertilisation, and reach a maximum during fry development and then decrease again (Lapkin et al., 1981). The rise in lethal temperature in the fry period points to maximisation of organism stability towards this factor. Consequently, in this period, the value of free energy imbalance is maximum, since, as stated above, it determines stability and the number of states of the system. However, the fact that free energy imbalance shows the extreme value in the fry period, does not mean, in our opinion, that this age category is optimal. Each age group is adapted for the fulfilment of definite biological function and reaches the optimum state at the selected temperatures appropriate to this age. Hence the absolute values of the optimisation indices at first increase, then decrease, whilst the relationships between these parameters are similar to those found on study of the statistical pattern of the temperature optimum. Such behaviour of the organism over time permits one to consider animal ontogenesis as a relaxation process, i. e. the tendency of the system to return to the primary state following some disturbance.

Possible applications of the optimisation criteria to transient processes in the living systems appear to be as follows. For a disturbance of small amplitude and short duration which does not lead the system to adaptive structure-function reconstruction, the minimum time required for the system to return to the initial state can be taken as the optimisation criterion. When the disturbance results in substantial structure-function reconstruction in the system, i. e. the system comes to a new stationary level, the optimal transient process is, according to Rabinovich (1982) that which is realised with the minimum intensity of free energy expenditure. We believe that both these optimisation criteria must be minimised in the finally selected, or neutral, temperature of the organism.

Optimal Heat Insulation of Homeotherms

M. A. Khanin, O. G. Bat'

Some of the general mechanisms of heat insulation in homeotherms are well known. For example, a relation between heat insulation and climatic conditions is obvious. At the same time, the available data indicate an essential difference in heat insulation of different weights for individuals under the same climatic conditions. The main parameter in heat insulation is the heat insulation coefficient which is the inverse of the heat transfer coefficient and varies in value from 0.1 m^2 to 1.3 m^2 °C/W. In studies of heat insulation in homeotherms attention has been given mainly to the dependence of the heat insulation coefficient on the fur layer thickness (Scholander et al., 1950 b; Hart, 1956), (Fig. 32 a, b). The almost monotonous behaviour of this dependence and the absence of a noticeable difference in heat insulation coefficients at fixed fur layer thickness (with some exceptions) suggests that the thermal conductivity of fur is similar for most species.

It is of interest to determine, using a corresponding mathematical model, how the principal heat insulation parameters depend on ecological and specific parameters.

The mathematical model of heat insulation, proposed for homeotherms, is based on the biological optimisation principle, which has been successfully applied as a method of phenomenological description in the structural and functional organisation of biological systems (Hess, 1914; Murray, 1926; Cohen, 1955; Rosen, 1967; Horsfield, 1967; Cohen, 1969; Milsum, 1968; Hämäläinen. 1978: Insarov, Semenov, 1982; Semevskaya, Semvskii, 1977; Khanin et al., 1976; Khanin, Bukharov, 1980; Chernousko, 1980).

Thermodynamics and Regulation of Biological and Processes

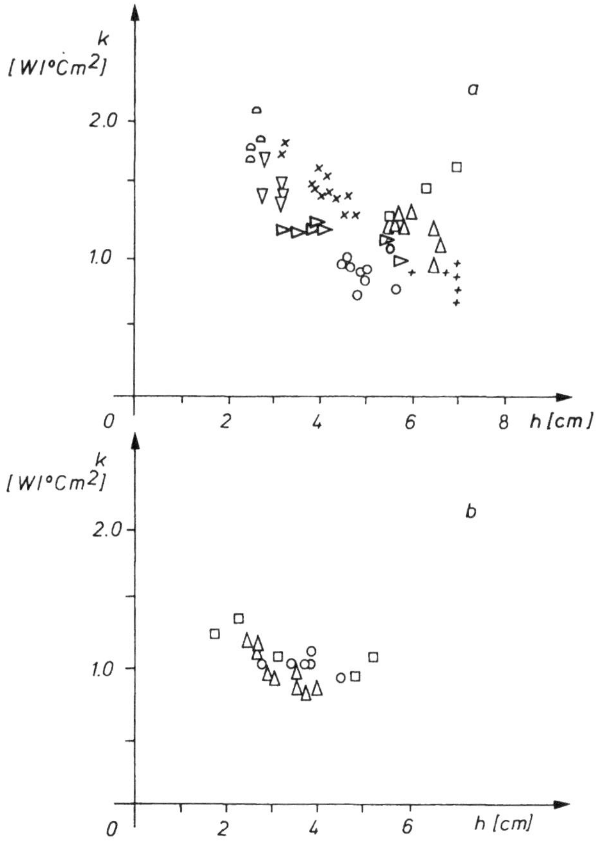

Fig. 32 Dependence of heat transfer coefficient (k) on fur layer thickness (h) of animals of northern hemisphere

- ◠ marten
- ▽ rabbit
- × dog, fox
- ▷ reindeer caribu
- ○ polar fox, glutton
- □ white bear
- △ wolf, bear
- + dall ram

(a - according to Scholander et al., 1950 c; b - according to Hart, 1956).

Rosen (1967) emphasised that the most important state in applying the optimisation principle to biology is to formulate the optimisation criterion.

In considering heat insulation it appears natural to proceed from an energetic optimisation criterion. In other words, in the course of natural selection species whose heat insulation provides minimum energy losses would be expected to survive.

The criterion of minimum energy expenditure must, of course, be accompanied by additional conditions following from the necessity to provide normal functioning of physiological systems, reproduction and persistence.

When formulating this criterion and its application to the case under consideration, it is necessary to consider separately the components of the energy balance of an individual which are related to specificity in heat insulation.

The most important energetic factor which depends on heat insulation intensity is the power expended to maintain body temperature when the environmental temperature is below the lower critical temperature. It is obvious that the presence of heat insulation makes it possible to reduce power losses at low environmental temperatures.

On the other hand, heat insulation is related to a certain additional energy expenditure. When moving, an individual expends additional power because of the increased weight associated with the fur coat. In addition, a certain energy is required to regenerate the heat insulating layer after moulting.

The components indicated in the energy balance must be included in the energetic criterion for heat insulation optimisation

$$W_T + W_M + W_r = \min \qquad (1)$$

where W_T is the additional power necessary to maintain the body temperature at environmental temperature below the lower critical temperature, W_M is the power expended by an individual in carrying the heat insulating layer, W_r is the power expended in growth of the heat insulating layer.

In accordance with the energetic optimisation hypothesis, the heat insulation parameters which provide a minimum in function (1) must be established during the course of evolution.

Let us consider, qualitatively, the dependence between energetic factors in criterion (1) and fur layer thickness; this dependence leads to an energetically optimal fur layer thickness. As the fur layer increases, energy losses to the environment due to heat transfer decrease, but power expenditure in motion and fur regeneration after moulting increase. Competition among these factors must determine the energetically optimal heat transfer coefficient and, for a given thermal conductivity, its optimal thickness.

To solve this problem, it is necessary to express the quantities W_T, W_M, and W_r through species-specific biological parameters, and quantities which characterise the environment.

The additional power necessary for an individual to maintain body temperature outside the thermoneutral zone, if the individual is at rest, is determined by the expression

$$W_T = k S (T_S - T_e) - W_o \qquad (2)$$

where k is the heat transfer coefficient; T_S is the body temperature of the individual; T_e is the environmental temperature; W_o is the basal metabolism of the individual; S is the surface area of the body.

During exercise W_T is less than the value determined by (2), and, during moderate exercise falls to zero.

Expression (2) determines the instantaneous value of the quantity W_T for a resting individual at an environmental temperature T_e. The optimisation criterion (1) must involve the average annual W_T taking into account exercise and seasonal changes in environmental temperature.

If we assume, approximately, that during exercise $W_T = 0$, the average annual W_T at constant environmental temperature takes the form

$$\bar{W}_T = \delta (k S (T_S - T_e) - W_o) \tag{3}$$

where δ is the fraction of time over which the individual is at rest.

Generally speaking, in (3) T_e should be understood as an average, annual, environmental temperature. Winter temperatures are the most influential in adaptation of heat insulating layer. In summer time, heat insulation decreases (moulting). Therefore, when determining the average annual W_T, one should consider T_e as an average winter temperature T_e^* and introduce a factor which takes into account the relative winter duration.

Thus, (3) takes the form

$$\bar{W}_T = \gamma \delta (k S (T_S - T_e^*) - W_o) \tag{4}$$

where γ is the relative winter duration. Using (4), we have assumed approximately that for all seasons except winter $W_T = 0$.

Now let us turn to the second term in criterion (1). According to zoological data, the power expenditure of an individual in satisfaying its fur requirements, when averaged over a sufficiently long period, amounts to 60 to 80 % of basal metabolism, i. e.

$$W_M = \alpha_M W_o \tag{5}$$

where $\alpha_M = 0.6$ to 0.8.

On the other hand, the mechanical power developed during movement is proportional to the total weight of the load carried, i. e. to the body weight

$$W_{mech} = \omega (P + P_M) \tag{6}$$

where P is the body weight of the individual, P_M is the fur weight, ω is a species-specific coefficient.

Equating (5) and (6), we find coefficient ω as:

$$\omega = \alpha W_0 / (P + P_M). \tag{7}$$

Now, using (7), we can find the additional power expended in carrying fur

$$W_M = \frac{\omega}{\eta} P_M = \frac{\alpha W_0 P_M}{\eta (P + P_M)} \tag{8}$$

where η is the skeletal muscle efficiency. Since $P_M \ll P$, P_M in the denominator of (8) can be neglected with respect to P. Expression (8) takes the form

$$W_M = \alpha W_0 P_M / \eta P. \tag{9}$$

(9) determines the power required by an individual to support fur.

Now let us express fur weight P_M in terms of other parameters of the heat insulation layer

$$P_M = S h \rho_M \tag{10}$$

where h is the fur thickness, ρ_M is the specific weight of the fur.

Taking into account (9) and (10), we find

$$W_M = \alpha W_0 S h \rho_M / \eta P. \tag{11}$$

(11) determines the power expended in carrying the heat insulation layer when all other forms of heat insulation, except for the fur coat, are neglected. In other words, we neglect the thermo-insulating properties of the hypodermal and fat layers in comparison with the heat insulation of the fur.

The last term in criterion (1) is determined by the expression

$$W_r = C_M P_M \varphi / T \eta_M$$

or (12)

$$W_r = c_M S h \varrho_M \varphi / \eta_M$$

where c_M is the specific heat capacity of the fur; η_M is the coefficient that accounts for the energetic efficiency of growth of the heat insulating layer; φ is the coefficient which accounts for the proportion of the fur lost on moulting; T is the fur coat growth period (T = 1 year).

Using (4), (11) and (12) and taking into account the relation between the heat transfer coefficient and the thermal conductivity,

$$k = \kappa / h \qquad (13)$$

we represent the optimisation criterion (1) in the following, explicit, form

$$\gamma \delta (k S (T_s - T_e) - W_o) + S \frac{\kappa}{k} \varrho (\varphi c_M / T \eta_M + \alpha W_o / \eta P) = \min$$

or, introducing dimensionless variables,

$$\gamma \delta (\tilde{k} (1 - T_e / T_s) - 1 + (A + B) / \tilde{k} = \min \qquad (14)$$

where

$$\tilde{k} = k / k_o; \quad k_o = W_o / S T_s$$

$$A = S^2 \kappa T_s \varrho \alpha / W_o \eta P; \quad B = S^2 \kappa T_s \varrho \varphi c_M / W_o^2 \eta_M T.$$

Differentiating (14) with respect to $\tilde{k}$ and equating the derivative to zero, we find the expression for the optimal heat transfer coefficient

$$\tilde{k}_{exp} = \left(\frac{A + B}{\gamma \delta (1 - T_e / T_s)} \right)^{1/2}. \qquad (15)$$

Knowing the optimal heat transfer coefficient, we can calculate the optimal thickness of the fur layer

$$h_{exp} = \varkappa / k_{exp} \quad (k_{exp} = \tilde{k}_{exp} W_o / ST_s). \tag{16}$$

One can also find the optimal, lower, critical temperature

$$k_{exp} S (T_s - T_1) = W_o. \tag{17}$$

Hence,

$$T_1 = T_s - W_o / S k_{exp}. \tag{18}$$

(15) determines the optimal heat transfer coefficient if the species-specific parameters in this expression, and the average winter environmental temperature, are known. (15), (16), and (18) can be used on consideration of their limits of application which result from the simplifying assumptions made in their derivation. Let us consider these limitations.

Equations (15), (16), (18) involve the basal metabolism of the individual, which is supposed to be the species-specific parameter independent of the season. For hibernating, large, animals the magnitude of the basal metabolism requires a corresponding correction. Special consideration should also be given to those cases where the lower critical temperature is equal to, or less than, the average winter environmental temperature. Under such conditions $W_T = 0$. Since the condition $T_1 \ll T_e^*$ holds for large animals, the mathematical model proposed can be applied in the study of the optimal heat insulation of individuals with a relatively low body weight ($P < 5$ kg). The mathematical model can be further extended to eliminate this limitation.

Water based mammals have specific heat insulation mechanisms. These mechanisms should be considered separately as should amphibians and birds.

Unfortunately, not all the coefficients in (15) are known accurately enough. Let us discuss the available data on the quantities in (15) and the ways of eliminating them.

Specific weight of fur

Unfortunately, in the literature available we have not found direct, quantitative, data on the specific weight of fur. But there are indirect data which enable the specific weight of fur to be estimated to an accuracy sufficient for our purpose. Thus, the fur weight per unit area for the white bear, in winter, is 11.10^{-3} kg/dm^2 (Frisch et al., 1974). Using results for the average thickness of bear fur (5 to 6 cm) (Scholander et al., 1950; Hart, 1956), one can estimate the specific weight of bear fur, ρ, to be about 20 kg/m^3. Similar data obtained for the wild Australian dog , dingo, (Schield, 1972; Scholander et al., 1950) allows an estimate of specific fur weight as 20 kg/m^3 too. Geptner et al. (1967) present some data on hide weight, fur and flesh thickness which make it possible to estimate the specific fur weight of the polar fox and polar wolf as about 20 kg/m^3.

The coefficient of fur thermal conductivity

Using the available data (Scholander et al., 1950; Hart, 1956) on the dependence of fur heat insulation on the thickness of the coat, one can find the thermal conductivity of fur (Table 5).

Thermal conductivities for the same species differ approximately by a factor of 1.5. This is due to differences in measures of fur thickness, whereas the heat transfer coefficients presented by Scholander and Hart were very similar. This discrepancy cannot be explained by differences in environmental temperature since both authors consider species inhabiting areas that are similar in climatic conditions. The discrepancy is likely to be

Table 5 Specific thermal data for some northern animals

Species	Area	Thermal conductivity W/M^2 °C	Heat transfer coefficient W/m^2 °C	Fur thickness cm	Source
wolf	Banks island	0.036	0.99	3.6	Hart, 1956
wolf	Alaska	0.062	0.90	6.9	Scholander et al.,1950
polar fox	Alaska	0.041	0.88	4.7	"
fox	Alaska	0.047	1.18	4.0	"
glutton	North East of Canada	0.038	1.00	3.8	Hart, 1956
white bear	Banks island	0.047	0.94	5.0	"
white bear	Alaska	0.081	1.35	6.0	Scholander et al.,1950

due to the accuracy of measurement of fur thickness. Acknowledging that Hart payed special attention to measurement of fur layer thickness, it seems reasonable to use Hart's results as a basis for calculation of the thermal conductivity of fur. The mean value of thermal conductivity calculated is in good agreement with that presented by Hammel (1955): $\bar{k}$ = 0.038 W/m^2 °C. For our calculations we have taken the value $\bar{k}$ = 0.038 W/m^2 °C.

Proportion of fur lost during moulting

The results of studies of winter, and summer, white bear fur (Frish et al., 1974) make it possible to estimate the proportion of fur weight loss on moulting to be 2/3. Unfortunately, for other species such data are not available.

Specific heat capacity

The fur of animals consists of the protein keratin. When estimating the heat capacity of fur, we used the known protein heat capacity value of 23.8 kJ/g.

Relative winter duration can be determined on the basis of the climatic characteristics of the area (Konyukova et al., 1971; Berry et al., 1945).

According to Asmussen et al. (1965), the efficiency of the skeletal muscles is within the range 20 to 25 %. In the following calculations we have assumed $\eta = 0.22$.

Because of the lack of corresponding data, the energetic efficiency of fur synthesis is taken as unity.

Comparison of theoretical and experimental results

In calculations we have taken the following numerical values for the parameters in equation (15):
$\alpha_M = 0.7$; $\delta = 0.5$; $\gamma = 0.25$; $\eta = 0.22$; $\eta_M = 1$; $C_M = 2.4\ 10^7$ J/kg; $\varphi = 2/3$; $T_S = 38$ °C; $\bar{\kappa} = 0.038$ W/m² °C; T = 1 year.

The basal metabolism W_o is determined by the empirical formula

$$W_o = f\,P^{3/4} \tag{19}$$

where f = 3.4 W/kg. The results presented below appear in two forms. The first of them (index 'a') refers to a relatively cold climatic conditions ($\gamma = 1/3$), the second (index 'b') to milder climatic conditions ($\gamma = 1/4$).

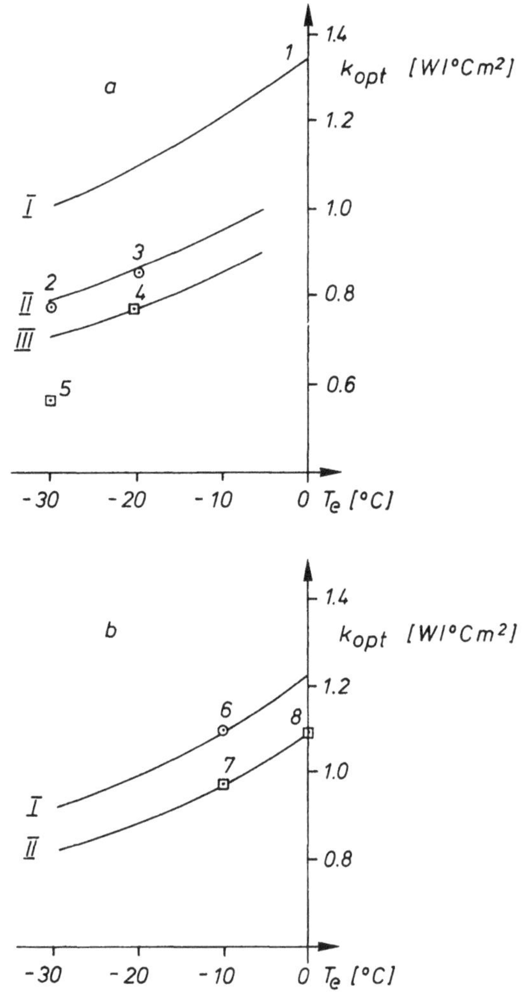

Fig. 33 Relationship between optimal heat transfer coefficient and environmental temperature Θ_e. Body weight of the individual serves as parameter.
a relative length of winter period ϑ = 1/3
I P = 0.3 kg; II P = 2kg; III I = 5 kg; 1 lemming, Alaska (Scholander et al., 1950 a); 2 hare, Alaska (Hart, 1971); 3 polar hare (Olnyanskaya, Slonim, 1947); 4 arctic fox, Alaska (Irving et al., 1955); 5 polar fox, Alaska (Scholander et al., 1950 a).
b relative length of winter period ϑ = 1/4.
I P = 2 kg; II P = 5 kg; 6 hare, European part of the USSR (Olnyanskaya, Slonim, 1947); 7 red fox, European part of the USSR (Olnyanskaya, Slonim, 1947); 8 fox from Middle Asia (Olnyanskaya, Slnonim, 1947).

Figures 33 a, b show the theoretical dependence of the optimal heat transfer coefficient value on the mean winter environmental temperature and the corresponding data obtained by measurement. The observed heat transfer coefficient value was determined from the dependence of the heat power of an individual at the environmental temperature

$$k = W_0 / S(T_S - T_1) \quad (20)$$

where T_1 is the lower critical temperature.

In Figures 33 a, b the parameter shown is the body weight of an individual.

As would be expected, optimal heat transfer decreases as the environmental temperature is lowered. There is also a strong dependence of the optimal heat transfer coefficient on the body weight of an individual. A decrease in heat transfer coefficient with increase in body weight is due to a reduction in heat flux through the body surface and also to a relatively small contribution of fur layer weight to the energetics of motion.

As is seen from Figure 33, the theoretical results are in a quite good agreement with oservation.

Figures 34 a, b show the dependence of the lower critical temperature on body weight. The situation is different with the polar fox. In this case, the divergence is likely to be due to a considerable difference between the heat transfer coefficient for polar fox fur, and the mean value $\kappa = 0.038\ W/m^2\ ^\circ C$ taken in these calculations.

The model proposed is approximate. The factors disregarded in this model are the following

1. Fur heat conduction non-uniformity-due to the presence of downy, coarse and directional hair.
2. The presence of heat insulation other than fur (the adipose layer, etc.).
3. Variation in basal metabolism in some species in winter (e. g. hiber-

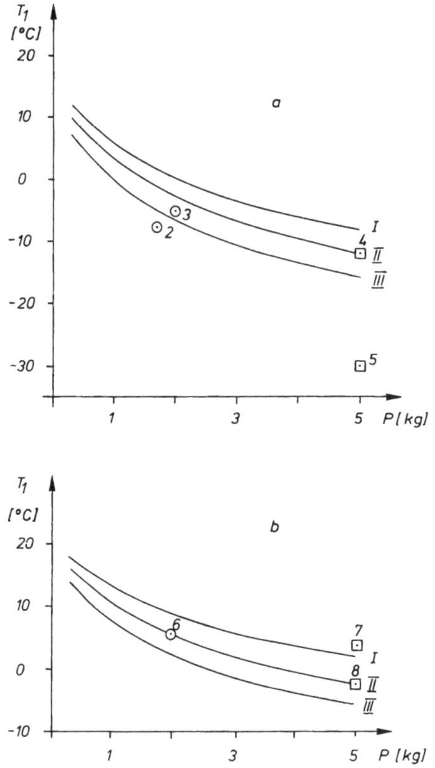

Fig. 34 Optimal temperature relationship of the lower border of thermoneutral zone and body weight P(P < 5 kg). Winter temperature of the environment T serves as a parameter.
a relative length of the winter period γ = 1/3.
I T_e = -10 °C; II T_e = -20 °C; III T_e = -30 °C.
b relative length of the winter period γ = 1/4.
I T_c = = °C; II T_c = -10 °C; III T_c = -10 °C.

nation in bears).

4. The necessity for an additional cutaneous integument in species with thicker fur.

5. Crowding (mutual heating) in collective animals.

There is nevertheless a sufficiently good agreement between theoretical and experimental results, suggesting that the model takes into account the most essential factors determining the heat insulating properties of homeotherms. It should also be noted that the model can easily be modified to take into account the factors mentioned above.

The good agreement between the theoretical and observed values supports the hypothesis of evolutionary pressure leading to the development of optimal energetic insulation.

Thermodynamics of Processes and the Evolution of Simple Biothermodynamic Populations

J. U. Keller

Introduction

The thermodynamics of irreversible processes has been applied successfully to several phenomena and problems in biology (Morowitz, 1960; Fritzsch, 1970; Bertalanffy, 1977; Lamprecht, 1978; Lamprecht, Zotin, 1978; Zotin, 1978; Zotin et al., 1978; Wieser, Gnaiger, 1980). Most applications deal with either a single biological system, for example a cell, or a fixed number of such systems. Often global phenomena are considered, i. e. phenomena which can easily be described by phenomenological thermodynamic terms as for example the heat exchange between a biological system and its surroundings or the evolution process of the system etc. (Schaarschmidt, Lamprecht, 1978; Wieser, Gnaiger, 1980; Zotin, Lamprecht, 1982).

Considering sets of biological systems existing together within a certain environment, one of the basic observations is that at almost any instant of time new biosystems are created and emerge whereas others are destroyed and vanish. We argue that, no matter how complex the interactions of biosystems with their surroundings, especially among each other, the whole evolution process of the set and especially the creation and destruction process is subject to the Second Law of Thermodynamics, which profoundly influences not only the evolution of a single system (Morowitz, 1960; Weizsäcker, 1976; Bertalanffy, 1977), but also that of the whole set or the biological population considered.

Thermodynamics and Regulation of Biological and Processes

In this paper we will derive, from the Second Law, within the framework of thermodynamics of irreversible processes an integral equation which allows, in principle, calculation of the total number of any species of biological systems present (in different states) in the set at any time t, and in addition the total number of all systems within the set considered can be calculated as function of time.

To our knowledge systems including a varying number of subsystems have neither been considered by the classical thermodynamics of irreversible processes nor by other phenomenological theories of processes in continuous matter (Pacault, 1971; Scharf, 1977, 1979; Lamprecht, 1978; Walter, Lamprecht, 1978). Therefore we have to use several new concepts and ideas which will be described in detail below.

Approaching the problem stated above we define the basic concept of a biothermodynamic system and a biothermodynamic population in the first section. Then, we give the basic equations of the thermodynamics of processes of a single biothermodynamic system. They are complemented by the conservation equations and constitutive relations of the entire system considered in the subsequent part. In the next section we derive the integral equation for the time derivative of a quasi-population. Then we give a special form of this equation holding for sets including only one type or species of biothermodynamic systems. This equation is discussed through use of an example in the last section describing the growth of the population of a very special and simple type of biothermodynamic system. It is thus demonstrated that the constitutive relations of a single system, describing the exchange of mass and energy with its surroundings, profoundly influence the growth rate of the total number of systems included in a biothermodynamic population at a given instant of time.

Populations of biothermodynamic systems

Let us consider a biothermodynamic system Σ_τ . By this we understand (Schottky et al., 1973) a certain piece of matter, distinguishable from its surroundings by any physical and/or chemical property, which

I) may exchange heat, mechanical work and mass with its surroundings and

II) may reproduce itself or another kind of system.

The system has to have clearly defined boundaries which divide its interior from its surroundings or environmental system Σ^* . One may distinguish between external and internal exchange of heat, work and mass, the latter occurring between any two parts of the system.

Biothermodynamic systems start their evolution in a certain, thermodynamically well defined state and they may either approach a final equilibrium state, characterised by vanishing internal exchanges of heat, work and mass, or dissolve in their vicinity to become undistinguishable from it*. This may also happen in purely thermodynamic systems such as Bénard's cells which occur in a liquid heated from beneath.

Any set of biothermodynamic systems Σ_τ, $\tau = 1, 2, 3\_$ a so-called biothermodynamic population, together with its surroundings Σ^* forms an entire system

$$\Sigma = \Sigma^* + \sum_\tau \Sigma_\tau , \qquad (1)$$

which again may exchange heat, work and mass with its surroundings or environmental system Σ_0 .

*) The thermodynamics of processes does not require knowledge of the nature of the 'birth' - or the 'death-process' of a biothermodynamic system, because we assume its concepts, especially the First and the Second Law, to hold irrespective of the length of the 'life-time' of a system.

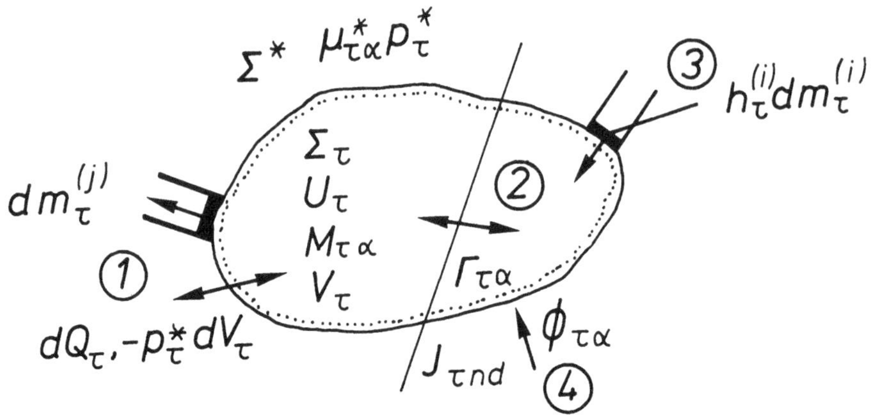

Fig. 35 Biothermodynamic system Σ_τ.
1: external exchange of heat and work; 2: internal exchange of heat and work; 3: convective exchange of mass and energy; 4: mass exchange by diffusion.

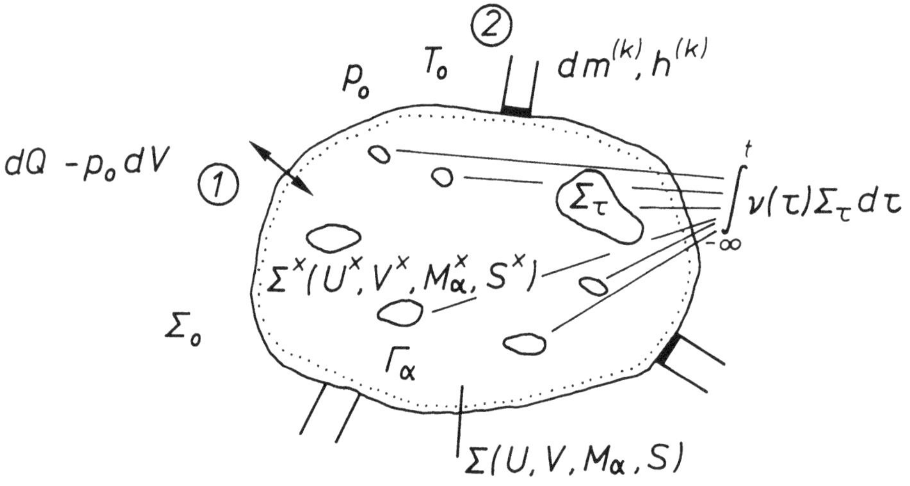

Fig. 36 The entire system Σ and the biothermodynamic population $\int d\tau\, \nu(\tau)\Sigma_\tau$.
1: exchange of heat and work; 2: quasi static convective exchange of mass and energy.

In what follows, we assume the symbol ' τ ' of any member system Σ_τ of a population not to be a natural number, but to indicate the instant of time, at which it has come into being. By this we understand that the system has reached a well defined 'initial state', typical for the species considered and indicating it's own existence as an individual system with respect to the thermodynamic exchange processes considered. Prior to time τ , the system Σ_τ may exist in a physical sense, but it is not considered to be an individual system on it's own, but rather to be part of any other system or of the environment of the set of thermodynamic systems considered. Assuming that systems are created at times τ_i, i = 1, 2, 3..., the total number [n] of systems existing at time t can be written as

$$[n(t)] = \int_{-\infty}^{t} d\tau \sum_i \delta(\tau - \tau_i), \tag{2}$$

where δ indicates Dirac's delta-function. Obviously [n(t)] is a step function (Fig. 37). However, there are arbitrarily many quasi-population functions n(t) having continuous time derivatives ν (t), and which obey the relations

$$n(t) = [n(t)] \quad \text{any } t = \tau_1, \tau_2, \ldots \tag{3}$$

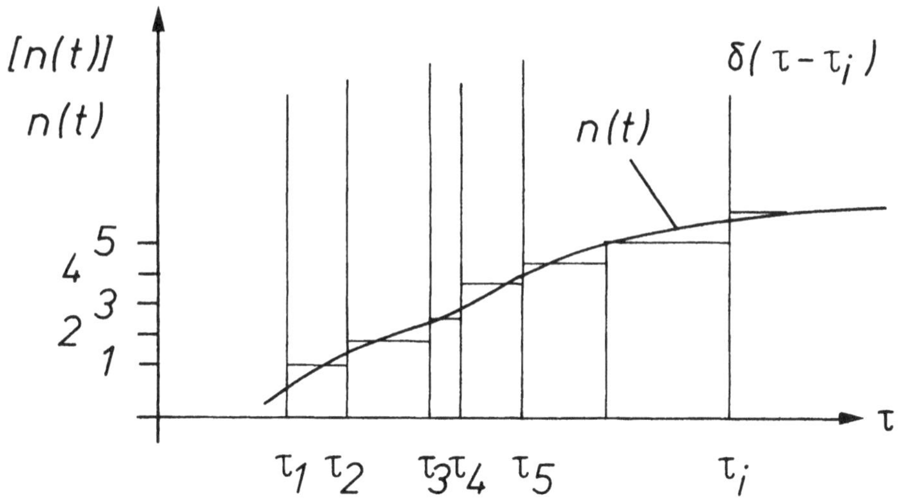

Fig. 37 The quasi-population function n(t) of a set of biothermodynamic systems Σ_{τ_i} starting their evolutions at times t = $\tau_1, \tau_2, \tau_3 \ldots$.

$$n(t) = \int_{-\infty}^{t} \nu(\tau)\, d\tau . \quad (4)$$

Using the quasi-density function $\nu(t)$, (1) formally or symbolically can be written as

$$\Sigma = \Sigma^*(t) + \int_{-\infty}^{t} \nu(\tau)\, \Sigma_\tau(t)\, d\tau . \quad (5)$$

Obviously this relation holds in an ordinary sense for any extensive quantity like the volume, the mass or the energy of the respective systems.

Thermodynamics of processes of a single biothermodynamic system Σ_τ

In this section we consider the evolution of a single biothermodynamic system Σ_τ and formulate from a macroscopic point of view the equations which govern the exchange of heat, mechanical work and mass between the system and its surroundings.

To avoid certain physical difficulties (Meixner, 1968, 1969) we here use a general thermodynamics of processes (Meixner, 1969; Keller, 1975, 1976) which includes (linear and non-linear) classical thermodynamics of irreversible processes as a special case (Meixner, Reitz, 1959; Päsler, 1975; Keller, 1977)*.

Assuming the system Σ_τ to contain the masses $M_{\tau\alpha}$, $\alpha = 1 \ldots N_\tau$ of N_τ many (chemically different) components which due to chemical reactions are produced at rates $T_{\tau\alpha}$ and introduced to Σ_τ either by diffusion processes (overall diffusion flux of component α : $\Phi_{\tau\alpha}$) or by convective massfluxes $\dot{m}^{(i)}_{\tau\alpha} = \gamma^{(i)}_{\tau\alpha}\, \dot{m}^{(i)}_{\tau}$, $i = 1, 2, \ldots, J_\tau$ with mass concentrations $\gamma^{(i)}_{\tau\alpha}$, the mass balance of any component is (Fig. 35)

*)This special case can, mathematically, be characterised by the assumption that all phenomenological laws are of 'differential type of order one'.

$$\dot{M}_{\tau\alpha} = \sum_{i=1}^{J_\tau} \gamma_{\tau\alpha}^{(i)} \dot{m}_\tau^{(i)} + \Phi_{\tau\alpha} + \Gamma_{\tau\alpha}, \quad \alpha = 1,..,N_\tau \tag{6}$$

The overall balance is

$$\dot{M}_\tau = \sum_{i=1}^{J_\tau} \dot{m}_\tau^{(i)} + \Phi_\tau, \tag{7}$$

$$M_\tau = \sum_{\alpha}^{N_\tau} M_{\tau\alpha}, \tag{8}$$

with

$$\sum_{\alpha}^{N_\tau} \gamma_{\tau\alpha}^{(i)} = 1 \qquad i = 1,2,..,J_\tau \tag{9}$$

$$\sum_{\alpha}^{N_\tau} \Gamma_{\tau\alpha} = 0, \tag{10}$$

the last relation expressing the law of mass conservation of classical chemistry.

Assuming that energy is added to Σ_τ by ordinary heat conduction, slow or quasistatic exchange of mass, mechanical work against the pressure outside the system and diffusion processes, the energy balance is

$$\dot{U}_\tau = \dot{Q}_\tau + \sum_{i}^{J_\tau} h_\tau^{(i)} \dot{m}_\tau^{(i)} - p_\tau^* \dot{V}_\tau + J_{\tau ud} \tag{11}$$

Here U_τ is the internal energy of Σ_τ, $\dot{Q}_\tau$ the rate of the conductive heat added to the system, $h_\tau^{(i)}$, $i = 1,2,...,J_\tau$ the specific enthalpy of the mass flux $\dot{m}_\tau^{(i)}$, p_τ^* the pressure outside of Σ_τ, V_τ the system's volume and $J_{\tau ud}$ the energy flux associated with diffusion processes between Σ_τ and its environment Σ^*. The last quantity is related to the diffusion fluxes $\Phi_{\tau\alpha}$ by the relation

$$J_{\tau ud} = \sum_{\alpha}^{N_\tau} \mu_{\tau\alpha}^* \Phi_{\tau\alpha}, \tag{12}$$

where $\mu_{\tau\alpha}^*$ is the chemical potential (per unit of mass) of component of Σ_τ (see Fig. 35).

Similarly the balance of the accompanying thermostatic entropy of any equilibrium or non-equilibrium state of Σ_τ is (Meixner, 1968, 1969; Keller, 1977)*.

$$\dot{S}_\tau = \frac{\dot{Q}_\tau}{T_\tau^*} + \sum_{i}^{J_\tau} s_\tau^{(i)} \dot{m}_\tau^{(i)} + J_{\tau sd} + P_{\tau s}. \qquad (13)$$

Here T_τ^* is the temperature outside of Σ_τ, $s_\tau^{(i)}$ the specific entropy of the convective mass flux $\dot{m}_\tau^{(i)}$, $J_{\tau sd}$ the entropy flux associated with diffusion processes between Σ_τ and Σ^* and $P_{\tau s}$ the entropy production within Σ_τ. The diffusive entropy flux $J_{\tau sd}$ is associated with the diffusion fluxes $\phi_{\tau\alpha}$ by the relation

$$J_{\tau sd} = \left(\frac{1}{T_\tau^*} - \frac{1}{T_\tau}\right) J_{\tau ud} = \left(\frac{1}{T_\tau^*} - \frac{1}{T_\tau}\right) \sum_{\alpha}^{N_\tau} \mu_{\tau\alpha} \phi_{\tau\alpha} \qquad (14)$$

The entropy production $P_{\tau s}$ is very often, but not necessarily always, positive (Meixner, 1968, 1969; Keller, 1975; Päsler, 1975). However, the overall production, i. e. the production during the whole evolution process of Σ_τ from its initial state at time τ to infinity is always non-negative, i. e. either positive or zero:

$$\Sigma_\tau : \quad \int_{t=\tau}^{\infty} P_{\tau s}(t)\, dt \geq 0. \qquad (15)$$

This is the Second Law of Thermodynamics for a biothermodynamic system Σ_τ. This formulation is similar to that given originally by R. Clausius for thermodynamic systems (see Meixner, 1968, 1969; Keller, 1975; 1977). The inequality

$$P_{\tau s}(t) \geq 0, \qquad (16)$$

often used in the literature does not hold for all irreversible processes actually occurring in nature. For example, it does not hold for heat con-

* 'Accompanying thermostatic parameters' of Σ_τ are, for example, its entropy S_τ, chemical potentials $\mu_{\tau\alpha}$, internal temperature T_τ or pressure p_τ and are defined as parameters, that would be realised in the system if it were in a thermodynamic equilibrium state with the actual extensive parameters, i. e. the internal energy U_τ, volume V_τ and masses $M_{\tau\alpha}$. The parameters S_τ, $\mu_{\tau\alpha}$ ... can be calculated from the fundamental equation $S_\tau = S_\tau(U_\tau, V_\tau, M_{\tau\alpha})$ of the system (Callen, 1960).

duction processes with finite velocity of propagation (Päsler, 1975; Keller, 1977). Generally speaking (16) holds only, if the consitutive equations of Σ_τ have a special mathematical structure, namely differential type of order one, which are exactly the equations used in linear and non-linear classical thermodynamics of irreversible processes. In this case (16) can be derived rigorously from the overall condition (15) (Meixner, 1969).

We now rearrange inequality (15) using (13) and the Gibbs' equation for Σ_τ

$$dS_\tau = \frac{1}{T_\tau} dU_\tau + \frac{P_\tau}{T_\tau} dV_\tau - \sum_{\alpha=1}^{N_\tau} \frac{\mu_{\tau\alpha}}{T_\tau} dM_{\tau\alpha} \tag{17}$$

where T_τ , P_τ , $\mu_{\tau\alpha}$ are the so-called accompanying temperature, pressure and chemical potentials of the system. We then get from (15) the semi-passivity inequality of Σ_τ:

$$\begin{aligned}\int_\tau^\infty dt \Big\{ & \left(\frac{1}{T_\tau} - \frac{1}{T_\tau^*}\right)(\dot{Q} + J_{\tau ud}) + \sum_i^{J_\tau}\left(\frac{1}{T_\tau} - \frac{1}{T_\tau^{(i)}}\right) h_\tau^{(i)} \dot{m}_\tau^{(i)} + \\ & + \sum_i^{J_\tau} \sum_\alpha^{N_\tau} \left(\frac{\mu_{\tau\alpha}^{(i)}}{T_\tau^{(i)}} - \frac{\mu_{\tau\alpha}}{T_\tau}\right) \dot{m}_{\tau\alpha}^{(i)} + \frac{1}{T_\tau} \sum_\alpha^{N_\tau} (\mu_{\tau\alpha}^* - \mu_{\tau\alpha}) \phi_{\tau\alpha} + \\ & + \frac{1}{T_\tau}(P_\tau - P_\tau^*) \dot{V}_\tau + \sum_{\ell=1}^{R_\tau} \left(-\frac{A_{\tau\ell}}{T_\tau}\right) \Gamma_{\tau\ell} \Big\} \geq 0 .\end{aligned} \tag{18}$$

In deriving this inequality we have used the conservation laws (6, 11), the relations (12), (14), and the Gibbs-Duhem-relation

$$g_\tau^{(i)} = \sum_\alpha^{N_\tau} \mu_{\tau\alpha}^{(i)} \gamma_{\tau\alpha}^{(i)} \qquad i = 1, 2, .., J_\tau . \tag{19}$$

In addition we have assumed that among the N_τ components included in Σ_τ, there are $N_\tau - R_\tau > 0$ many linear independent chemical reactions where $1 < R_\tau < N_\tau$ is the number of components whose production rates $\Gamma_1, .., \Gamma_{R\tau}$ can be chosen independently from each other (Päsler, 1975; Keller, 1977). The remaining production rates $\Gamma_{\tau R_\tau+1}, \dots, \Gamma_{\tau N_\tau}$ can be represented as linear functions of $\Gamma_{\tau 1} \dots \Gamma_{\tau R_\tau}$

$$\Gamma_{\tau j} = \sum_{\ell=1}^{R_\tau} \nu_{\tau j \ell} \Gamma_{\tau \ell} \qquad j = R_\tau + 1, \dots, N_\tau \tag{20}$$

which can be calculated from the stoichiometric equations of the system. Using these relations, de Donder's chemical affinities can be introduced

$$A_{\tau\ell} = \mu_{\tau\ell} + \sum_{j=R_\tau+1}^{N_\tau} \mu_{\tau j} \nu_{\tau j \ell} \qquad \ell = 1, \dots, R_\tau \tag{21}$$

leading to the identity

$$-\sum_{\alpha=1}^{N_\tau} \frac{\mu_{\tau\alpha}}{T_\tau} \Gamma_{\tau\alpha} = \sum_{\ell=1}^{R_\tau} \left(-\frac{A_{\tau\ell}}{T_\tau}\right) \Gamma_{\tau\ell} . \tag{22}$$

Contrary to the chemical potentials $\mu_{\tau\alpha}$, the chemical affinities vanish in equilibrium states. Therefore, the condition for chemical equilibrium in is

$$A_{\tau\ell} = 0 \qquad \ell = 1, \ldots, R_\tau \tag{23}$$

$$\Gamma_{\tau\ell} = 0 \qquad \ell = 1, \ldots, R_\tau \tag{24}$$

which due to (20) implies that

$$\Gamma_{\tau j} = 0 \qquad j = R_\tau + 1, \ldots, N_\tau \tag{25}$$

The inequality (18) includes several terms which represent different irreversible processes namely, conductive and diffusive heat exchange, quasi-static convective heat exchange, convective exchange of mass, diffusion, exchange of mechanical work and chemical reactions. The inequality sign in (18) indicates that the (even and odd) factors of each term are not independent from each other. We take this into account by assuming constitutive relations to hold as follows:

Flux representation — Force representation

$$\dot{Q} + J_{\tau ud} = \mathfrak{A}_{W\tau}(\cdot|\cdot) \qquad \frac{1}{T_\tau} - \frac{1}{T_\tau^*} = \mathfrak{B}_{W\tau}(\div) \tag{26}$$

$$h_\tau^{(i)} \dot{m}_\tau^{(i)} = \mathfrak{A}_{Wh\tau}(\cdot|\cdot) \qquad \frac{1}{T_\tau} - \frac{1}{T_\tau^{(i)}} = \mathfrak{B}_{Wh\tau}(\div) \tag{27}$$

$$\dot{m}_{\tau\alpha}^{(i)} = \mathfrak{A}_{W\tau\alpha}(\cdot|\cdot) \qquad \frac{\mu_{\tau\alpha}^{(i)}}{T_\tau^{(i)}} - \frac{\mu_{\tau\alpha}}{T_\tau} = \mathfrak{B}_{W\tau\alpha}(\div) \tag{28}$$

$$\Phi_{\tau\alpha} = \mathfrak{A}_{d\tau\alpha}(\cdot|\cdot) \qquad \frac{1}{T_\tau}(\mu^*_{\tau\alpha} - \mu_{\tau\alpha}) = \mathfrak{B}_{d\tau\alpha}(\div) \tag{29}$$

$$\dot{V}_\tau = \mathfrak{A}_\tau(\cdot|\cdot) \qquad \frac{1}{T_\tau}(p_\tau - p^*_\tau) = \mathfrak{B}_\tau(\div) \tag{30}$$

$$\Gamma_{\tau\ell} = \mathfrak{A}_{c\tau\ell}(\cdot|\cdot) \qquad -\frac{A_{\tau\ell}}{T_\tau} = \mathfrak{B}_{c\tau\ell}(\div) \tag{31}$$

$$(\cdot|\cdot) = \left(\frac{1}{T_\tau} - \frac{1}{T^*_\tau} \cdots - \frac{A_{\tau\ell}}{T_\tau}\right) \qquad (\div) = (\dot{Q}_\tau + J_{\tau u\alpha} \cdots \Gamma_{\tau\ell})$$

$$i = 1, 2, \ldots, J_\tau \qquad \alpha = 1, 2, \ldots, N_\tau \qquad \ell = 1, 2, \ldots, R_\tau$$

The symbols $\mathfrak{A}_{\ldots}, \mathfrak{B}_{\ldots}$ indicate generally non-linear functionals of the thermodynamic forces and fluxes respectively. These functionals cannot be chosen arbitrarily but are subject to several conditions which restrict their mathematical structure considerably (Meixner, 1969; Keller, 1975, 1976). We here restrict ourselves to mentioning that they vanish for vanishing arguments, i. e.

$$\mathfrak{A}_{\ldots}(0 \ldots 0) = 0 \qquad \mathfrak{B}_{\ldots}(0 \ldots 0) = 0 \tag{32}$$

If the functionals $\mathfrak{A}_{\ldots}, \mathfrak{B}_{\ldots}$ reduce to (linear or non-linear) functions of their arguments, (26) to (31) reduce to the well-known phenomenological equations of classical thermodynamics of irreversible processes (Meixner, 1969; Keller, 1975).

The constitutive equations (26) to (31) and the thermostatic equations of state of Σ_τ form a system of equations from which, in principle, all thermodynamic quantities of Σ_τ can be calculated as functions of time, given the external parameters T^*_τ, p^*_τ ... and an appropriate set of initial conditions.

Thermodynamics of processes of the entire system

The entire system Σ including a biothermodynamic population ($\Sigma_\tau, \tau = \tau_1, \tau_2, \ldots$), (see (5) and Fig. 36) may exchange with its surroundings Σ_o , heat, mechanical work and mass. In this section we will formulate the basic equations which these processes can be described by.

The mass balance is

$$\dot{M}_\alpha = \sum_{k=1}^{K} \gamma_\alpha^{(k)} \dot{m}^{(k)} + \Gamma_\alpha \qquad \alpha = 1, 2, \ldots, N \quad (33)$$

Here M_α is the total mass of component α in Σ , $\dot{m}^{(k)}$, $k = 1, 2 \ldots K$, the convective mass flows entering Σ , $\gamma_\alpha^{(k)}$ the mass concentration of component α in $\dot{m}^{(k)}$ and Γ_α the chemical production rate of component α . Summing all relations (33) we get the overall mass balance equation

$$\dot{M} = \sum_{k=1}^{K} \dot{m}^{(k)}, \quad (34)$$

with

$$M = \sum_{\alpha=1}^{N} M_\alpha \quad (35)$$

$$\sum_{\alpha=1}^{N} \gamma_\alpha^{(k)} = 1 \quad (36)$$

$$\sum_{\alpha=1}^{N} \Gamma_\alpha = 0 \quad (37)$$

The conservation of energy law is

$$\dot{U} = \dot{Q} + \sum_{k=1}^{K} h^{(k)} \dot{m}^{(k)} - p_o \dot{V}, \quad (38)$$

where U is the internal energy of Σ , $\dot{Q}$ the rate by which conductive heat is added to the system, $h^{(k)}$ the specific enthalpy of the mass flux $\dot{m}^{(k)}$, V the volume of the system and p_o the outside pressure (Fig. 36).

The balance equation for the accompanying thermostatic entropy is

$$\dot{S} = \dot{Q}/T_0 + \sum_{k=1}^{K} s^{(k)} \dot{m}^{(k)} + P_s , \tag{39}$$

where T_0 is the temperature outside of Σ , $s^{(k)}$ the specific entropy of the convective mass flux $\dot{m}^{(h)}$ and P_s the entropy production.

Gibbs' equation for the accompanying thermostatic parameters of Σ is (see footnote p. 312)

$$dS = dU/T + P/T . dV - \sum_{\alpha=1}^{N} \frac{\mu_\alpha}{T} dM_\alpha . \tag{40}$$

From the Second Law of Thermodynamics

$$\Sigma : \qquad \int_{-\infty}^{+\infty} P_s \, dt \geq 0 , \tag{41}$$

we derive, similar to the procedure carried out above, a semipassivity inequality of the system Σ :

$$\begin{aligned} \int_{-\infty}^{+\infty} dt \Big\{ & \left(\frac{1}{T} - \frac{1}{T_0}\right)\dot{Q} + \sum_{k=1}^{K}\left(\frac{1}{T} - \frac{1}{T^{(k)}}\right) h^{(k)} \dot{m}^{(k)} + \\ & + \sum_{k=1}^{K}\sum_{\alpha=1}^{N}\left(\frac{\mu_\alpha^{(k)}}{T^{(k)}} - \frac{\mu_\alpha}{T}\right)\dot{m}^{(k)} + \\ & + \frac{1}{T}(p - p_0)\dot{V} + \\ & + \sum_{l=1}^{R}\left(-\frac{A_l}{T}\right)\Gamma_l \Big\} \geq 0 . \end{aligned} \tag{42}$$

Here we have used the Gibbs-Duhem relations

$$g^{(k)} = \sum_{\alpha=1}^{N} \mu_\alpha^{(k)} \gamma_\alpha^{(k)} \qquad k = 1, 2, \dots, K \tag{43}$$

and introduced de Donder's affinities via the relations

$$- \sum_{\alpha=1}^{N} \frac{\mu_\alpha}{T} \Gamma_\alpha = \sum_{l=1}^{R} \left(-\frac{A_l}{T}\right) \Gamma_l , \tag{44}$$

assuming that there are N-R chemically independent reactions among the N components occurring in Σ .

Again each term in (42) describes a special irreversible process, namely ordinary heat conduction, convective heat exchange, quasistatic convective exchange of mass, the exchange of mechanical work and chemical reactions. The phenomenological equations of these processes can be deduced from the inequality (42) to be:

Flux representation	Force representation	
$\dot{Q} = \mathfrak{A}_W(\cdot|\cdot)$	$\frac{1}{T} - \frac{1}{T_0} = \mathfrak{B}_W(\div)$	(45)
$h^{(k)}\dot{m}^{(k)} = \mathfrak{A}_{Wk}(\cdot|\cdot)$	$\frac{1}{T} - \frac{1}{T^{(k)}} = \mathfrak{B}_{Wk}(\div)$	(46)
$\dot{m}_\alpha^{(k)} = \mathfrak{A}_{m\alpha}^{(k)}(\cdot|\cdot)$	$\frac{\mu_\alpha^{(k)}}{T^{(k)}} - \frac{\mu_\alpha}{T} = \mathfrak{B}_{m\alpha}^{(k)}(\div)$	(47)
$\dot{V} = \mathfrak{A}(\cdot|\cdot)$	$\frac{1}{T}(p - p_0) = \mathfrak{B}(\div)$	(48)
$\Gamma_l = \mathfrak{A}_{cl}(\cdot|\cdot)$	$-\frac{A_l}{T} = \mathfrak{B}_{cl}(\div)$	(49)

$$(\cdot|\cdot) = \left(\frac{1}{T} - \frac{1}{T_0} \ldots - \frac{A_l}{T}\right) \qquad (\div) = (\dot{Q} \ldots \Gamma_l)$$

$$k = 1, 2, \ldots, K \qquad l = 1, 2, \ldots, R \qquad \alpha = 1, 2, \ldots, N$$

The symbols k = 1 ... K, l = 1 ...R,   = 1 ... N respectively represent generally non-linear functionals of the thermodynamic forces and fluxes. The phenomenological equations together with the conservation laws (33), (38) and the 'accompanying' thermostatic equations of state - which are not mentioned explicitly here, but are included in Gibbs' equation (40) - form a complete set of equations from which all thermodynamic parameters of Σ can, in principle, be calculated as functions of time given the external parameters T_0, p_0... and a set of initial conditions.

An integral equations for the growth rate of a biothermodynamic population

The thermodynamic processes, i. e. exchange of heat, work, mass, chemical reactions etc. occurring within a member system Σ_τ of a biothermodynamic population are strongly associated with the respective processes occurring in the overall, or entire system, Σ . This is a consequence of (5) which becomes more evident if the relations between all extensive quantities of the systems Σ_τ , Σ^* and Σ are written down explicitly. Thus we have for the masses, the volumes, the internal energies, the entropies and the chemical production rates of the respective systems

$$M_\alpha(t) = \int_{-\infty}^{t} \nu(\tau) M_{\tau\alpha}(t)\, d\tau + M_\alpha^*(t) \qquad \alpha = 1, 2, .., N \tag{50}$$

$$V(t) = \int_{-\infty}^{t} \nu(\tau) V_\tau(t)\, d\tau + V^*(t), \tag{51}$$

$$\mathcal{U}(t) = \int_{-\infty}^{t} \nu(\tau) \mathcal{U}_\tau(t)\, d\tau + \mathcal{U}^*(t), \tag{52}$$

$$S(t) = \int_{-\infty}^{t} \nu(\tau) S_\tau(t)\, d\tau + S^*(t) \tag{53}$$

$$\Gamma_\alpha(t) = \int_{-\infty}^{t} \nu(\tau) \Gamma_{\tau\alpha}(t)\, d\tau + \Gamma_\alpha^*(t) \qquad \alpha = 1, 2, \ldots, N .$$

Here $\nu(t) = \dot{n}(t)$ is the time derivative of the quasipopulation function n(t) (see (3), (4)) and M_α^* ...S^* are the masses of components $\alpha = 1...N...$ and the (accompanying thermostatic) entropy of the environmental system Σ^* (see Fig. 36).

By taking time derivatives of (50) to (53) we get

$$\dot{M}_\alpha(t) = \nu(t) M_{t\alpha}(t) + \int_{-\infty}^{t} \nu(\tau) M_{\tau\alpha}(t)\, d\tau + \dot{M}_\alpha^*(t) \tag{54}$$

$$\alpha = 1, 2, \ldots, N$$

$$\dot{V}(t) = \nu(t) V_t(t) + \int_{-\infty}^{t} \nu(\tau) V_\tau(t)\, d\tau + \dot{V}^*(t) \tag{55}$$

$$\dot{\mathcal{U}}(t) = \nu(t)\mathcal{U}_t(t) + \int_{-\infty}^{t} \nu(\tau)\dot{\mathcal{U}}_\tau(t)\,d\tau + \dot{\mathcal{U}}^*(t) \tag{56}$$

$$\dot{S}(t) = \nu(t)\,S_t(t) + \int_{-\infty}^{t} \nu(\tau)\,\dot{S}_\tau(t)\,d\tau + \dot{S}^*(t) \tag{57}$$

where $m_{t\alpha}(t) \ldots S_t(t)$ are the masses of components $\alpha = 1 \ldots N \ldots$ and the entropy of the biothermodynamic system Σ_t in its initial state at time t, i. e. at the beginning of its evolution or in its 'status nascendi'.

Let us now consider Gibbs' equation for the environmental system

$$\Sigma^*: \qquad dS^* = \frac{d\mathcal{U}^*}{T^*} + \frac{p^*}{T^*}\,dV^* - \sum_{\alpha=1}^{N} \frac{\mu_\alpha^*}{T^*}\,dM_\alpha^* \tag{58}$$

where T^*, p^*, μ_α^* are so-called accompanying thermostatic temperature, pressure and chemical potentials respectively (see footnote on p. 312). Inserting in this equation for $\dot{M}_\alpha^* \ldots \dot{S}^*$ expressions from (54) to (57) we get

$$\begin{aligned} &\dot{S} - \frac{1}{T^*}\dot{\mathcal{U}} + \sum_\alpha \frac{\mu_\alpha^*}{T^*}\dot{M}_\alpha = \\ &= \nu\Big(S_t - \frac{1}{T^*}\mathcal{U}_t + \sum_\alpha \frac{\mu_\alpha^*}{T^*} M_{t\alpha} - \frac{p^*}{T^*} V_t(t) + \\ &+ \int_{-\infty}^{t} \nu(\tau)\Big(\dot{S}_\tau(t) - \frac{1}{T^*}\dot{\mathcal{U}}(t) - \frac{p^*}{T^*}\dot{V}_\tau(t) + \\ &+ \sum_\alpha \frac{\mu_\alpha^*}{T^*}\dot{M}_{\tau\alpha}(t)\Big)\,d\tau \end{aligned} \tag{59}$$

Using the Gibbs-Duhem equation for Σ_t in its initial state at time t,

$$S_t(t) = \frac{1}{T_t}\mathcal{U}_t(t) + \frac{p_t}{T_t}V_t(t) - \sum_\alpha \frac{\mu_{t\alpha}}{T_t} M_{t\alpha} \tag{60}$$

and Gibbs' equations (17) and (40) we get from (59)

$$\lambda(t)\nu(t) = \tilde{L}(t) + \int_{-\infty}^{t} d\tau\,\nu(\tau)\,\tilde{K}_\tau(t) \tag{61}$$

with

$$\lambda(t) = \left(\frac{1}{T^*(t)} - \frac{1}{T_t(t)}\right) U_t(t) + \left(\frac{P^*}{T^*} - \frac{P_t}{T_t}\right) V_t(t) \qquad (62)$$

$$- \sum_{\alpha=1}^{N} \left(\frac{\mu_\alpha^*}{T^*} - \frac{\mu_{t\alpha}}{T_t}\right) M_{t\alpha}(t)$$

$$\tilde{L}(t) = \left(\frac{1}{T^*} - \frac{1}{T}\right) \dot{U} + \left(\frac{P^*}{T^*} - \frac{P}{T}\right) \dot{V} \qquad (63)$$

$$- \sum_{\alpha=1}^{N} \left(\frac{\mu_\alpha^*}{T^*} - \frac{\mu_\alpha}{T}\right) \dot{M}_\alpha$$

$$\tilde{K}_\tau(t) = \left(\frac{1}{T_\tau} - \frac{1}{T^*}\right) \dot{U}_\tau + \left(\frac{P_\tau}{T_\tau} - \frac{P^*}{T^*}\right) \dot{V}_\tau - \sum_{\alpha=1}^{N} \left(\frac{\mu_{\tau\alpha}}{T_\tau} - \frac{\mu_\alpha^*}{T^*}\right) \dot{M}_{\tau\alpha} . \qquad (64)$$

We rearrange the r.h.s of (63) and (64) by inserting the conservation laws of Σ_τ , (6), (11) and of Σ , (33), (38). We then get from (61)

$$\lambda(t) v(t) = L(t) + \int_{-\infty}^{t} d\tau\, v(\tau) K_\tau(t) \qquad (65)$$

with

$$L(t) = \left(\frac{1}{T^*} - \frac{1}{T}\right) \left(\dot{Q} - p_0 \dot{V} + \sum_k h^{(k)} \dot{m}^{(k)}\right) +$$

$$+ \left(\frac{P^*}{T^*} - \frac{P}{T}\right) \dot{V} - \qquad (66)$$

$$- \sum_{\alpha=1}^{N} \left(\frac{\mu_\alpha^*}{T^*} - \frac{\mu_\alpha}{T}\right) \Big(\underbrace{\sum_{k=1}^{K} \gamma_\alpha^{(k)} \dot{m}^{(k)}}_{\dot{m}_\alpha^{(k)}} + \Gamma_\alpha^* \Big)$$

$$K_\tau(t) = \left(\frac{1}{T_\tau} - \frac{1}{T^*}\right)\left(\dot{Q}_\tau + J_{\tau u \alpha} + \sum_i^{J_\tau} h_\tau^{(i)} \dot{m}_\tau^{(i)}\right) + \left(\frac{P_\tau}{T_\tau} - \frac{P^*}{T^*}\right)\dot{V}_\tau - \sum_\alpha^{N > N_\tau}\left(\frac{\mu_{\tau\alpha}}{T_\tau} - \frac{\mu_\alpha}{T}\right)\Gamma_{\tau\alpha} - \sum_\alpha^{N_\tau}\left(\frac{\mu_{\tau\alpha}}{T_\tau} - \frac{\mu_\alpha^*}{T^*}\right)\left(\sum_i^{J_\tau} \dot{m}_{\tau\alpha}^{(i)} + \phi_{\tau\alpha}\right) \tag{67}$$

Assuming the constitutive relations of Σ_τ, ((26) to (31)) and of ((36) to (49)) inserted into (66) and (67), the relation (65) becomes a linear integral equation of the Volterra type for the time derivative $\nu(t)$ of the quasi-population function n(t)!

The parameter $\lambda(t)$, defined by (62), is a characteristic quantity of the initial state of any type of biothermodynamic system. The inhomogeneous term L(t) in (65) includes the present state and the processes occurring in the environmental system Σ^* and in the entire system Σ as a whole. Therefore, this term takes into account the influence of the environment Σ has on the production - and the annihilation process of the biothermodynamic systems. The kernel function $K_\tau(t)$ includes thermodynamic parameters describing the exchange processes of heat, work and mass of the biothermodynamic system Σ_τ with its environment Σ^*. Obviously these processes should have a profound impact on the production and annihilation of new biothermodynamic systems.

If all biothermodynamic systems Σ_τ in a population are of the same kind, i. e. have the same initial state and the same constitutive laws and if the environmental system Σ^* is very large and has constant intensive parameters T^*, p^*, μ_α^*, the parameter λ = const and the kernel function $K_\tau(t)$ depends on the difference $(t-\tau)$ of its time parameters only*. Therefore (65) reduces to a simple convolution equation

* This does not imply that all systems within the population are at any time in the same state of their own evolution. It only implies that the evolution of two member systems of the population $\Sigma_{\tau 1}$, $\Sigma_{\tau 2}$, given the same environmental conditions, can be transformed into each other by a simple shift of time, i. e. by the transformation $t \rightarrow t-(\tau_2 - \tau_1)$.

$$\lambda v(t) = L(t) + \int_{-\infty}^{t} d\tau \, v(\tau) K(t-\tau) \tag{68}$$

having with the initial conditions $v(t) = 0 \ldots t < 0$ the Laplace-transform solution

$$v(t) = \frac{1}{2\pi i} \int_{c-i\infty}^{c+i\infty} \frac{\hat{L}(p)}{\hat{K}(p)-\lambda} e^{pt} \, dp \ldots \quad t \geq 0 \tag{69}$$

where the integration in the complex p-plane has to be performed such that $c > \mathrm{Max}\,(\mathrm{Re}\,p_i)_i$ where $\mathrm{Re}\,p_i$ indicates the real part of any singularity p_i of the integrand in (69).

For isolated systems Σ ($\dot{U} = 0$, $\dot{V} = 0$, $\dot{m}^{(k)} = 0$) of the above type, $L(t) = 0$ and (68) has only the solution

$$v(t) = 0 \qquad n(t) = n_0 = \mathrm{const.} \tag{70}$$

In such systems biothermodynamic systems can neither be created nor annihilated, instead the total number of systems Σ_τ is constant! This is a consequence of a statement of Fredholm on (in this case homogeneous) integral equations of the Volterra type (Duschek, 1961).

A simplified version of the growth rate equation

Let us consider a somewhat simplified version of the growth rate of *population equation* (65). We assume the intensive parameters $T^*_\tau \ldots$ outside a biothermodynamic system Σ_τ, not to be too different from the respective parameters $T^* \ldots$ of the environmental system Σ^*. Assuming this, the constitutive relations of Σ_τ in the force representation ((26) to (31)) can be introduced in the kernel-function (67) to give

$$K_\tau = \mathcal{B}_{W\tau}(\dot{Q}_\tau + J_{\tau ud}) \ldots \qquad T^*_\tau \cong T^*$$

$$+ \sum_i^{J_\tau} \mathcal{B}^{(i)}_{Wk\tau} \dot{m}^{(i)} h^{(i)}_\tau \qquad T^{(i)}_\tau \cong T^*$$

$$
\begin{aligned}
&+ \mathcal{B}_\tau \dot{V}_\tau && T_\tau \cong T^* \\
&+ \sum_\alpha^{N_\tau} \sum_i^{J_\tau} \mathcal{B}^{(i)}_{m\tau\alpha} \dot{m}^{(i)}_{\tau\alpha} && \mu^{(i)}_{\tau\alpha} \cong \mu^*_\alpha \\
&+ \sum_\alpha \mathcal{B}_{d\tau\alpha} \Phi_{\tau\alpha} && \mu^*_{\tau\alpha} \cong \mu^*_\alpha \qquad (71) \\
&+ \sum \mathcal{B}_{c\tau\ell} \Gamma_{\tau\ell} \quad \ldots && T_\tau \cong T^* \\
&= P_{\tau s}(t) .
\end{aligned}
$$

The approximations used for each term are noted in the respective lines. In addition, we have used the relation

$$
\begin{aligned}
\sum_{\alpha=1}^{N>N_\tau} \frac{\mu_\alpha}{T} \Gamma_{\tau\alpha} &= \sum_{\alpha=1}^{N_\tau} \frac{\mu_\alpha \Gamma_{\tau\alpha}}{T} + \sum_{\beta=N_\tau+1}^{N} \frac{\mu_\beta}{T} \Gamma_{\tau\beta} = \\
&= \sum_{\ell=1}^{R_\tau} \frac{A_\ell^{(\tau)}}{T} \Gamma_{\tau\ell} = 0 .
\end{aligned} \qquad (72)
$$

Here the index β indicates those chemical components not inherent in Σ_τ, but possibly present anywhere in Σ, i. e. $\Gamma_{\tau\beta} = 0$, and $A_\ell^{(\tau)}$ are the thermostatic affinities of the (linearly independent) components $\ell = 1 \ldots R_\tau$ which however vanish by definition in any state of chemical equilibrium, i. e. $A_\ell^{(\tau)} = 0$. Thus under the restrictions mentioned in (71) the kernel-function K_τ is identical with the entropy production in the system Σ_τ, which, as mentioned earlier is always positive for exchange processes of differential type of order one (16).

Consider the case that the environmental system Σ^* is very large compared to all biothermodynamic systems $\sum_\tau \Sigma_\tau$ existing at any time t. We can then expect the intensive parameters of Σ^* to be not too different from that of the total system Σ, viz.

$$
\begin{aligned}
T^* &= T - \Delta T && (73) \\
p^* &= p + \Delta p \\
\mu^*_\alpha &= \mu_\alpha + \Delta\mu_\alpha && \alpha = 1, 2, \ldots, N .
\end{aligned}
$$

Then we have

$$L(t) = \frac{\Delta T}{T^{*2}}\dot{u} + \left(\frac{P^*}{T^{*2}}\Delta T + \frac{\Delta P}{T^*}\right)\dot{v} - \sum\left(\frac{\mu_\alpha^*}{T^*}\Delta T + \Delta\mu_\alpha\right)\times \\ \times \frac{\dot{M}_\alpha'}{T^*} + \mathcal{O}(2) = L^*(t) + \mathcal{O}(2), \tag{74}$$

where by definition

$$\dot{M}_\alpha' = \sum_{k=1}^{K} \gamma_\alpha^{(k)} \dot{m}^{(k)} + \Gamma_\alpha^* \qquad \alpha = 1, 2, \ldots, N \tag{75}$$

and 0(2) indicates terms vanishing of second order if $\Sigma^* \to \Sigma$.

Similarly, if we assume the intensive parameters of a biothermodynamic system in its initial state to be noted too different from the respective parameters in its surroundings Σ^* , viz.

$$T_t = T^*(t) - \Delta T_t \qquad p_t = p^*(t) - \Delta p_t \\ \mu_{t\alpha} = \mu_\alpha^*(t) - \Delta\mu_{t\alpha} \qquad \alpha = 1, 2, \ldots, N_t \tag{76}$$

the eigenvalue function $\lambda(t)$ can be written as

$$\lambda(t) = \frac{\Delta T_t}{T^*(t)^2} U_t(t) + \left(\frac{P^*}{T^*}\Delta T_t + \Delta p_t\right)\frac{V_t(t)}{T^*} - \\ - \sum_\alpha^{N_t}\left(\frac{\mu_\alpha^*}{T^*}\Delta T_t + \Delta\mu_{t\alpha}\right)\frac{M_{t\alpha}(t)}{T^*} + \mathcal{O}(2) \\ = \lambda^*(t) + \mathcal{O}(2). \tag{77}$$

Inserting (71), (74) and (77) into the population equation (65) and neglecting terms of order two we get

$$\lambda^*(t)\, v(t) = L^*(t) + \int_{-\infty}^{t} v(\tau)\, P_{\tau s}(t)\, d\tau . \tag{78}$$

This is an approximation of the rigorous population equation (65). The functions λ^* and L^* (t) are defined by (74) and (77) respectively. The kernel-function s the entropy production of a single biothermodynamic system Σ_τ , the mathematical and physical structure of which depends decisively on the constitutive equations of the system. Equation (78) should be valid for biothermodynamic populations including systems whose initial state is not too far away from a state which would be in thermodynamic equilibrium with its environmental system Σ^* , which itself is assumed to be very large compared to all biothermodynamic systems included in the population.

An example for the evolution of a simple species of biothermodynamic systems

Let us consider the evolution process of a population consisting of very simple biothermodynamic systems of the same kind, which however at any arbitrary instant time will be in different states of their individual evolution processes. A member system Σ_τ of the population is assumed to consist of a certain amount of solid mass $M_{\tau 1}$ of a single chemical component '1'. It is embedded in an environment Σ^* containing two different components '0' and '1'. The only exchange process between Σ_τ and Σ^* is assumed to be isothermal diffusion or adsorption-desorption of component '1'. Chemical reactions may neither occur in Σ_τ nor in Σ^*. Thus the conservation laws and constitutive equations are:

(6) $$M_{\tau 0} = 0 \tag{79}$$

$$\dot{M}_{\tau 1} = \Phi_{\tau 1} \tag{80}$$

(11) $$\dot{U}_\tau = J_{\tau u d} = \mu^*_{\tau 1} \Phi_{\tau 1} \tag{81}$$

(29) $$\Phi_{\tau 1} = \mathfrak{A}_{d\tau 1}\left(\frac{1}{T}\left(\mu^*_{\tau 1} - \mu_{\tau 1}\right)\right) \tag{82}$$

$$\frac{1}{T}\left(\mu^*_{\tau 1} - \mu_{\tau 1}\right) = \mathfrak{B}_{d\tau 1}\left(\Phi_{\tau 1}\right). \tag{83}$$

We assume that due to appropriate mass-supply to the entire system Σ, the chemical potential $\mu^*_{\tau 1}$ of component '1' in the vicinity of Σ_τ equals the accompanying thermostatic chemical potential of this component in the environmental system Σ^*, which itself is assumed to be constant

$$\mu^*_{\tau 1} = \mu^*_1 = \text{const}. \tag{84}$$

and we restrict consideration to linear diffusion laws, i. e.

$$\mathcal{B}_{d\tau 1}(\Phi_{\tau 1}) = \frac{\delta}{T}\Phi_{\tau 1} , \tag{85}$$

where $\delta > 0$ is a reciprocal diffusion constant, and consider only isothermal situations, that is we assume

$$T_\tau = T^* = T = const \tag{86}$$

then the evolution of Σ_τ is completely determined. Indeed, from (80), (83), (84) and (85) we infer that

$$\dot{M}_{\tau 1} = \Phi_{\tau 1} = \frac{1}{\delta T}(\mu_1^* - \mu_{\tau 1}) = const \tag{87}$$

and hence that

$$M_{\tau 1}(t) = M_{\tau 1}(\tau) + \Phi_{\tau 1}(t-\tau) \qquad t \geq \tau \tag{88}$$

indicating that the mass of a single system which started its evolution at time τ with the initial value $M_{\tau 1}(\tau)$ increases proportional to its age $(t-\tau)$. Of course this result can only hold within a limited period of time, i. e. assumption (84) has to be replaced by a more realistic one after a certain juvenile period of the system Σ_τ.

The entire system Σ is assumed to contain a fixed amount of component '0', and to be supplied with component '1' by an isothermal convective mass flow $\dot{m}_1$.

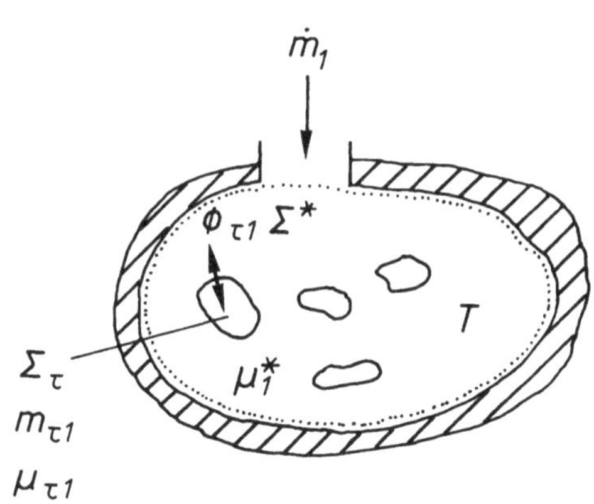

Fig. 38 Evolution of a simple biothermodynamic population by isothermal mass exchange.

The conservation laws for mass and energy are

(33) $$\dot{M}_0 = 0 \tag{89}$$

$$\dot{M}_1 = \dot{m}_1 \tag{90}$$

(38) $$\dot{U} = h_1 \dot{m}_1 - p_0 \dot{V}. \tag{91}$$

Assuming for the initial state of any system Σ_t

$$T_t = T^* = T = \text{const} \tag{92}$$

$$p_t = p^* = p = \text{const} \tag{93}$$

we get from (62)

$$\lambda = -\frac{1}{T}(\mu_1^* - \mu_{t1}) M_{t1} \doteq \frac{\lambda_1}{T} = \text{const} \tag{94}$$

and similarly from (66), (67) and (85)

$$L(t) = -\frac{1}{T}(\mu_1^* - \mu_1)\dot{m}_1 \doteq \frac{1}{T}\Delta\mu \dot{m}_1(t) \tag{95}$$

$$K = \mathcal{B}_{d\tau 1} \Phi_{\tau 1} = \frac{\delta}{T}\Phi_{\tau 1}^2 = \text{const}. \tag{96}$$

Thus the growth-rate equation (65) for the population is

$$\lambda_1 v(t) = \Delta\mu \dot{m}_1(t) + \delta\Phi_{\tau 1}^2 \int_{-\infty}^{t} d\tau\, v(\tau). \tag{97}$$

Introducing the initial condition

$$\left.\begin{aligned} v(t) &= 0 \qquad (98) \\ n(t) &= n_0 = \text{const} \qquad (99) \end{aligned}\right\} \quad t \leq 0$$

its solution can be written as

$$v(t) = \frac{1}{\lambda_1} \int_0^t \Delta\mu(t')\, \dot{m}_1(t')\, e^{(t-t')/t_r}\, dt' \tag{100}$$

where the characteristic time t_r is defined by

$$t_r = \lambda_1 / \delta\Phi_{\tau 1}^2 \gtrless 0 . \tag{101}$$

This can be either positive or negative, indicating a growing or a decreasing number

$$n(t) = n_0 + \frac{1}{\lambda_1} \int_0^t dt' \int_0^{t'} dt'' \, \Delta\mu(t'')\, \dot{m}_1(t'')\, e^{(t'-t'')/t_r} \tag{102}$$

of member systems Σ_τ in the population. Obviously $n(t) = n_o$ = const if there is no mass supply at all, i. e. if $\dot{m}_1 = 0$ for $t \geq 0$. In addition, the number of systems Σ_τ increases (decreases) the faster, the smaller the absolute value of the 'initial parameter' $\lambda_1 \gtrless 0$ is and the larger the mass mass flow $\Phi_{\tau 1}$ between Σ_τ and its environmental system Σ^* is. A more detailed discussion of (97), (100) and (102) will be given in another publication (Keller, 1984).

Acknowledgement

Discussions on the role of thermodynamics of irreversible processes in biology with Prof. I. Lamprecht, Berlin (West) are gratefully acknowledged.

Synergetics – The Problem of Selforganisation

Self-Organisation Processes and Their Modelling

V. A. Vasiliev, Yu. M. Romanovskij

The development of theories of biological evolution, morphogenesis, and embryogenesis made it possible to approach problems of the origin of life and development of the organism. Mathematical modelling, although not yet yielding quantitative predictions, put the study of these biological processes on the same level with other self-organisation in organic and inorganic systems which are far from thermodynamic equilibrium.

The term self-organisation has been used widely only recently. This term has resulted from an attempt to include, into a unique system of concepts, the common aspects of these phenomena in developing objects of different origin, namely, astrophysical, geological, physical, chemical, biological, and even sociological (see, e. g. Haken, 1974; Nicolis, Prigogine, 1977; Eigen, 1971; Eigen, Schuster, 1977, 1978 a, b; Romanovskij, 1981; Ebeling, Feistel, 1982). Sometimes, self-organisation theory is regarded as a special scientific field, called synergetics (Haken, 1978). Expedience in the identification of this field will be seen later, but this fact itself, already indicates the general scientific significance and pressure of the self-organisation problem, its complexity and specificity. Self-organisation is studied within a mathematical branch of science, in cybernetics. A notable feature of the synergetic approach is the physical method: from knowledge and hypotheses concerning elementary (microscopic) processes leading to an understanding of cooperative phenomena and formatting of the macrostructure of a system.

The concept of self-organisation, as well as structure, should be referred to primary, initial states and these are defined through concepts which are also very complicated. So, one can consider self-organisation (Haken, 1978), for example, as a spontaneous appearance of space and time structure order out of disorder (chaos) in a system. Although such an interpretation

Thermodynamics and Regulation of Biological and Processes

of self-organisation cannot be considered rigorous, at an intuitive level it is rather satisfactory, and we shall bear it in mind in what follows.

Self-organisation is possible only in a thermodynamically open system, i. e. one which can exchange energy or substance with an external medium. The deviation from thermodynamic equilibrium must also be sufficiently large and exceed critical values. In this case for processes there must exist essentially irreversible stages and chains of positive feedback. In mathematical modelling of the dynamics of self-organisation one has to deal with non-linear systems of equations (Nicolis, Prigogine, 1977; Ebeling, 1976).

Stipulating that self-organisation processes at different evolutionary stages in organic, and inorganic, systems can be divided into two classes (Romanovskij, 1981, 1982) the first class consists of systems, where there appear no new elements composing these systems. The variety of elements is not large here, and they can change only in quantity, redistribution in space and time. In the second class of self-organised systems, selection and multiplication of a few elements out of a very large number of newly formed and interconverting elements is possible.

As an example of processes of the first type we take auto-oscillations and auto-waves and, in particular, the case of stationary, dissipative structures important in an understanding of form-building. The second class includes processes of pre-biological evolution of macromolecules, formation of species, chemical isotopes, the appearance and development of human languages. These processes are a result of multiplication and concurrent selection.

The mathematical models and methods of investigation are similar in each of the two classes of self-organisation for objects of quite different nature. The differences are not due to mathematical specificity in the models, but to specificity in the problems posed in each type of self-organisation. Let us consider some of them.

Interesting examples can be found in auto-wave systems (Vasliliev et al., 1979), whose mathematical modeis can be represented in the form

$$\frac{\partial}{\partial t} x_i = f_i(x_1, x_2, \ldots, x_n, P) + D_i \Delta x_i \qquad i = 1, 2, \ldots, n \quad (1)$$

where t is time, x(r, t) are variables (component concentrations. potentials, etc.), D_i are diffusion coefficients, f_i are functions describing quasi-local point interactions in the system. The presence of chains of positive feed-back in point kinetics permits self-organisation of waves in the system (1).

Dependent on the type of space-time organisation systems of auto-oscillating processes are divided into the following groups:

- standing, self-generated waves
- stationary, space-inhomogeneous states (stationary dissipative structures)
- restarting regimes (echo)
- leading centres (stable localised wave sources)
- quasi-stochastic auto-oscillations (chemical 'turbulence' or, in other words, space desynchronisation of systems)
- reverberators (spiral waves).

These phenomena are possible in one and the same system at different parameter values. Sometimes the type of the process is determined by the initial state of the system. Consider some characteristic features of self-organisation-structure formation in an example of stationary dissipative structures (Vasiliev et al., 1982).

First of all notice that the system of equations (1) is equivalent to a system of differential equations with an infinite number of variables. Indeed, any solution of this system satisfying the boundary conditions, for example, Neiman's solution (impermeable ends at $r = \pm L/2$) can be represented as

$$x_i(t, r) = \sum_{j=-\infty}^{\infty} A_{ij}(t) \cos(\pi j r / L) \qquad (2)$$

Substitute (2) into (1), multiply by cos $(\pi j r / L)$ and integrate over dr and as a result, we obtain the equation for mode amplitudes

$$\frac{d}{dt} A_{ij} = F(A_{11}, A_{12}, \ldots) \qquad j \in (-\infty, \infty),\ i = 1..n \quad (3)$$

Stationary solutions of system (3), where there exist $A_{ij} \neq 0$ at $j = 0$ describe dissipative structures. Near the first point of bifurcation in parameters (P or D_1, or D_2...) there appear two states $A_{ij} = A_{ik}(j = -k)$ corresponding to structures of different polarity. At subsequent bifurcation points there appear new pairs of states with different space periods. These points can be stable at one and the same parameter values, which implies multistationary states of the system.

The number of stationary pairs (3) (M) can be, arbitrarily, large. For example, in the case $n = 2$ $M \to \infty$ at $D_1/D_2 \to \infty$, where D_2 is the diffusion coefficient of the auto-catalytic component. In such situations parametric system control and a parametric record of information in the final state appears impossible, and a specific dynamic system control is necessary, which uses, of course, much more information. Indeed, a parametric record may contain only information on the system components which determine both kinetic parameters (P) and D_i. Dynamic control also implies storage of information about the manner of development of the object as a whole. From the biological point of view it is obvious that a parametric information record is more advantageous since it requires much less genetic material for storage or organism structure parameters. It would not be out of place to recollect that a number of limitations on the application of Turing's ideas (1952) were conditioned by the possibility of realising different structures (Waddington, 1982). The following circumstance was, though, disregarded. Regions of attraction to different, finite structures have, as a rule, essentially different volumes. Figure 39 shows the ground attracting state for the model of the Belousov-Zhabotinskii reaction (Vasiliev et al., 1979) at $D_1/D_2 \gg 1$ (Fig. 39 b, c, d). Competing states (there are about 100) are stable and small and have smaller amplitudes and periods. For quantitative determination of the attraction region volume it is convenient to use the variables of representation (3):

$$V_k = \int_{\Omega_k} \cdots \int \prod_{i=1,2} \prod_{j \in S} dA_{ij} \qquad (4)$$

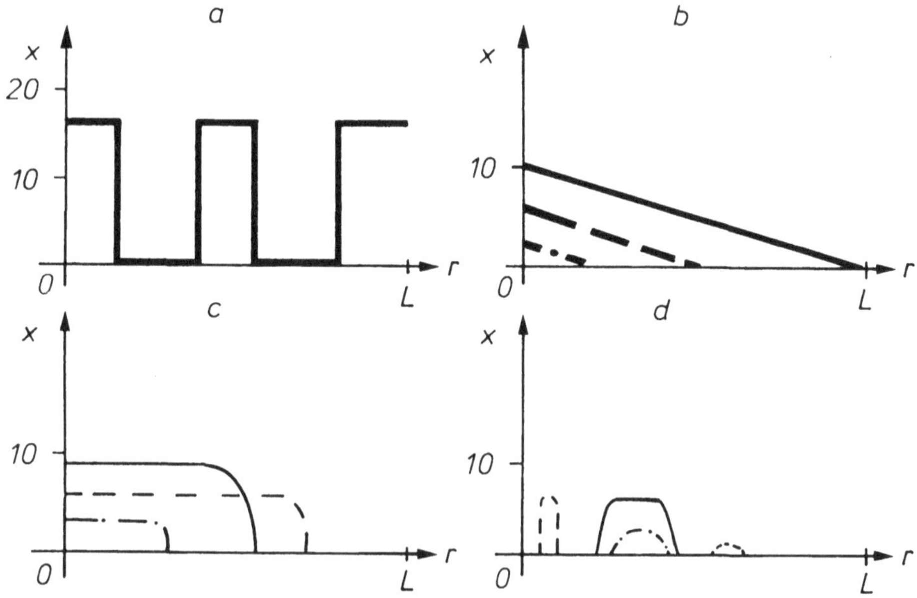

Fig. 39 Formation of dissipation structure. Explanations in the text.

where V_k is the attraction region volume of the k-th state, S is a set of modes for which the Turing's instability condition is valid, Ω_k is the set of the initial conditions of the k-th state.

To calculate values of V_k, we carry out a numerical experiment using the Monte-Carlo method. A random-number generator generates independent values A_{ij} $(A_{ij} \leq max\{x(r_i t)\})$, and then system (1) is solved. Note that as a rule the process of numerical integration should rapidly approach the limit. The point is that the initial state rapidly (for time of the order of the characteristic time of auto-catalysis) passes over to a certain state with, typically, a one-peak distribution form (Fig. 39 d), from which, by means of self-completion, there then forms the whole main structure.

The relative volumes of the attraction regions of competing states at $D_1/D_2 \gg 1$ appears to be less than 0.1 %. They appear only when the initial conditions have period and form close to the corresponding characteristics of the final state. At $D_1/D_2 \sim 1$ some cases are possible where there are several V_k of the same order. However, even in these cases small fluctuations in the dynamic variables lead to the fact that a single state can be realised.

It should be noted that there remains degeneracy under the replacement of all j by -j, i. e. structural polarity. This is removed in the presence of system inhomogeneities, for example, small gradients in the parameters, small fluxes through boundaries, etc.

Thus, although structure formation models of type (1) possess multistationary character, the structure formation can be characterised only by information about the system parameters (P, D_i). The quantitative value of the ratio between essential dynamic and parametric information is determined from $\sum V_j / V_k$ $(j \neq k)$. The smaller this ratio, the smaller the probability of formation of erroneous structure.

In a number of cases which model efficient self-organisation processes, multistationary character and dynamic control over the system state are of primary importance, for example, in operon systems describing differentiation (Sendov, 1976; Romanovskij et al., 1974). Multistationary states of elements play an essential role in the appearance of an ordered variety of functional properties in the system, and its stability as a whole. Although the theory of such systems is still embryonic, a number of interesting models have already been developed. Differentiation and formation are shown to be possible on the basis of genetic triggers. Two ways of modelling exist: one is modelling developing objects on the basis of the theory of finite-state automation (Sendov, 1976). The other is via dynamic models (Romanovskij et al., 1974; Solganik, Chernavskij, 1980). The greatest insight is perhaps gained by a model combining these two approaches.

Mathematical modelling of second-class self-organisation processes has much in common with what has been said above. Consider some problems on the example of concurrent hypercycle selection (Eigen, Schuster, 1977; 1978 a, b), following the procedures of Feistel et al. (1980). Let, in a homogeneous solution, exist two closed cycles of reactions consisting of substances ${}^1C_i$ and ${}^2C_i$ $(i=1,2,\ldots,n)$ which interact according to equations

$$\frac{d^{1,2}}{dt} C_i = \alpha_{1,2}\, {}^{1,2}C_i\, {}^{1,2}C_i - \Phi^{1,2}C_i\,, \quad {}^{1,2}C_0 = {}^{1,2}C_n\,. \tag{5}$$

In this case the conditions of selective limitation hold:

$$\sum_i {}^1C_i + \sum_i {}^2C_i = B_1 + B_2 = B = \text{const}$$

It can be shown that at $\alpha_1 \approx \alpha_2 \gg 1$ competition between hypercycles is realised only by their summed compositions B_1 and B_2:

$$\frac{d}{dt} B_i = B_i(\alpha B_i/n - \tilde{\Phi}) \qquad i = 1,2 \tag{6}$$

where

$$\tilde{\Phi} = (\alpha_1 B_1 + \alpha_2 B_2)/Bn\,.$$

The system of equations (6) is multistationary and is dynamically controlled. This means that the hypercycle which is kinetically more complete, i. e. the one with the higher value of α_i/n_i, will not overcome the less complete hypercycle if its initial concentrations are sufficiently large.

As has been stressed by Oparin (1968), drops may form in solutions of organic molecules with relatively large concentrations of substances called coacervates. Assuming, that in coacervates there exist hypercycles, which, on reaching a certain volume, divide and interact with one another, one can obtain equations describing the dynamics of the change in the volume of the coacervate for the k-th hypercycle type ($\sum V_k = \text{const}$):

$$\frac{d}{dt} V_k = V_k(E_k - \langle E \rangle) \tag{7}$$

where

$$E_k = \alpha_k B_k / n_k\,.$$

Equations (7) describe a system with parametric control over the final state, in this system each kinetically more complete hypercycle (that with a maximum value of α_k/n_k) succeeds, irrespective of the initial state of the system.

From the evolutionary standpoint the interpretation of the above results is very simple. The appearance of new substances due to mutations in systems of type (5) or (7) causes the formation of new elements in the system, i. e. hypercycles, such that the variety of system elements is, in principle, very large. In the case of processes of type (5) competition among elements can guarantee only a random choice, which depends on the initial state. On the other hand, systems of type (7) provide an element development, and selection which is the best at a given moment; either with the best catalytic characteristics (α_i) or with shorter reaction chains (n_i). In the evolution of life both types of less efficient self-organisation processes took place. An example of random choice may be the fixing of a unique chirality for biological molecules; selection of the best version is the basic factor of evolution.

Self-organisation models, both in the first and in the second class, contain kinetic equations which are obtained by averaging. Many authors have emphasised the role of internal fluctuations which, as has already been said, are of particular importance in understanding transitions from one state to another. Some authors even speak of phase transitions, order parameters, etc. near bifurcation points which determine these transitions (Nicolis, Prigogine, 1977; Haken, 1978). On the other hand, systems of kinetic equations of an order higher than three (this is, of course, the case in hypercycle theory) may also give, in addition to the expected multi-stationary states, dynamic quasi-stochastic solutions in the form of strange attractors. Quasi-stochastic regimes act, on the one hand, against ordering and on the other hand, can accelerate evolution and, by removing finite values, turn the system from the attraction region of one stationary state into the attraction region of another and, thus, intensify the search for better versions. In this case the class of models describing concurrent selection widens, but not the random choices of a given state.

With the words 'the best version' we mean that the process of evolution has a positive direction, i. e. progressive evolution takes place. It is rather difficult to give a rigorous definition of progressive evolution although some intuitive criteria of progress have already been chosen. Mathematical definitions of progressive evolution in biology, although suggested by a number of authors, for example, by Shnol (1979), cannot be valid in all cases. This circumstance hampers to a great extent the choice of criteria of the quality of models of the second-class self-organisation processes. In all probability, a precise definition can be made only by fixing the concept of biological information, its quantity, and value (Volkenstein, Chernavskij, 1979).

Let us make the following statement: the velocity of biological and, may be, even pre-biological evolution, in selection processes depends on the ways individual elements memorise information and their methods of storage. The more perfect these mechanisms, the easier the value of information is realised in the adaptation of objects to varying external conditions, and hence, the quicker evolution. It is most probable that in inorganic nature evolution velocity depends on the capacity of systems to store and accumulate negentropy which can be estimated, on the one hand, in the language of the information theory and, on the other hand, through the thermodynamics of non-equilibrium processes.

Most general and complete self-organisation models must give non-stationary solutions, on the basis of which one can estimate the characteristic times of the evolutionary processes. This concerns, especially, second-class processes, in particular, selection. This problem is only just being formulated. In the end, it is simply the real times of the evolutionary process as a whole and at separate stages that may appear as the test of this or that model. In models of first-class self-organisation processes non-stationary behaviour is sometimes also of primary importance in determining the applicability of models, for example, in an analysis of peaking regimes in non-linear active media (Kurdyumov, 1982).

In conclusion we would like to emphasise that the study of both classes of self-organisation processes is closely associated not only with mathematical, formal aspects, but also with contents. The properties of one system predetermines the characteristics of subsequent systems. The knowledge of the general, self-organisation laws will indicate new trends in the investigation of biological phenomena and will make it possible to control efficiently complex ecological systems.

Stochastic Dissipative Structures

A. S. Pikovsky, M. I. Rabinovich

In recent years researchers have shown an increasing interest in those processes proceeding in non-linear non-equilibrium systems and media. This happens because these are wide-spread systems (they are encountered in many fields of physics, chemistry, biology, etc.) and the phenomena observed in them are diversified and beautiful. The analysis of such phenomena is based on simplified models described, as a rule, by differential equations in ordinary or partial derivatives. An important role in these models is played by non-linear equations with diffusion

$$\frac{\partial u(\vec{r},t)}{\partial t} = f_i(u_1,\ldots,u_n) + D_i \Delta u_i(\vec{r},t), \quad i=1,\ldots,n \qquad (1)$$

where D_i are diffusion coefficients.

Equations of the type (1) describe distributed chemical reactions, electronic devices, combustion processes, neuron structures, population of organisms, etc. (Vasiliev et al., 1979; Zeldovich, Malomed, 1982). In systems with diffusion, different types of behaviour can be observed, namely, stationary homogeneous regimes, oscillations periodic in space and (or) in time, propagating excitation waves, and, finally, irregular stochastic regimes, both homogeneous and inhomogeneous. The present paper is aimed at the analysis of some types of stochastic behaviour in systems of type (1).

New concepts as to the nature of stochastic processes in non-equilibrium, dissipative systems with aggregated parameters have appeared and been formulated recently. They are associated with the discovery of strange attractors - nontrivial attracting sets in the phase space, on which the solutions to the equations behave in such a tangled and confuse manner that they cannot be distinguished from a random signal, whatever meaning we impart to this term. Such stochastic proccesses appear in systems without

Thermodynamics and Regulation of Biological and Processes

noise and fluctuation and are determined by an exponential instability of individual solutions to small disturbances of the initial conditions (Rabinovich, 1978).

At present most detailed consideration has been paid to chaotic oscillations in the systems described by ordinary, low-order, differential equations as well as to those described by difference equations. There exist descriptions, for example, of interrupted spatially homogeneous regimes in chemical reactions (Turner et al., 1981), ecological populations, etc.. At the same time, spatially inhomogeneous chaotic regimes are also possible due to the distributed character of such systems. In this paper we describe three types of spatially, inhomogeneous, chaotic regimes in systems with diffusion (1). The first type are stationary in time, chaotic, dissipative structures, the second type are slightly spatially inhomogeneous regimes associated with stochastic time dependence, and, finally, the third type are auto-wave order-disorder transition.

Stationary dissipative structures

It is known that in system (1) the stationary homogeneous state, $_i$ = const, can be unstable (if the diffusion coefficients D_i are different), as a result of which there form spatially inhomogeneous structures (1). Stationary structures are described by equations following from (1) if we put $\partial u_i / \partial t = 0$:

$$\frac{d^2 u_i}{dx^2} = - f_i(u_1, \ldots, u_n) \tag{2}$$

System (2) may have chaotic solutions, which corresponds to stationary-in-time and stochastic-in-space structures. We shall show this through an example of the following system

$$\begin{aligned} \frac{\partial u}{\partial t} &= \beta u - \gamma v + D_1 u_{xx} \\ \frac{\partial v}{\partial t} &= - \nu v + g(u) + D_2 v_{xx} . \end{aligned} \tag{3}$$

These equations describe the one-dimensional version of the known predator (v) - prey (u) model, where a predominant factor limiting the population number is a non-linear sating of the predator.

In the stationary case ($\partial/\partial t \equiv 0$) from (3) there follows a fourth-order equation

$$D_1 D_2 u^{IV} + (D_2\beta - D_1 \nu)u'' - \nu\beta u + \gamma g(u) = 0. \tag{4}$$

The stationary, spatially homogeneous state, $u = u_0$, $v = \beta u_0/\gamma$ is determined from the relation

$$g(u_0) = \gamma^{-1} \nu\beta u_0 . \tag{5}$$

By expanding g in a series and restricting ourselves to the quadratic term, we obtain for small deviations from the homogeneous state, $y = u - u_0$,

$$D_1 D_2 y^{IV} + (D_2\beta - D_1\nu)y'' + (\gamma g'(u_0) - \nu\beta)y + \frac{1}{2} g''(u_0) y^2 = 0 \tag{6}$$

It is convenient to transform to dimensionless variables

$$\begin{aligned} y &= [2(\gamma g'(u_0) - \nu\beta)/\gamma g''(u_0)]\,\omega \\ x &= (D_1 D_2/(\gamma g'(0) - \nu\beta)^{1/4})\,\xi \end{aligned} \tag{7}$$

and then (6) is reduced to the form

$$\omega^{IV} + \varepsilon\omega'' + \omega - \omega^2 = 0 \tag{8}$$

where the dimensionless parameter ε is equal to

$$\varepsilon = (D_2\beta - D_1\nu)/(D_1 D_2 (\gamma g'(u_0) - \nu\beta))^{1/2} \tag{9}$$

(8), considered as an equation of motion in four-dimensional phase space, ($\omega, \omega', \omega'', \omega'''$) has the integral

$$E = 6\omega'\omega''' - 3(\omega'')^2 + 3\varepsilon\omega'^2 + 3\omega^2 - 2\omega^3 = \text{const} \tag{10}$$

Accordingly, the solutions (8) lie on the three-dimensional surfaces E E = const. Real values of the integral E are determined by the boundary conditions. Let us assume that the structures appear in an unlimited region

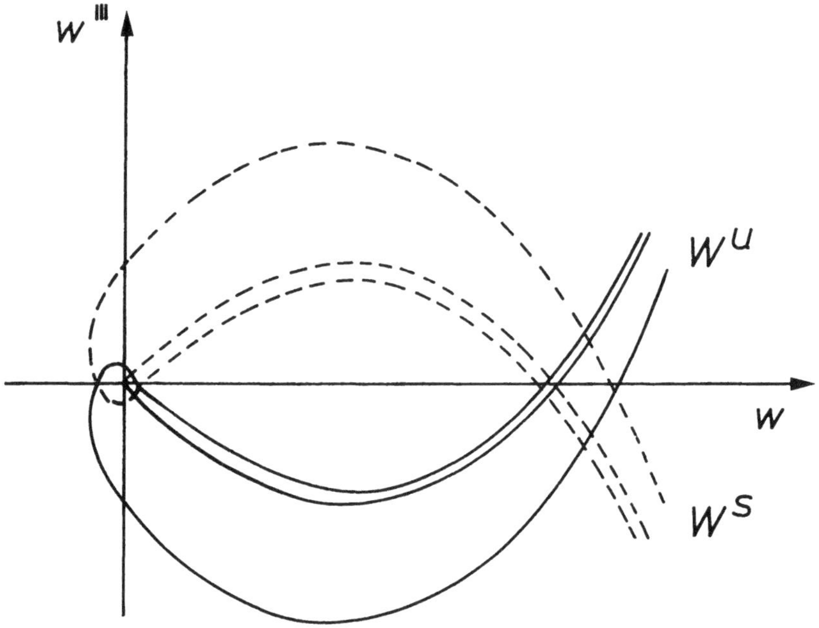

Fig. 40 Explanation in the text.

$-\infty < \xi < +\infty$, and be localised, i. e. $\omega', \omega'', \omega''', \omega \to 0, \xi \to \pm\infty$. For such structures E = 0.

Localised structures in phase space (8) are described by homoclinic trajectories which tend to the origin at $\xi \to +\infty$ and at $\varepsilon \to -\infty$. Homoclinic trajectories are lines of intersection of an unstable, (W^u), and stable, (W^s), manifold of zero equilibrium, at $|\varepsilon| < 2$. These manifolds are two-dimensional.

These trajectories cannot be determined analytically, and thus the problem has been solved through computation. The cross-sections W^u and W^s of the surface $\omega' = 0$, $\omega'' = -(\omega^2 - 2\omega^3/3)^{1/2}$ at $\varepsilon = 1$ are presented in Fig. 40. W^u and W^s are seen to be intersected transversally (at a non-zero angle), and therefore the theorem (Devaney, 1976) of the existence of a countable set of homoclinic trajectories, is applicable. All possible types of localised structures corresponding to these trajectories can be described using the methods of symbolic dynamics (Bowen, 1979). Some solutions are presented in Fig. 41. These solutions can be coupled, in an arbitrary manner, which fact is responsible for the stochastic character of spatially inhomogeneous regimes. From the elementary structures shown in Fig. 41 one can construct periodic solutions, the period being arbitrarily large, and the internal arrangement arbitrarily complicated.

Although the above stochastic structures occur in a dissipative system, (6) describing these structures is conservative. Hence, stochastic, dissipative structures are analogous to stationary solutions of the corresponding conservative equations. In particular, to equation (6) there is associated the problem of stationary waves via a modified equation (Gorshkov et al., 1979). Similar stationary solutions also exist in the problem of spatial structures in a chain of particles with non-linear interaction. A discrete mapping is here an analog of (6).

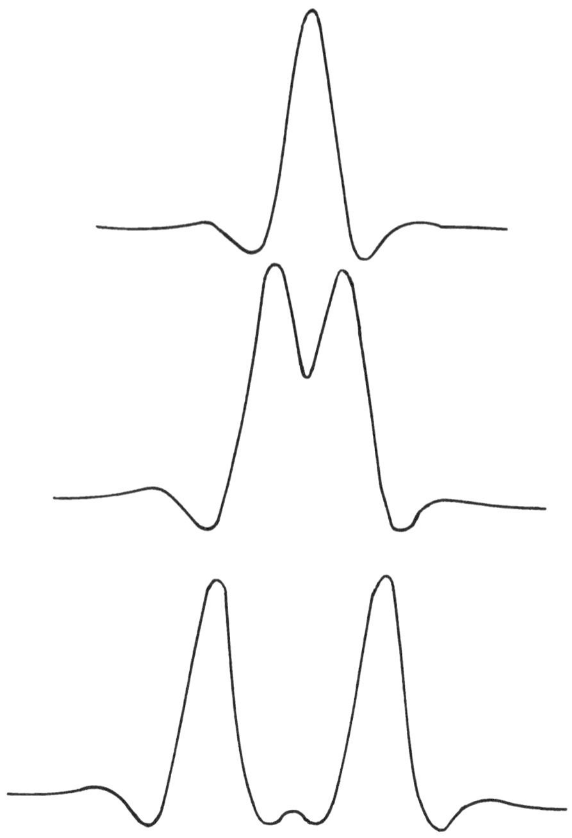

Fig. 41 Explanation in the text.

The appearance of inhomogeneous regimes of stochastic oscillations

Let us consider the system described by (1) with the boundary condition of the absence of flux:

$$\frac{\partial u_i(\vec{r},t)}{\partial \vec{l}} = 0 , \quad \vec{r} \in \partial G \tag{12}$$

where $\vec{l}$ is the vector of the normal to the surface of a closed region G. We assume, that in this system, there exist spatially homogeneous stochastic oscillations $u_i(\vec{r},t) = v_i(t)$, where v_i satisfy the system or ordinary differential equation

$$\frac{\partial v_i}{\partial t} \quad f_i(v_1,\ldots,v_n,t) \qquad i=1,2,\ldots,n \tag{13}$$

Let us consider the dynamics of spatially-inhomogeneous perturbations $w_i(\vec{r},t)$ of this solution. On linearising (1), we obtain

$$\frac{\partial w_i}{\partial t} = A_{ij}(t) w_i(\vec{r},t) + D_i \Delta w_i(\vec{r},t) \tag{14}$$

where $A_{ij} = \partial f_i(v_1,\ldots,v_n,t)/\partial v_j$.

Let us expand the solution ($w_1,\ldots,w_n$) in eigenfunctions of the auxiliary problem

$$\Delta y(\vec{r}) + \nu y(\vec{r}) = 0 \qquad \vec{r} \in G \tag{15}$$

with the boundary conditions (12). As is known, the eigenfunctions (15) $\{y_k\}$ form a whole set, and the eigenvalues are positive $0 = \nu_0 < \nu_1 \leq \nu_2 < \ldots$ (Tricomi, 1954). Writing the expansion in the form

$$w_i(\vec{r},t) = \sum_k W_{ik}(t) y_k(\vec{r}) \tag{16}$$

and substituting into (14), we have

$$\frac{dW_{ik}(t)}{dt} = A_{ij}(t)W_{jk}(t) - D_i v_k W_{ik}(t). \tag{17}$$

In the general case of arbitrary coefficients, D_i, it is impossible to analyse the solutions (17). Thus we first consider equal diffusion coefficients $D_I = D_2 = \ldots = D_n = D$. Then, making in (17) the substitution of variables

$$W_{ik} = \exp(-D v_k t) Z_i(t) \tag{18}$$

we obtain for

$$dZ_i/dt = A_{ij}(t) Z_j. \tag{19}$$

System (19) exactly coincides with the equations derived in linearising (13) near the solution $\{ v(t) \}$. As is known, the criterion of stochastic oscillation is an exponential increase in disturbances (Rabinovich, 1978), a and at long times the solutions (19) behave like exp (λt), where λ is the Lyapunov characteristic exponent (Ruelle, 1979). Accordingly,

$$W_{ik} \sim \exp[(\lambda - D v_k)t].$$

Thus, a disturbance corresponding to $v_0 = 0$ always increases, but this does not lead to spatially inhomogeneous regimes since y_0 = const. Stability toward spatially inhomogeneous disturbances is determined by the eigenvalue v_1. Instability occurs if

$$\lambda > v_1 D. \tag{20}$$

Otherwise, the regime of spatially homogeneous stochastic oscillations is stable.

In the case of different diffusion coefficients D_i this method does not hold, but one can expect, however, that if the diffusion coefficients differ only a little, the stability is determined, as before, by criterion (20), where D is some mean diffusion coefficient. Since stochastic oscillations often have the form of a succession of groups of increasing oscillations (Devaney, 1976; Rabinovich, 1978), we consider as a model the linear system

$$\dot{u}_1 = \alpha_1 u_1 + \beta_1 u_2 + D_1 \Delta u_1 \tag{21}$$

$$\dot{u}_2 = \beta_2 u_1 + \alpha_2 u_2 + D_2 \Delta u_2 \tag{22}$$

At $\beta_1\beta_2 - \alpha_1\alpha_2 < 0$, $\alpha_1 + \alpha_2 > 0$ (22) has a uniform solution in the form of oscillations with an exponentially increasing amplitude. An analog of the Lyapunov characteristic exponent is here, the increment $\lambda = 0.5(\alpha_1 + \alpha_2)$. Investigation of stability yields the following result: a perturbation corresponding to the mode, y_k, increases if,

$$\lambda > \nu_k (D_1 + D_2)/2 \tag{23}$$

or

$$D_1 D_2 \nu_k^2 - \nu_k(\alpha_1 D_2 + \alpha_2 D_1) + \alpha_1\alpha_2 - \beta_1\beta_2 < 0. \tag{24}$$

Condition (23) coincides with (20) if, for the effective coefficient D, one takes arithmetic mean. Condition (24) corresponds to a purely diffusion instability mechanism. To fulfill this condition, the diffusion coefficients must differ sufficiently.

'Order-disorder' transformation waves

The presence of diffusion in a non-equilibrium medium is known to cause various instabilities whose development may result in the establishment of stochastic regimes. Among the different possibilities for establishment of such regimes, we consider, here, stationary auto-waves which describe transition from a static (unexcited) state of the medium to a regime of stochastic space-time pulsations, and also, transformation waves to which there corresponds a boundary, moving at a constant velocity, and separating in space, the regions of regular stochastic oscillations. As a basic model we consider a two-component onedimensional medium described by equations of the form (1)

$$\frac{\partial u}{\partial t} = u + u^3 - \alpha u w + D_1 \partial^2 u/\partial x^2 \qquad (25)$$
$$\frac{\partial w}{\partial t} = \beta w - \delta w^3 - \gamma u^2 + D_2 \partial w/\partial x^2$$

The stationary transformation waves in question depend only on the moving coordinate $\xi = x - v t$, and they satisfy the equations

$$\frac{d^2 u}{d\xi^2} + \frac{v}{D_1}\frac{du}{d\xi} + \frac{1}{D_1}(1 + u^2 - \alpha w)u = 0 \qquad (26)$$
$$\frac{d^2 w}{d\xi^2} + \frac{v}{D_2}\frac{dw}{d\xi} + \frac{1}{D_2}(\beta w - \gamma u^2 - \delta w^3) = 0$$

These equations describe, simultaneously, a system of two related non-linear oscillators with damping. At $\xi \to 0$ all the solutions (26) tend to zero equilibrium. If v is small, the character of motion is close to stochastic oscillations which exist in (26) at $v = 0$. Thus, (26) describe the stationary 'order-disorder' transition wave.

Concluding remarks

The above examples confirm that in non-equilibrium, distributed systems, dissipative stochastic structures are possible which are in no way connected with random inhomogeneities or non-stationary states of the medium. In one-dimensional media parts of such random structures can be located between periodic structures or can be in agreement with unexcited regions of non-equilibirum medium. The question of spatial evolution of chaotic region in two-dimensional media seems to be very interesting. One of the simplest possibilities here is stochastic, cylindrical or spiral, waves induced by a source of such waves working in a stochastic regime, i. e. by a leading centre or reverberator.

Corresponding point models which refer to chemical kinetics and biology have been thoroughly studied.

We would like to mention here another class of stochastic dissipative structures whose dynamics is determined by an external field. An external field can change local parameters of the medium, for example, the velocity of increase of individual components or the magnitude of coupling between components. These changes manifest themselves in the appearance of stochastic patterns.

Symmetry Breaking Instabilities in Chemical Systems

G. Dewel, P. Borckmans, D. Walgraef

Introduction

Perhaps the most fascinating and intriguing aspect of natural phenomena is that complex systems may undergo symmetry breaking instabilities leading to pattern formation or coherent temporal behaviour over macroscopic space and time scales (Nicolis, Prigogine, 1977).

Therefore the understanding of why order may appear spontaneously in open systems far from equilibrium and which symmetries are selected among a large manifold of possibilities has become a major theme of research both theoretically and experimentally. Although such instabilities have been shown to occur in most branches of physics (Swinney, Gollub, 1981), excitingly, such structures have also been suggested to provide a basis for many problems of biological nature. Indeed spontaneous spatial ordering in open systems has many analogies with phenomena such as mitosis, embryonic development, and morphogenesis, whereas the concept of coherent temporal behaviour is important for the description of slime mould aggregation, fungal growth, and cardiac arrythmia (Winfree, 1980).

It is now a well accepted working hypothesis (sustained by some experimental evidence) that, at a molecular level, many such biological structures may arise from appropriate metabolic circuits involving catalytic reactions controlled by feedback regulation.

Therefore reaction-diffusion systems form an important class of models capable of accounting for major features of some biological organisations.

Thermodynamics and Regulation of Biological and Processes

Consider, therefore, open systems in which a set of coupled chemical reactions is occurring. For the sake of simplicity isothermal and mechanical equilibrium are assumed.

The phenomenological rate equations for the local concentrations (reaction-diffusion equations) of the intermediate species $\{x_i\}$ then take the form

$$\partial_t x_i = f_i(\{x_j\}, \lambda) + D_i \nabla^2 x_i \tag{1}$$

where $\{D_i\}$ are diffusion coefficients, $\{f_i\}$ the rates of change of $\{x_i\}$ due to chemical reactions (usually non-linear functions of the $\{x_i\}$), and λ represents a set of parameters describing the external constraints (concentration of a buffer product). The system is also provided with adequate boundary conditions.

When the external constraints are increased beyond the linear regime around thermal equilibrium, which corresponds, for the homogeneous steady state $\{X_i^o\}$, to the law of mass action, various instabilities may arise which are located by testing the linear stability of this reference state.

These are located by analysing the linear stability of the reference state $\{X_i^o\}$. Linearising (1) one obtains

$$\partial_t x_q^i(t) = L_{ij}(q^2, \lambda)\, x_q^j(t) \tag{2}$$

where

$$x_q^i(t) = X_q^i(t) - X_i^o = \frac{1}{L^{d/2}} \int d^d r\, e^{iqr} x^i(r)$$

with

$$L_{ij}(q^2, \lambda) = (\partial f_i / \partial x_j)^o - q^2 D_i .$$

The stationary state is asymptotically stable if all the eigenvalues of the matrix L_{ij} have negative, real parts.

Various instabilities may develop when λ is varied, for example:

1. L_{ij} has one eigenvalue $\omega_q(\lambda)$ for $|q| = 0$ and $\lambda = \lambda_c$ corresponding to transitions between homogeneous steady states.

2. L_{ij} has one eigenvalue $\omega_q(\lambda)=0$ for $|q|=q_c\neq 0$ and $\lambda=\lambda_c$, this soft-mode instability induces the formation of stationary structures (Turing instability).
3. L_{ij} has two complex conjugate eigenvalues with zero real value when $\lambda=\lambda_c$, $q=0$. This hard-mode instability may lead to time-periodic solutions of the limit-cycle type.

In the region where the instability develops many authors have stressed the need to include local fluctuations in the description of systems in which a coherent behaviour takes place on the macroscopic level. Following the Landau-Lifshitz theory of fluctuating hydrodynamics (Landau, Lifshitz, 1959), we add appropriate Langevin noise terms $\eta_i(\underline{r},t)$ to the balance equations (1). Because they take care of the rapid small-scale effects, we assume that

$$\langle \eta_i(\underline{r},t)\,\eta_j(\underline{r}',t')\rangle = 2\left[\Gamma^c_{ij}(\underline{r}) + \nabla_i^2\,\Gamma^d_{ij}(\underline{r})\right]\delta(\underline{r}-\underline{r}')\,\delta(t-t')$$

and also that the system remains in a state of local equilibrium throughout the entire transition region. The coefficients Γ^c and Γ^d may be determined by a local fluctuation dissipation theorem. Indeed, the instabilities involve macroscopic disturbances that do not modify the fluctuations on a microscopic scale where the short-scale equilibrating processes remain efficient in maintaining local equilibrium.

Although solution of equation (1) in its full generality seems improbable, it turns out that in the vicinity of some critical values of the control parameters (instability points) the dynamics of the system can be described by a few variables, the order parameters (σ_q), whereas all other variables may be eliminated because of the slaving principle (Haken, 1978), as is usually done, resulting from the separation of length and time scales near the instability points. Therefore, in these regimes the dynamics are purely those of the order parameters. The fluctuations of these modes in a great many situations are shown to obey equations of motion which are generalisations of the so-called time-dependent Ginzburg-Landau equations (TDGL) originally introduced to deal with equilibrium phase transitions (Patashinskii, Prokowskii, 1979). They have the following structure

$$\partial_t \sigma_{\underline{k}} = -\Gamma \frac{\delta \mathcal{F}}{\delta \sigma^*_{-\underline{k}}} + V_{\underline{k}} + \eta_{\underline{k}} . \tag{4}$$

The first term in the right hand side describes all contributions exhibiting a gradient structure and therefore leading to a generalised Landau potential function which has the following typical structure (d = space dimension)

$$\mathcal{F} = \frac{1}{\Gamma} \Big\{ \sum_{\underline{k}} \omega_{\underline{k}} |\sigma_{\underline{k}}|^2 + \sum_{\underline{k},\underline{k}'} v(\underline{k},\underline{k}') \sigma_{\underline{k}} \sigma_{\underline{k}'} \sigma_{-\underline{k}-\underline{k}'} + \tag{5}$$
$$+ \sum_{\underline{k},\underline{k}',\underline{k}''} u(\underline{k},\underline{k}',\underline{k}'') \sigma_{\underline{k}} \sigma_{\underline{k}'} \sigma_{\underline{k}''} \sigma_{-\underline{k}-\underline{k}'-\underline{k}''}$$

which determines the stationary behaviour of the system. On the other hand the divergence free current $V_{\underline{k}}$ which is such that

$$\sum_{\underline{k}} (\partial_{\sigma_{\underline{k}}} V_{\underline{k}} + \partial_{\sigma^*_{\underline{k}}} V^*_{\underline{k}}) e^{-\mathcal{F}} = 0 . \tag{6}$$

The noise term $\eta_{\underline{k}}$ represents the contributions of all the slaved modes.

Because of the pronounced analogies (critical slowing down, long range critical fluctuations, symmetry breaking ...) between instabilities far from equilibrium and equilibrium phase transitions that are examplified by the structure of equations (4) and (5), the latter concepts may be used to shed some new light on behaviour in the vicinity of instabilities. For instance, the theory of the RNG has furnished techniques to assess the importance of inhomogeneous fluctuations near critical points (Patashinskii, Pokrovskii, 1979). Besides providing computational techniques for the evaluation of critical indices, it leads to a deeper understanding of the concept of universality and shows that the degree to which order is modified by the fluctuations depends strongly on the spatial dimension of the system. The properties of degenerate, ordered systems exhibiting breaking of continuous symmetry have also been clarified, and the concept of spontaneous symmetry breaking has allowed the derivation of general relations for the correlations functions independently of any approximation (Patashinskii, Pokrovskii, 1979).

We are, however, aware that, although such analogies are helpful they are seldom complete and their eventual limitations must be discussed. On the other hand, non-equilibrium phase transitions, as the instabilities have become known, are much richer because they include formation of limit-cycles, motion on tori and even chaos.

Here, we can again establish universality classes which may however also depend on geometry and size (boundaries).

Now it is well known that as one lowers the dimensionality of the system, the influence of fluctuations increases dramatically. As a consequence not only is long range order in general impossible in 1 dimensional, but this also happens in some 2 dimensional systems undergoing spontaneous breaking of a continuous symmetry.

Many large-aspect ratio systems (thin layers) undergoing the instabilities we have considered may be thought as 2 dimensional. For such systems a new kind of order appears: topological order characterised by the power law behaviour of its correlation functions and the presence of defects such as dislocations and disclinations that play a paramount role in the fusion or desynchronisation of this order (Kosterlitz, Thouless, 1973; Halperin, 1979).

These situations are being increasingly studied both theoretically and experimentally. The situations most often analysed appear in Rayleigh-Bénard (Bergé, 1982; Gollub, 1981) and Bénard-Marangoni (Pantaloni et al., 1981) convection, nematic convective (Dubois-Violette et al., 1978) or elliptical shear flows (Dreyfus, Guyon, 1981). Instabilities in chemical (Showalter, 1980) or biochemical (Boiteux, Hess, 1980) systems also lead to analogous situations. The common features exhibited by these various systems can be traced to the generation of long range phase fluctuations by spontaneous breaking of a continuous symmetry group (Ivanitzky et al., 1981; Smoes, 1980).

This paper is organised as follows. The next section analyses the consequence of the Turing instability leading to a 'chemical crystal' whereas the following section studies the Hopf bifurcation to a limit cycle, in other words a 'chemical clock'. In both sections the influence of topological defects is analysed in the case of 2 dimensional systems.

Spatial structures

When they are driven sufficiently far from equilibrium many systems, wherein chemical reactions are coupled with diffusion, undergo a symmetry breaking transition leading to macroscopic, steady, spatial structures. Instability occurs when one of the roots of the characteristic equations, associated with the linearised kinetic equations, has a positive real part. The corresponding eigenvector σ_q (unstable mode) which is a linear combination of the concentrations of the various species plays the role of the order parameter. Linear stability analysis fixes the critical value, λ_c , of the bifurcation parameter and the critical wavelength, $2\pi/q_c$. Contrary to most hydrodynamic instabilities, this critical wavelength is determined by the intrinsic properties of the system, it is given in terms of the ratio of the two competing mechanisms (chemical rates and diffusion constants) the dynamic equilibrium of which leads to the structure. Systems with a dimensionality higher than one exhibit a great degeneracy. In isotropic media this degeneracy is twofold. First for a given wavelength there is a great variety of possible patterns (rolls, hexagons ...). Second, in unbounded media there is also an orientational degeneracy which can only be removed by lateral boundary conditions. In anisotropic media such as, for instance, in the mesophyses of liquid crystals an intrinsic mechanism raises the orientational degeneracy by inducing preferred directions for the wave vectors characterising the structure.

As pointed out in the introduction on can obtain, close to the instability, a reduced description of the system in terms only of the critical mode σ_q . The generalised potential, F, (Brazowskii's potential) takes the form:

$$F = \frac{1}{\Gamma} \left\{ \frac{1}{2} \sum_q \omega_q |\sigma_q|^2 + \frac{v}{3!} \sum_{q,q'} \sigma_q \sigma_{q'} \sigma_{-q-q'} + \right.$$
$$\left. + \sum_{\{q\}} \frac{v(q+q')}{4!} \sigma_q \sigma_{q'} \sigma_{q''} \sigma_{-q-q'-q''} \right. . \qquad (7)$$

The frequency, ω_q , tends to zero when approaching the instability (critical slowing down). In isotropic systems it depends only on the length of the wave vector: on the other hand in uniaxial materials we can have, for instance,

$$\omega_q = r_o + D(q^2 - q_c^2) + A^2 q^2 \sin^2 \varphi \qquad (8)$$

where $r_o = (\lambda_c - \lambda)/\lambda_c$

and φ is the angle of the wave vector, q , with the principal axis of the medium A is an anisotropy constant. D, u and v depend on the intrinsic properties of the system.

The functional F plays the role of a generalised free energy far from thermal equilibrium. Similar functionals have been derived near to Rayleigh-Bénard instability (Graham, 1974; Swift, Hohenberg, 1977) or from convective instabilities in nematic liquid crystals (Manneville, 1978). We now show that the non-linear terms (u, v in (7) provide a pattern selection mechanism already present in the mean field picture.

Landau theory

In mean field approximation the most probable configuration (which minimises F) can be characterised by m pairs of wave vectors $q_i, -q_i; q_i^2 = q_c^2$. (Walgraef et al., 1982). In space variables the order parameter becomes

$$\sigma(r) = 2 \sum_{i=1}^{m} |\sigma_i| \cos(q_c (1_i \cdot r) + \varphi_i) \qquad (9)$$

where the amplitudes, σ_i , satisfy the state equation:

$$\delta F / \delta \sigma_{q_i} = 0 .$$

For the sake of clarity we shall mainly consider here the isotropic case with u and v constant. Two classes of structures must be considered.

a) Structures appearing subcritically through a second order transition where the principle of exchange of stability holds.

These are characterised by m independent pairs of wave vectors; as a result for such cases the cubic invariant in F makes no contribution. The amplitudes are equal to

$$\sigma_m \begin{cases} = 0 & \lambda < \lambda_c \\ = [-2r_0/(2m-1)u]^{1/2} & \lambda > \lambda_c . \end{cases} \tag{10}$$

The relative stability of the various patterns may be calculated by comparing the corresponding contribution to the potential, where we find

$$F_m = - \frac{m}{2m-1} \left(\frac{r_0^2}{u} \right) . \tag{11}$$

Since $F_{m-1} < F_m$, the most stable phase corresponds to m = 1; a roll structure. It consists of a periodic variation of the concentration of the intermediate species in the direction conjugate to q_i :

$$\sigma(x) = \left[-\frac{2r_0}{u}\right]^{1/2} \cos q_c x + c_2[-r_0] \cos 2q_c x + c_3[-r_0]^{3/2} \cos 3q_c x \tag{12}$$

(c_1 and c_2 are constants).

The second and the third term in (12) are the contributions due to the harmonics generated by non-linear terms. Moreover, a stability analysis when compared with a homogeneous perturbation approach shows that this roll pattern is the only stable structure in this class.

b) Structures appearing subcritically; the transition is then of the first order type because of the contribution due to the cubic terms in the potential. The patterns are constructed on m pairs of vectors arranged to form equilateral triangles. The most stable pattern of this class is described by the following order parameter

$$\sigma_3(r) = 2a_3 [\cos q_c x + \cos \tfrac{q_c}{2}(x + \sqrt{3}y) + \cos \tfrac{q_c}{2}(x - \sqrt{3}y) \quad (13)$$

+ harmonics

where a_3 obeys the following state equation

$$r_0 + v a_3 + \frac{5}{2} u a_3^2 = 0 . \quad (14)$$

Depending on the sign of the cubic term, the maxima of concentration respectively define a triangular, ($v > 0$), or a honeycomb, lattice ($v < 0$). This pattern is reminiscent of the Bénard l or g hexagonal structures which appear in non-Boussinesq fluids.

In conclusion, in isotropic systems the hexagonal (triangular) pattern is the first to appear; then, following further increasing in the bifurcation parameter, ('structural') transition to the roll structure takes place.

Of course anisotropies greatly modify pattern selection. In most anisotropic systems the first structure to appear is a roll pattern making a well defined angle with the principal axes of the medium (cf. convective instabilities in nematics). For instance in the example given in (8) the rolls are parallel to the principal axis of the system. Moreover for a given anisotropy this angle does not vary with the distance from the threshold in agreement with experiments on nematic systems subjected to an elliptical shear.

The angular dependence of the non-linear coupling terms in (7) (e. g. $u(q + q')$) also plays an important role in the pattern selection mechanism. To illustrate this point we consider the case where the cubic term in (8) is equal to zero. The amplitude of the square pattern is then given by

$$\sigma_3^2 = -r_0/(u_0/2 + u_{ND}) \tag{15}$$

where u_D is the coupling constant between parallel wave vectors and u_{ND} between perpendicular wave vectors. A stability analysis around the state (15) shows that the squares are stable when $u_D - 2u_{ND} > 0$ whereas the rolls are stable when $u_D - 2u_{ND} < 0$. As a result the angular dependence of the quartic coupling furnishes in this case a selection mechanism.

For $\lambda > \lambda_c$, linear stability around the homogeneous states predicts that a range, $\delta_{\parallel}$, of wave numbers in the neighbourhood of q_c correspond to unstable modes which are thus amplified: $q = q_c \pm \delta_{\parallel}$ where $\delta_{\parallel} = (-r_0/4Dq_c^2)^{1/2}$.

However only a part of this range leads to layered roll solutions which can be realised physically. It is thus important to check the stability of these layered structures with $|q_i| \neq q_c$ against inhomogeneous perturbations. In the case of Brazowskii's model two types of disturbances must be considered (isotropic case, u = constant):

I. The zig-zag instability which induces a wavy distortion of rolls takes place when

$$q_i^2 \leq q_c^2 \quad \text{for} \quad \lambda \to \lambda_c . \tag{16}$$

II. The Eckhauss instability which corresponds to a longitudinal compression of the structure occurs when q lies in one of the two ranges

$$[q_c - \delta_{\parallel}, q_c - \delta_{\parallel}/\sqrt{3}] \quad \text{or} \quad [q_c + \delta_{\parallel}, q_c + \delta_{\parallel}/\sqrt{3}] \tag{17}$$

Indeed in our model the oblique roll instability (corresponding to perturbation with a wave vector making a finite angle with q_i) is less dangerous than the Eckhauss or the zig-zag instability because it appears when $q > q_c + \delta_{\parallel}/\sqrt{2}$.

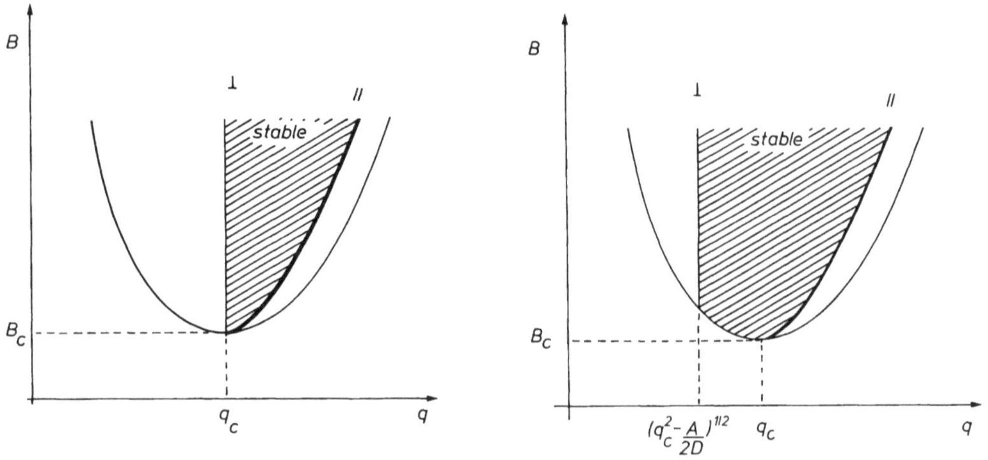

Fig. 42 Sketch of the stability diagram for the structure periodic in one dimension.
a isotropic system
b anisotropic system.

Here also the anisotropy A (cf. 8) modifies these results. For instance the borderline for the onset of the zig-zag instability is shifted to lower values of $|q_i|$ given by Fig. 42.

$$(q_i^2 - q_c^2) = -A/2D$$

Phase fluctuations

As in equilibrium the breakdown of translational and rotational symmetry at the Turing instability point generates long range fluctuation in the ordered phase. These 'dissipative Goldstone modes' (Patashinskii, Pokrovskii, 1979) are responsible for the nonexistence of symmetry breaking instabilities in infinite two dimensional systems. However this destruction of spatial order is effective only in very large systems (of the size of the universe!). As a result most finite systems can display a kind of order which is, nevertheless, greatly modified by these long range fluctuations. Because they are not subjected to restoring forces contrary to amplitude perturbations, phase fluctuations dominate at long wavelengths. The corresponding kinetic equations can be obtained by substituting in the TDGL equations the following expression

$$\sigma(\underline{r}) = \sum_{i=1}^{m} [\Psi_i + \Psi_i^*]$$

with

$$\Psi_i(\underline{r}) = R(\underline{r}) \exp [i q_i (\underline{r} + \underline{u}(\underline{r}))]$$
$$= R(\underline{r}) \exp i (q_i \underline{r} + \varphi_i(\underline{r}))$$

and then performing the adiabatic elimination of the amplitude (R(r).

For perfectly periodic structures $\varphi_i = q_i \cdot u$ is constant, whereas deformations such as compressions or torsions induce a spatio-temporal variation of φ_i .

In the case of layered structures the relaxation of a deformation with a spatial period large with respect to λ_c obeys the following diffusion equation (Pomeau, Manneville, 1979)

$$\partial_t \varphi = D_{\parallel} \partial_x^2 \varphi + D_{\perp} \partial_y^2 \varphi - K \partial_y^4 \varphi \qquad (19)$$

($\underline{1}_y$ parallel to the roll axis).

The coefficients $D_{\parallel}$, $D_{\perp}$ and K corresponding respectively to longitudinal and transverse modes depend on the amplitude and the wave vector of the rolls. For the lowest order in r_0 and $(q_i^2 - q_c^2)$, one finds for $D_{\parallel}$ and $D_{\perp}$

$$D_{\parallel} = 4 D q_c^2 + 6 D (q_i^2 - q_c^2) - \frac{16 D^2 q_i^2 (q_i^2 - q_c^2)^2}{u \sigma_1^2} \qquad (20)$$
$$D_{\perp} = 2 D (q_i^2 - q_c^2) + A .$$

Experimental values for these coefficients have been obtained in the case of convective structures (Croquette, Schosseler, 1982; Guazzelli et al., 1981). In isotropic systems one finds $D_\perp \ll D_\parallel$ moreover $D_\perp$ is expected to vanish at the threshold whereas in anisotropic cases it remains finite and positive even near the threshold in agreement with equation (20) (Guazzelli, 1981) where $D_\perp$ is related directly to the anisotropy constant. In the (λ , q) plane the curves $D_\parallel = 0$; $D_\perp = 0$ delineate the region where the non-equilibrium pattern is unstable against longitudinal compression ($D_\parallel = 0$) or modulations of the structure ($D_\perp = 0$). Using expressions (19) and (20) the criteria derived above for the Eckhauss ($D_\parallel = 0$) and the zig-zag instabilities ($D_\perp = 0$) can be deduced.

Including a Gaussian noise term into (19) describing short scale amplitude fluctuations, then the phase fluctuations are described in the steady state, by the following potential:

$$V_1 = \frac{\sigma_1^2}{2\Gamma} \int dx \int dy \, [D_\parallel (\partial_x \varphi)^2 + D_\perp (\partial_y \varphi)^2 + K (\partial_y^2 \varphi)^2] . \tag{21}$$

From (21) the order parameter correlation function can be calculated as:

$$C(\underline{r}) = \langle \Psi(\underline{r}) \Psi^*(0) \rangle \sim \left(\frac{D_\parallel}{D_\perp} x^2 + y^2 \right)^{-\eta} \tag{22}$$

where $\eta = \Gamma / 4\pi \sigma_1^2 (D_\perp D_\parallel)^{1/2}$.

This algebraic decay is an indication of quasi long range order which is characteristic of two-dimensional systems (Kosterlitz, Thouless, 1973; Halperin, 1979).

Similarly in the case of hexagonal or triangular patterns the displacement vector $\underline{u}$ obeys the following equation (Walsgraef et al., 1982)

$$\partial_t \underline{u} = \mu a_3^{-2} [\nabla^2 \underline{u} + 2 \,\mathrm{grad\,div}\, \underline{u}] + \mu a_3^{-2} (\nabla^2)^2 \underline{u} \tag{23}$$

where $\mu = 3 D q_c^4$ plays the role of a rigidity modulus and $\mu = \mu / 2q_c^2$ is a second order elasticity coefficient.

These general diffusion equations (19) and (23) are characteristic of dissipative systems. They are to be contrasted with the usual equations of motion in the theory of elasticity derived from Newtonian dynamics. As a result Goldstone modes in non-equilibrium structures are diffusive modes; contrary to propagating phonon modes in equilibrium crystals. The potential for the phase fluctuations now take the form

$$V_3 = \frac{a_3^2}{\Gamma} \int d^2r \Big\{ \mu \Big([\partial_x u_x + \partial_y u_y]^2 + \frac{1}{2} [\partial_x u_x - \partial_y u_y]^2 + \frac{1}{2} [\partial_x u_y + \partial_y u_x]^2 \Big) + \mu \big((\nabla^2 u_x)^2 + (\nabla^2 u_y)^2 \big) \Big\} \tag{24}$$

and this is analogous to the elastic Hamiltonian of an isotropic solid. In this case one also obtains, from (25), a power law decay for the order parameter correlation function.

$$C_i(\underline{r}) = \langle \psi_i(\underline{r}) \psi_i^*(0) \rangle \simeq |\mathbf{r}|^{-\eta}$$

with

$$\eta = \Gamma q_c^2 / 3\pi \mu a_3^2 .$$

By analogy with the equilibrium case, defects such as dislocations or disclinations are expected to play an important role also in 2dimensional non-equilibrium systems (Kosterlitz, Thouless, 1973; Halperin, 1979). A dislocation in two dimensions is a point defect in an otherwise perfect structure in which part of an extra minimum iso-concentration line is added. These are described by a Burgers vector b (defined by the geometric construction in Fig. 43). Disclination are also points defects around which the iso-concentration lines rotate by some definite angle related to the symmetry of the 'lattice' as illustrated in Fig. 44.

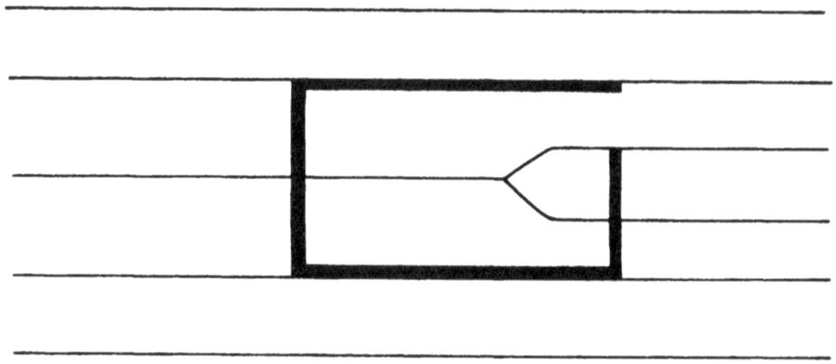

Fig. 43 A dislocation in a layered structure shown with its Burger's circuit.

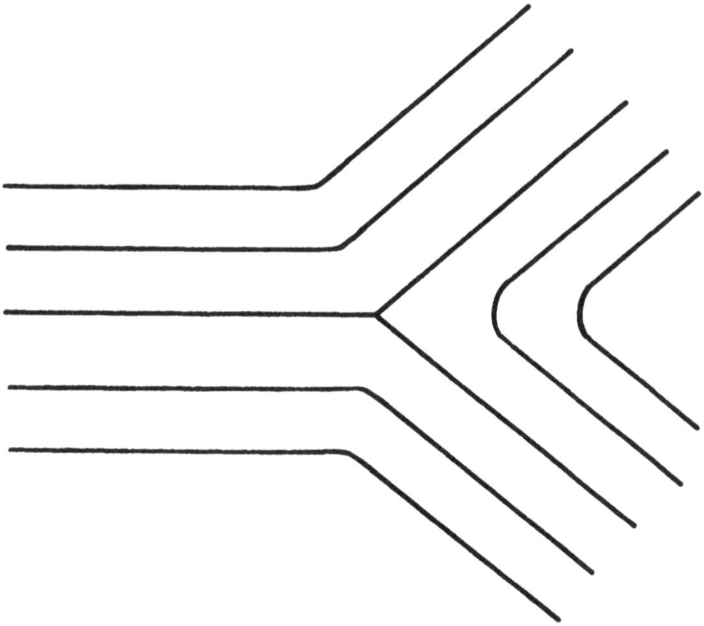

Fig. 44 A typical disclination in a layered structure.

In the case of a layered structure the Burgers vector is perpendicular to the roll axis and, for a contour around a dislocation we have:

$$\int \underline{\nabla}\varphi \cdot d\underline{\ell} = 2\pi m \qquad (m = \text{integer}) \tag{25}$$

The dislocation is associated with a phase singularity. Near the dislocation $(y < (K/D_{\parallel})^{1/2}$ the region of strong and small distortions are separated by a parabola of aperture, $\lambda = (K/D_{\parallel})^{1/2}$ for larger distances $(y > (K/D_{\perp})^{1/2}$ these regions are separated by two straight lines of slope $(D_{\parallel}/D_{\perp})^{1/2}$. The study of the deformation field induced by the defect can thus lead to a measure of the diffusion coefficients of the structure. From this deformation field, that decreases roughly as the inverse first power of the distance from the core, one may estimate the probability of occurrence of an isolated dislocation (Dewel et al., 1981):

$$P \sim \exp - \left[\frac{Rm^2}{2\Gamma} (D_{\parallel} D_{\perp})^{1/2} \ln \frac{L}{a_0} + E_c \right] . \tag{26}$$

(L is the linear size of the system, a_0 the core diameter and E_c the core energy).

Employing an argument developed by Kosterlitz and Thouless (1973) for equilibrium systems one sees that the probability associated with the number of possible positions of an isolated dislocation behaves as

$$P \sim L^2 \exp - \frac{Rm^2}{2\Gamma} (D_{\parallel} D_{\perp})^{1/2} \ln (L/a_0) . \tag{27}$$

Hence, for $m^2 (D_{\parallel} D_{\perp})^{1/2} > \frac{4\Gamma}{\pi}$ the probability for the creation of isolated dislocations is vanishingly small in large macroscopic systems. Only neutral arrays of dislocations may be present in the medium as their potential is

$$V_t = \frac{R}{2\Gamma} \sum_{i,j} m_i m_j (D_{\parallel} D_{\perp})^{1/2} \ln \left[\frac{D_{\parallel}}{D_{\perp}} (x_i - x_j)^2 + \right. \\ \left. + (y_i - y_j)^2 \right] \tag{28}$$

$(\sum_i m_i = 0, \ |r_i - r_j| > a_0,$ r_i is the centre of the dislocation of charge m_i) which leads to a finite probability. However, for $m^2 (D_{\parallel} D_{\perp})^{1/2} < 4\Gamma/\pi$ the probability for the creation of isolated dislocations is finite and they may spontaneously form, or, dislocation pairs may unbind and destroy the quasi-long-range translational order. The quasi-long-range orientational order may then be destroyed by the spontaneous unbinding of the disclinations. This argument may also be used in the case of polygonal structures.

Low constraint melting

This type of topological transition may, in principle, be responsible for the melting of non-equilibrium structures for low values of the bifurcation parameter.

From the melting criterion, (e. g. $m^2 (D_{\parallel} D_{\perp})^{1/2} = 4\Gamma/\pi$ for the layer), one deduces that this kind of transition takes place when $\sigma_1^2 (\sim - r_0)$ is comparable to the noise intensity which is expected to be very small.

Moreover this phase fluctuation driven transition may only occur when the size of the system is much larger than the amplitude correlation length $\xi \sim D^{1/2} / \sigma_1$ otherwise the separation between phase and amplitude is no longer justified and the transition is of the Landau-Hopf type. In the case of honeycomb or triangular patterns another restrictive condition appears as the amplitude a_3 has a lower bound $(v/5u)^{1/2}$. As a result a topological phase transition may only occur if $\Gamma \gg (v/5u)^2$ otherwise one has a first order transition of the mean field type. For instance, in the case of the transition to non-Boussinesq hexagonal cells in water near 4 °C (Dubois et al., 1978) one expects a first order transition describable by a Landau potential for the amplitude. On the other hand for Turing instability in the trimolecular chemical model may be varied continuously from positive to negative values. Hence a topological transition might occur in this model which, despite its idealised nature, may be used to describe spatio-temporal organisation in biochemical models such as glycolysis. Therefore the arguments developed here should be relevant to an interpretation of imperfect polygonal structures as obtained, for example, by Boiteux and Hess (1980).

At low values of the bifurcation parameter two types of melting mechanism may occur dependent on the noise intensity. For low intensities the transition is of the Landau-Hopf type while for higher intensities the transition will be of topological nature.

At higher values of the bifurcation parameter pairs, or arrays, of defects may appear spontaneously in the various structures described here and, since the characteristic wavelengths are macroscopic, they can be more easily observed then those in equilibrium systems.

High constraint melting

When the bifurcation parameter increases further, non-linear coupling between the harmonics of the spatial oscillations of the order parameter induces a smooth deformation of the structure (Walgraef et al., 1982). It looks more and more like a soliton lattice where homogeneous regions are separated by narrow bands and where the phase changes abruptly. The wavelength of this pattern increases very slowly. For sufficiently high values of the bifurcation parameter, this deformation may become sufficiently important to induce a screening effect on the dislocation deformation field leading to a finite probability for isolated dislocations. In the case of layers this effect leads to the melting of the pattern while for polygonal structures the formation of grain boundaries may be favoured. Hence in the latter case the destruction of translational quasi-order occurs via the formation of crystallites in the system. These crystallites maintain, however, the orientational quasi-order which is further destroyed via disclination unbinding. The scenario outlined here for the melting (Walgraef et al., 1982) of structures in large systems shows analogies with experimental results obtained in the case of convective instabilities in nematic systems submitted to elliptical shears (Dreyfus, Guyon, 1981) and with Bénard-Marangoni effects (Pantaloni et al., 1981).

To conclude this section we emphasise the fact that non-equilibrium structures which appear to be stable in a deterministic description may, however, be strongly affected by long range fluctuations which may develop spontaneously in the medium. Small systems will, in general, not be affected by such fluctuations since they are not compatible with the boundaries. In large systems however these fluctuations are able to induce topological defects which may be responsible, at sufficiently high values of the bifurcation parameter, for the melting of the structure. This melting corresponds to the spontaneous decorrelation of the system and cannot be obtained from a purely deterministic description based on the phenomenological equations of the active medium.

Chemical oscillations

The Hopf bifurcation is another example of symmetry breaking instability. Such a bifurcation occurs when two eigenvalues of the linear evolution matrix associated with the dynamic system under consideration become purely imaginary whilst the other ones have negative real parts. The corresponding eigenmodes are, hence, the two slow modes of the problem. The remaining fast modes may then be adiabatically eliminated and the evolution of the system is given, on the longest time scale of the problem, through its two component slow mode dynamics (Haken, 1978).

Hence, such systems may be described, in the vicinity of the Hopf bifurcation by the following Langevin equations for the slow modes (Σ, Σ^*) in their rotating frame ($\sigma = \Sigma \exp i\omega_0 t$) where $\pm i\omega_0$ are the imaginary eigenvalues of the linear evolution matrix at the bifurcation point), up to the first significant non-linear terms:

$$\partial_t \sigma(\underline{r}, t) = -[r - c\nabla^2]\sigma(\underline{r}, t) - u|\sigma|^2\sigma(\underline{r}, t) + \eta(\underline{r}, t). \qquad (29)$$

This equation may be derived explicitly from real models (Walgraef et al., 1982). In this case, σ and σ^* are well defined linear combinations of local concentrations while $r = r_1 + ir_2$, $c = c_1 + ic_2$, $u = u_1 + iu_2$ may be related to the various chemical reaction rates and diffusion constants of the problem. Moreover r depends explicitly on the bifurcation parameter λ ($r_1 \sim (\lambda_c - \lambda)/\lambda_c$, $r_2 \sim O(((\lambda_c - \lambda)/\lambda_c)^2)$) and vanishes at the bifurcation point. They may also be more generally obtained in the framework of the theory of normal forms (Arnold, 1980). Quadratic terms in σ do not appear near the instability point ($|r_1| \ll 1$) and are significant for times lower than, ω_0^{-1} ; the rotating wave approximation is valid and they oscillate too rapidly to give a non-vanishing contribution to the dynamics. As stated in the first section equation (29) is equivalent to a time dependent Ginzburg-Landau equation where the generalised potential, $\tilde{\mathcal{F}}$, and the divergence free current, $V_{\underline{k}}$, are such that:

$$\mathcal{F} = \frac{1}{\Gamma}\Big[\sum_{\underline{k}}(r_1 + c_1 k^2)|\sigma_{\underline{k}}|^2 + (2L^d)^{-1}\sum_{\underline{k},\underline{k}',\underline{k}''} u(\underline{k},\underline{k}',\underline{k}'')\sigma_{\underline{k}}\sigma^*_{\underline{k}'}\sigma_{\underline{k}''}\sigma^*_{-\underline{k}-\underline{k}'-\underline{k}''}$$

and

$$V_{\underline{k}} = i(r_2 + c_2 k^2)\sigma_{\underline{k}} + L^{-d}\sum_{\underline{k}',\underline{k}''} v(\underline{k},\underline{k}',\underline{k}'')\sigma_{\underline{k}-\underline{k}'-\underline{k}''}\sigma^*_{\underline{k}'}\sigma_{\underline{k}''} \qquad (30)$$

with

$$v(\underline{k},\underline{k}',\underline{k}'') = -iu_2 + i(c_1u_2 - c_2u_1)\beta(\underline{k},\underline{k}',\underline{k}'')\big[\alpha(\underline{k},\underline{k}',\underline{k}'' + ic_2\beta(\underline{k},\underline{k}',\underline{k}'')\big]^{-1}$$

$$\alpha(\underline{k},\underline{k}',\underline{k}'') = 4r_1 + c_1(k^2 + k'^2 + k''^2 + (k + k' + k'')^2)$$

$$\beta(\underline{k},\underline{k}',\underline{k}'') = k^2 - k'^2 + k''^2 - (k + k' + k'')^2 .$$

It may easily be checked that the divergence free condition is satisfied up to the sixth order in σ . This is consistent with the approximation scheme which allows one to neglect non-linearities of order higher than 3 in equation (29).

With the TDGL model described above we are able to investigate the behaviour of a system near its Hopf bifurcation and we will now mimic real experimental situations. We will first consider the case of well stirred reactors where the inhomogeneities are washed out and secondly the case of batch reactors where inhomogeneities may play an important role according to the space dimensionality of the system.

Oscillations in a well-stirred reactor

The stirring mechanism considered here is supposed to be sufficiently fast so as to be decoupled from the chemical kinetics. Its effect is equivalent to a considerable enhancement of the diffusion mechanism and is able to suppress all inhomogeneities on the time scales associated with the slow mode dynamics. Hence the TDGL description becomes $(\sigma_0 = L^{-d/2}\sigma_{q=0})$

$$\partial_t \sigma_0 = -\Gamma L^{-d}\frac{\delta \mathcal{F}_0}{\delta \sigma_0^*} - i[r_2\sigma_0 + u_2|\sigma_0|^2\sigma_0] + \eta_0$$

where

$$\mathcal{F}_0 = \frac{L^d}{\Gamma}\left[r_1|\sigma_0|^2 + \frac{u_1}{2}|\sigma_0|^4\right]. \tag{31}$$

The distribution function of the fluctuations is now

$$P \propto N \exp - \mathcal{F}_0 . \tag{32}$$

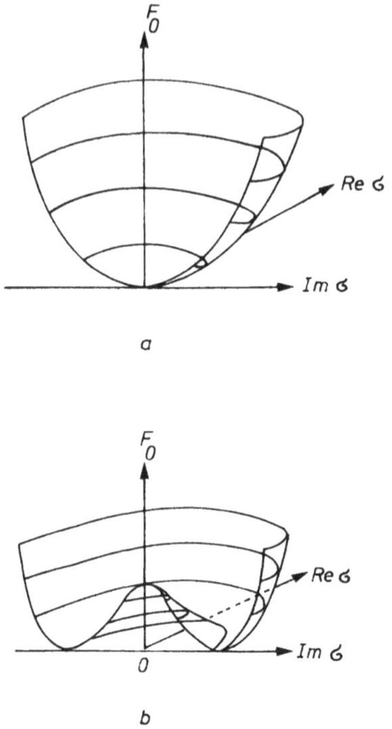

Fig. 45 Aspect of the potential in the homogeneous description:
a below the hard mode instability
b above the hard mode instability $(\lambda > \lambda_c)$.

$\mathcal{F}_o$ is the standard homogeneous potential for a system with a two component order parameter such as the XY model; the superfluid or the single mode laser. Its shape, a Gaussian well as long as r_1 is positive, becomes crater-like in the ordered region, $(r_1 < 0)$, as shown in Fig. 45. From the point of view of the theory of critical phenomena all these systems belong to the same universal class (Patashinskii, Prokrowskii, 1979) whereas their dynamics may be different (in this case, the specific aspects result from the divergence free current V_o).

Effectively, mean field analysis predicts a second order like transition at $r_1 = 0$ $(\lambda = \lambda_c)$ from a homogeneous steady state $-\Sigma_q = \Sigma_q^* = 0$ $(r_1 > 0, \lambda < \lambda_c)-$ to a limit cycle $-$ $\Sigma_{q \neq 0} = 0$, $\Sigma_o = R_o \exp i(\omega_o t + \phi)$ $-$. Its amplitude $R_o = (r_1/u_1)^{1/2} \sim (\lambda - \lambda_c)^{1/2}$ grows with the classical critical index 1/2 and $\phi = \phi_o - (r_2 + u_2 R_o^2)t$ where ϕ_o is an arbitrary phase. As the generalised potential, $\mathcal{F}_o$, depends only on the radius of the limit cycle, all the phases, ϕ_o , are equally probable and the transition occurs with the spontaneous breakdown of the rotational symmetry in the ($\mathrm{Re}\,\sigma$, $\mathrm{Im}\,\sigma$) plane. The study of the stability of the limit cycle shows that, in the mean field approximation, there is no symmetry restoring mechanism for the phase: any perturbation of the system induces a constant phase shift of the oscillation. Hence the fluctuations may neither be neglected in the vicinity of the transition nor in the ordered region. Their effect on the static (in the rotating frame) properties of the system may be evaluated from (32). They give rise to a rounding-off effect of the transition as for the single mode laser. As far as the dynamics is concerned the distribution function of the fluctuations of the order parameter $(\sigma_o = R \exp i\phi)$ satisfy the following Fokker-Planck equation:

$$\partial_t P(R,\phi,t) = \frac{1}{2}\Big[\frac{1}{R}\partial_R(r_1 R^2 + u_1 R^4) + \partial_\phi(r_2 + u_2 R^2)\Big] P(R,\phi,t) + \frac{\Gamma}{4L^d}\Big[\frac{1}{R}\partial_R R \partial_R + \frac{1}{R^2}\partial_\phi^2\Big] P(R,\phi,t). \qquad (33)$$

and the diffusion term of this equation leads to a symmetry-restoring mechanism, so-called phase diffusion. Indeed in the ordered regime the amplitude fluctuations relax rapidly in contrast to the phase fluctuations, and, on the longest time scale, the order parameter may be written as

$R_o \exp i\phi$ and its mean value behaves as

$$\langle \sigma(t) \rangle = R_o \langle \exp i\phi(t) \rangle = \\ = R_o \exp\left[i(\phi_o - (r_2 + u_2 R_o^2)t) - \frac{\Gamma t}{4R_o^2 L^d}\right]. \tag{34}$$

As a consequence the life time of the order parameter is proportional to the volume of the reactor, the squared amplitude of the oscillations, and inversely proportional to the noise intensity. Nevertheless, for sufficiently large systems $(L^d \gg \Gamma R_o^{-2})$, the phase may be considered, on experimental time scales, as locked.

From the discussion outlined here, two essential aspects of the dynamic behaviour of homogeneous systems deserve special attention. The fluctuations of the amplitude of the oscillations relax much more rapidly than the phase fluctuations. Furthermore these are able to desynchronise the system after sufficiently long times (phase diffusion). Hence one may also expect that long range phase inhomogeneities play an important role in batch reactors.

Oscillations in batch reactors and chemical auto-waves

In this case, while the homogeneous oscillations remain linearly stable, inhomogeneous fluctuations may develop. Their stationary distribution is given by the potential $\mathcal{F}$ which for $u_2 = c_2 = 0$ is exactly the Ginzburg-Landau potential for an equilibrium system with a two-component or order parameter. Moreover as the k dependent terms in $u\,(u_2, c_2 \neq 0)$ have a critical dimensionality of two, the static critical behaviour associated with $\mathcal{F}$ is that of the n = 2 universal class for systems of spatial dimensionality d > 2. For example, one obtains, from the renormalisation group approach of critical phenomena, that the order parameter behaves as $\sigma_o \sim (\lambda - \lambda^*_c)^\beta$ with $\beta = 1/2 - 3\epsilon/20 + \ldots$ for $\epsilon = 4 - d$ while the static correlation function goes like $\langle \sigma_k \sigma_k^* \rangle \sim$

$\sim \Gamma/(c_1 k^2 + \xi_1^{-2})$ where $\xi_1 \sim |\lambda - \lambda_c^*|^{-\nu_1}$ and $\nu_1 = \frac{1}{2} + \frac{\epsilon}{10} + \ldots$, λ_c^*

is the true instability point since the mean field instability point λ_c is shifted by the fluctuations (Patashinskii, Pokrovskii, 1979). In the disordered phase the dynamic correlation functions behave as (Hentschel, 1978)

$$\langle \sigma_{-k}^*(t)\, \sigma_k(0) \rangle \sim 2 \langle \sigma_{-k}^* \sigma_k \rangle \exp - \Gamma t / \tau(k) \quad (35)$$

with

$$\tau^{-1}(k) = c_1 k^2 + \xi_1^{-2} + i\,(c_2 k^2 + c_2 \xi_1^{-2} + \xi_2^{-2})$$

where

$$\xi_2 \sim |\lambda - \lambda_c^*|^{-\nu_2} \text{ and } \nu_2 = 1/2 .$$

As discussed elsewhere (Walgraef et al., 1981), these non classical effects whould only be observable in systems with low diffusion and high reaction rates.

Let us now investigate more specifically the consequences of spontaneous phase symmetry breaking associated with transition to the limit cycle behaviour. Despite this symmetry breaking, the order parameter may rotate homogeneously in the $(\mathrm{Re}\,\sigma, \mathrm{Im}\,\sigma)$ plane with no change in potential.

Consequently, small wavenumber fluctuations are important. The existence of such long range fluctuations in systems with broken continuous symmetries is well known in quantum field theory as the Goldstone theorem (Bogolubov, 1971). Using the standard methods to study the response of the system to a long wavelength phase shift we may derive the following Bogolubov inequality for the transverse correlation function (Walgraef et al., 1981)

$$g_\perp(k) = \langle |\sigma_k - \sigma_k^*|^2 \rangle \geq |\sigma_0|^2 / [\mathrm{coeff}]\, k^2 \tag{36}$$

which implies the destruction of long range order in d $\leq$ 2 infinite systems and hence the spontaneous desynchronisation of the oscillations.

Phase fluctuations

This effect could be of prime interest in many oscillating chemical or biological, non-linear media described by general reaction-diffusion equations and which may be considered as nearly two-dimensional.

In the ordered regime $(r_1 < 0)$ the Langevin equations (29) may be written in phase and amplitude variables $(\Sigma(\underline{r},t) = R(\underline{r},t) \times$ $\times \exp i[(\omega_0 - r_2 - u_2 R_0^2)t + \phi(\underline{r},t)])$:

$$\partial_t R = u_1 R(R_0^2 - R^2) + c_1(\nabla^2 R - R(\nabla\phi)^2) - c_2(2\nabla R \nabla\phi + R\nabla^2\phi) + \eta_R$$
$$R\partial_t\phi = u_2 R(R_0^2 - R^2) + c_2(\nabla^2 R - R(\nabla\phi)^2) + c_1(2\nabla R\nabla\phi + R\nabla^2\phi) + \eta_\phi \tag{37}$$

Linear stability analysis performed on this system, around the limit cycle defined by R = R_0, $\phi = \phi_0$, shows that when $u_1 c_1 + u_2 c_2 > 0$ it is stable against small radius and phase fluctuations while it is unstable when $u_1 c_1 + u_2 c_2 < 0$ (in this case phase turbulences may occur (Kuramoto, 1976). In the present analysis we consider only the first case. The eigenvalues of the linear evolution matrix are well separated when the range ℓ of the fluctuations is such that $\ell > L_0 \sim ((u_1 c_1 + u_2 c_2)/r_1 u_1)^{1/2}$ Hence inhomogeneities of range $\ell < L_0$ evolve on a common time scale and are strongly coupled while for $\ell > L_0$ amplitude and phase fluctuations evolve on different time scales and the amplitude, which is the fast variable, may be adiabatically eliminated. So, on the longest time scale, the dynamics of the oscillator is its phase dynamics given by the stochastic Burgers equation (Kuramoto, Yamada, 1976)

$$\partial_t \phi(\underline{r},t) = \mu \nabla^2 \phi + \nu (\underline{\nabla}\phi)^2 + \eta_\phi$$

where $$\mu = c_1 + u_2 c_2 / u_1 \,;\; \nu = (u_2 c_1 - c_1 c_2)/u_1 \tag{38}$$

$$\langle \eta_\phi(\underline{r},t)\eta_\phi(\underline{r}',t')\rangle = \frac{2\Gamma}{R_0^2}\,\delta(\underline{r}-\underline{r}')\,\delta(t-t').$$

Note that the validity of this equation is not limited to the vicinity of the instability (Neu, 1979) but far from it the shape of the limit cycle would affect the noise intensity. One furthermore deduces from (38) that the asymptotic distribution function of the phase fluctuations behaves as

$$P(\{\phi(\underline{r})\}) = N \exp\left[-\frac{\mu R_0^2}{2\Gamma}\int_{L_0} d\underline{r}\,(\underline{\nabla}\phi)^2\right] \tag{39}$$

(where N includes short range fluctuations). Consequently, large, low-dimensional systems (d = 1.2) do not exhibit true long range order. Effectively, the order parameter correlation function calculated with (39) becomes

$$|\langle \sigma(\underline{r}')\sigma^*(\underline{r})\rangle| \sim \begin{cases} R_0^2\left(\dfrac{|\underline{r}-\underline{r}'|}{L_0}\right)^{-\frac{2\Gamma}{R_0^2\mu}} & d = 2 \quad \text{(quasi long range order)} \\[2ex] R_0^2 \exp -\dfrac{2r\Gamma}{\mu R_0^2}|\underline{r}-\underline{r}'| & d = 1 \quad \text{(short range order)} \end{cases} \tag{40}$$

($L \gg |\underline{r}-\underline{r}'| \gg L_0$, L being the linear size of the reactor). Let us remark that L_0 plays the role of a critical desynchronisation length as systems of size $L < L_0$ still oscillate synchronously while larger systems may become partly desynchronised for asymptotically long times. However for chemical systems in batch reactors this asymptotic behaviour is scarcely ever reached and we will, rather, focus our attention on the early stages of their evolution after the interruption of the stirring which first maintained the homogeneity of the oscillations. With the stochastic description of equation (38) we are able to estimate the probability of occurrence of a phase inhomogeneity and to follow its further evolution through the Burgers equations. Consider first axis-symmetric fluctuations: as phase and amplitude are strongly coupled on short length scales, even small disturbances of range

$\ell < L_o$ may induce local frequency shifts of the oscillations. The regions where these shifts take place oscillate with a period shorter than that of the bulk and radiate concentric waves of isoconcentration lines (target patterns). From the stochastic analysis outlined above one finds that a fluctuation of amplitude α and range L_o occurs at $\underline{r} = \underline{r}_i$ and $t = t_o$ with probability

$$P \sim \exp[-v\mu(t)R_o^2\alpha^2/\Gamma]\;;\;\mu(t) = \mu\left[1 + \frac{\Gamma D}{\mu R_o^2}\cdot\left(\frac{L_o^2}{L_o^2 + 2\mu t}\right)^2\right] \qquad (41)$$

(D is the effective diffusion coefficient induced by the initial stirring mechanism). The frequency shift induced by this fluctuation gives rise to the following phase behaviour far from the centre (see e. g. Kuramoto, Yamada, 1976; Neu, 1979; Hagan, 1981):

$$\phi(\underline{r}, t) = \mathrm{Max}\,(0, v\varkappa^2(t - t_o) - \varkappa|\underline{r} - \underline{r}_i|) \qquad (42)$$

Where $\varkappa$, the wavenumber of the target pattern, is an increasing function of α $(\varkappa L_o \sim \exp - 2v/\alpha)$ for $\alpha > \alpha_o$ (where α is a function of the paramters of the bulk oscillation). These target patterns arise in various two-dimensional active media (Winfree, 1980) (e. g. aggregating *Dictyostelium discoideum*, cardiac muscles, nerve fibers, fungi mycelia, Belusov-Zhabotinskii (BZ) and chlorite oscillators). In the chemical case the explanation of their origin has led to a controversy about whether some 'catalytic particles' are nessary for their generation (see e. g. Winfree, 1978; Kopell, Howard, 1981). We conclude here that for self-oscillating systems they may be generated spontaneously by internal fluctuations. The centres are statistically distributed versus their wavelength (Fig. 46) and their total number may be related to the parameters of the oscillation (Fig. 47, 48). These effects can be tested experimentally and are, at present, in agreement with the qualitative experimental data obtained in the BZ reactions (Smoes, 1980; Zaikin, Zhabotinskii, 1970). However, a more refined quantitative investigation is needed to definitively check the results outlined above (Vidal, Pacault, 1983).

On the other hand, vortex-like fluctuations associated with phase singularities analogous to the vortices of 2 dimensional XY or superfluid systems (Kosterlitz, Thouless, 1973; Halperin, 1979) may also appear in

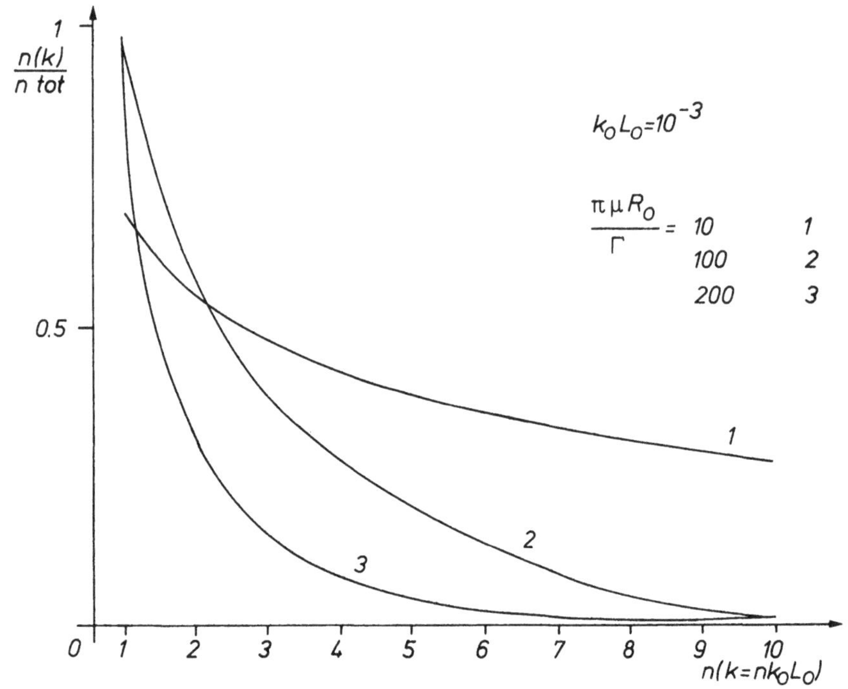

Fig. 46 Fraction of centres as a function of their wavenumber for different different values of the parameters.

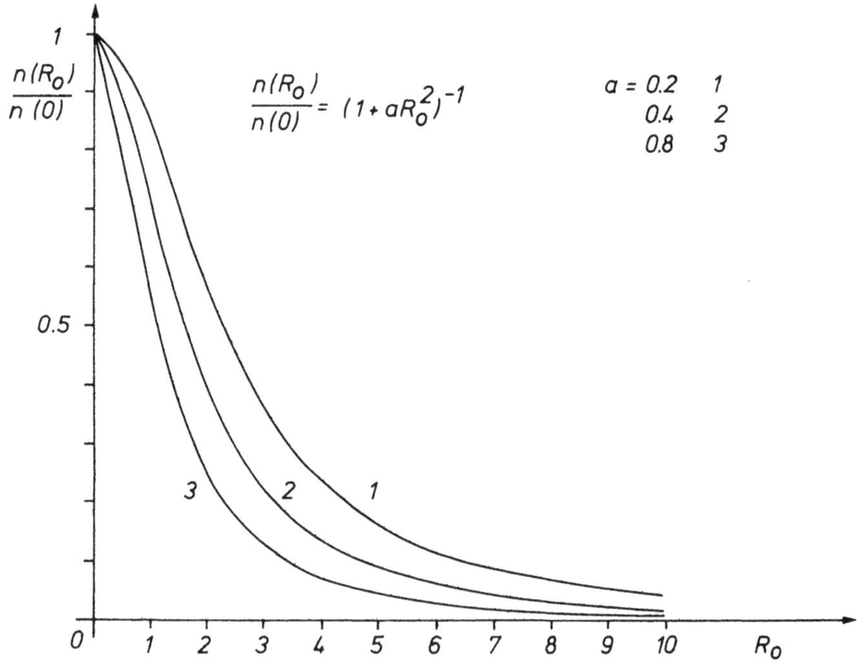

Fig. 47 Total number of centres as a function of the amplitude of the oscillations.

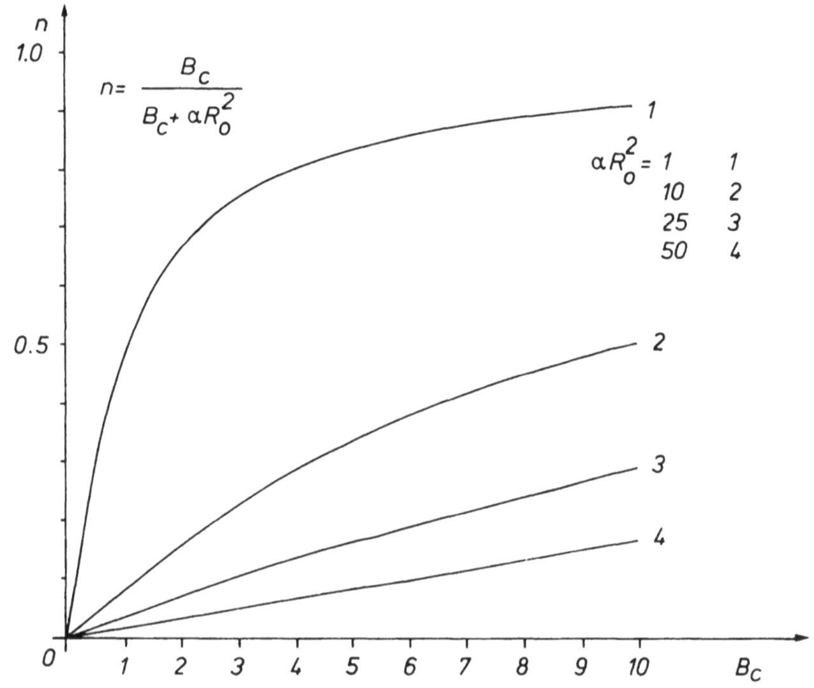

Fig. 48 Total number of centres as a function of the bifurcation parameter B_c.

oscillating chemical systems. These vortices have an integer topological charge, or winding number, N. Their long time evolution deduced from Burgers equation leads to rotating N-armed Archimedean spiral waves of iso-concentration lines (Walgraef et al., 1981; Kuramoto, Yamada, 1976; Cohen et al., 1978). The probability associated with such a spiral wave is vanishingly small in large reactors as is the case for isolated vortices in equilibrium systems (Kosterlitz, Thouless, 1973; Halperin, 1979). The probability of a pair of spirals of opposite winding number + N and - N, located at $\underline{r}_i$ and $\underline{r}_j$ behaves as $\exp[-2\pi N^2/\Gamma\, \mu(t) R_0^2 \ln|\underline{r}_i - \underline{r}_j|]$, $(|\underline{r}_i - \underline{r}_j| \gg L_0)$. N being an integer this probability also leads to a vanishingly small average number of spontaneous spiral pairs. Experimentally spiral waves may nevertheless appear under special circumstances (they may for example be induced by imperfect stirring mechanisms, inhomogeneities of the medium (Winfree, 1980; Agladze, Krinski, 1982).

No spontaneous appearance of isolated spirals which would destroy any order in the system, and hence suppress its oscillatory behaviour, has yet been reported. Let us however note that the possible relationship between the appearance of such waves and cardiac arrythmias (Ivanitsky et al., 1981; Winfree, 1981) suggests further analysis in this direction.

Dynamic Model of the Origin of Order in Controlled Macrosystems

V. S. Lerner

The conditions of the origin of ordering in non-equilibrium processes is determined by the properties of macrosystems and it is reasonable to study the laws relating to their function through use of dynamic macromodels (Trukhayev, Lerner, 1974; Lerner, Gusak, 1978). To describe regularities, we use variational principles as the functional of information entropy distribution as defined by the solutions of the controlled, stochastic equation (Lerner, 1981).

$$d\tilde{x}(t) = (a+u)dt + \sigma d\xi_t;\ \tilde{x} = \tilde{x}(s,t,u,\xi_t);\ \tilde{x}(s) = x(t_o), \qquad (1)$$

where $a = a(s, t, \tilde{x})$, $\sigma = \sigma(s, t, \tilde{x})$ are drift and diffusion matrices; ξ_t is a Wiener process; $u = u(\tilde{x}, s. t, \tau)$ are controls with feedback which minimise the final functional by selecting a control law

$$\Delta S = M_{\tilde{x},s}[\omega_s^{\tau}];\quad \omega_s^{\tau} = \omega(\tilde{x}(s), t, u, a, \sigma, \tau, \xi_t) \qquad (2)$$

Here ω_s^{τ} is an additive diffusion process functional which takes into account multiplicatively, the probability of interactions on the trajectories, (1), which are approximated by portions of the extremals of the functional ΔS. Equations of extremal lead to information analogs of controlled equations of non-equilibrium processes (Lerner, 1981)

$$\begin{aligned} &\frac{\partial x}{\partial t} = L(x,u)X(x,u);\ X = \text{grad}\,\Delta S' \\ &\Delta S' = \Delta S_o - \int H dt;\ H = -\partial \Delta S/\partial t;\ u = (u_1,..,u_n) \end{aligned} \qquad (3)$$

with the Hamiltonian $H = H(x(\tau), u(s, \tau))$ discretely varying at the points of union of extremals, and also, give rise to ordinary differential equations

$$\begin{aligned} &\frac{\partial x}{\partial t} = A^t(x,t)x + u^t;\ \frac{\partial x}{\partial l} = A^l(x,l)x + u^l; \\ &x(t_o) = \tilde{x}(S);\ x(l_o) = \tilde{x}(l_o(S)) \end{aligned} \qquad (4)$$

Thermodynamics and Regulation of Biological and Processes

and equations in partial derivatives

$$\frac{\partial x}{\partial t} = A(x,t,l)\frac{\partial x}{\partial l} + u; \quad A(x,t,l) = C;$$
$$x(0) = x(t_0, l_0); \quad x = (x_1, \ldots, x_n) \qquad (5)$$

as dynamic models of concentrated, and distributed, systems;

$$l = l(l_x, l_y, l_z) \qquad (6)$$

are coordinates of the external normal to the surface

$$\Delta S_0 = inv; \quad u = (u^t(s,\tau), u^l(l_0,\mu)) \qquad (7)$$

are optimal controls which change respectively the time and space solutions (4); (τ, μ) are space and time intervals of (7) which separate the parts of extremals under consideration, in particular, the condition

$$M_{\tilde{x},s}[d\tilde{x}_i(\tau)\tilde{x}_k(\tau)] = M_{\tilde{x},s}[d\tilde{x}_k(\tau)\tilde{x}_i(\tau)]$$
$$= \frac{1}{2}\frac{\partial r_{ik}}{\partial \tau}d\tau \qquad i,k = 1,2,\ldots,n. \qquad (8)$$

Random-field equations form a micro-level, whereas equations of non-equilibrium processes form a macro-level of the dynamic model of a macro-system closed at intervals (4). For (3) ΔS is a proper functional which defines (4), (5) as Euler-Ostrogradskii equations which are discrete-continuous at the intervals (8) (Lerner, 1972, 1980): The minimum of this proper functional provides invariance in the total, non-equilibrium entropy of a thermodynamic system, i. e. fulfilment of the conservation laws and for the processes connected by a chain of successive transformations it matches the kinetics and maximises their useful effects. Taking into account the connection of (2) with the transition probability on the trajectories (1), (3), the above-mentioned criterion characterises the maximum of the probability of transition into a given set, which reflects the statistical meaning of the conservation laws. Equality of the transition probabilities at the moments (8) gives rise to stochastically equivalent states $\tilde{x}_i(\tau) \sim \tilde{x}_k(\tau)$, in which the drift and diffusion matrices are related as

$$a^0(\tau) = \sigma\sigma^T(\tau)\left[\int_s^\tau \sigma\sigma^T dt + r(s)\right]^{-1}; \quad a^0(\tau)\tilde{x}(\tau) = a(\tau) \qquad (9)$$

and can be determined from the derivative of the conventional covariation

$$\sigma\sigma^T = dr/dt\,;\quad r = M_{\tilde{x},s}[\tilde{x}\,\tilde{x}^T] \tag{10}$$

where

$$\sigma = \sigma(t, \tilde{x})$$

characterise the operator of diffusion and electron conductivity in physico-chemical processes (Lerner, 1969). A direct measurement of the above functions makes it possible to identify the kinetic operator of the macro-model

$$L(\tau) = \frac{1}{2}\sigma\sigma^T(\tau)\,;\quad L(\tau)r^{-1}(\tau) = A^t(\tau)\,;\quad A^t = \| \lambda_i^t \|_{i-1}^{n} \tag{11}$$

the drift operator (9) and the Hamiltonian of the proper functional

$$H(\tau) = -\frac{\partial \Delta S'}{\partial t}(\tau) = \frac{1}{2}\,Sp\left[\frac{\partial r}{\partial t}(\tau)\,r^{-1}(\tau)\right] = Sp\,A^t(\tau). \tag{12}$$

Condition (9) connects the parameters of the regular and random components of solutions (3) and leads to equalising the relative phase velocities of the interacting processes

$$\frac{\partial x_i}{x_i \partial t}(\tau_j) = \lambda_i^t(\tau_j) = \lambda_k^t(\tau_j) = \frac{\partial x_k}{x_k \partial t}(\tau_j)\,;\quad i, k, j = 1, .., n \tag{13}$$

which corresponds to the Onsager theorem for controlled processes in the form

$$L_{ik}(u_k(\tau_j)) = \frac{1}{2}\frac{\partial r_{ik}}{\partial t}(\tau_j) = \frac{1}{2}\frac{\partial r_{ki}}{\partial t}(\tau_j) = L_{ki}(u_i(\tau_j))\,; \tag{14}$$

at the moments τ_j, i. e. 'opening' of the Hamiltonian system closed at intervals (8). The conditions (9) to (14), as well as drawing the process to the extremal, are realised by means of the optimal, reduced controls $v = (A^t)^{-1} u$ (synthesised simultaneously with identification (11), (12)) which take fixed values along portions of the extremals and change via a jump in value at their break points, according the law

$$v(\tau) = -2x(\tau)$$

where τ is to be found from (8) and (13). The set of samples (15) forms an optimal code determined by the dimensionality and properties of the operator (3). Fulfilment of the consistency conditions in the field of extremal ΔS requires application of 'needle' controls

$$\delta v = v(\tau) - v(\tau + o(\tau)). \tag{16}$$

The optimal controls introduced in (1):

$$u_i(\tau_i)dt = -2d\tilde{x}_i(\tau) = 2\lambda_i^t(\tau_i)\tilde{x}(\tau_i)$$

give rise to an equivalent drift, with $a_i = -\lambda_i^t \tilde{x}_i$; to condition (9), noting (10) and (11), there corresponds

$$\begin{aligned} &\tfrac{1}{2}(\sigma\sigma^T)_i(\tau_i)r_i^{-1}(\tau_i)\tilde{x}(\tau_i)dt = d\tilde{x}_i(\tau_i) \\ &\tfrac{1}{2}\, dr_i(\tau_i)r_i^{-1}(\tau_i) = -\lambda_i^t(\tau_i)dt \end{aligned} \tag{17}$$

and (15), (16) are originally involved in the functions of the initial values of the state vector (1), even for unknown model parameters, whose identification (11), (12), (17) takes place in the process of optimal motion under the action (15), (16). Relations (8) to (14) are generalised for spacially distributed systems (Lerner, 1980). Control (15) moving in space at the velocity, C, (5), enables the optimal motion and the identification to coincide in space and time. As the optimal motion stops, the controlled random field is transformed into space-time oscillations which determine dissipative wave processes.

Needle controls realise space cooperation of equally probable states (13) which are accompanied by a jump in the velocity $\dot{x}_i(\tau + o(\tau)) \to 0$ at the moment of cooperation. At the time of equalisation of eigenvalues (13), (14), the state vector and the operator (3) successively decrease dimensionality, and this corresponds to consolidating the states of a dynamic model. The ordering mechanism consists of successive equalisation of relative phase velocities of interacting processes, a decrease of the velocity up to zero at the interaction moment and a continuation of the process at equal relative velocities. The optimal interactions that give rise to ordering provide total entropy stabilisation in a transition process and a decrease of $\Delta S_0(\tau)$ at the end of the process. The increase of internal entropy over time macro-intervals is in agreement with its minimum external increase provided by controls in the process of formation of equi-probable states (τ) and their cooperation $(O(\tau))$. This guarantees stationary character and stability in the ordering increase mechanism, characteristic of biological systems. In the process of optimal motion a multi-level, space-time structure is formed which reflects a consistent succession of cooperating interacting macro-states. Consolidation gives rise to hierarchical models formed by controlled stochastic micro-level processes. The controlled transformations in question are mixing trans-formations, and that is why the set of macro-states

$$x(\tau) = M_{\tilde{x}(\tau)}[\tilde{x}(\tau)] \text{ at } M[\Delta S_0(\tau)] = \frac{h}{2} \quad (18)$$

being an analog of non-equilibrium stationary states, possesses ergodicity properties. At the moments (8), corresponding to the information entropy extremum, the condition of uncoupling of mutual and time correlations is fulfilled; dynamic and diffuse processes are equally probable. The control of the process at these moments provides, within the interval (τ), an optimal filtration of a random process without loss of information, and a direct measurement of the electron conductivity operator (10) replaces the statistical averaging technique in (8) to (14). The information model built on the basis of agreement between variational principles and statistics reflects such macrosystem properties as

- the existence of processes on micro- and macro-levels;
- consolidation of states;
- continuous kinetics and discrete variability of the structure and proper-properties;
- reversibility of motion within quantisation intervals and irreversibility of processes within larger intervals (when a closed system is opened);
- uniqueness in estimating information for the processes of observation;
- bio-physico-chemical kinetics;
- control.

Modelling such systems combines indentification of parameters with re-construction of a multi-level structure and optimal control which minimises non-equilibrium entropy. Identification and modelling of natural processes, for example, in organ regeneration (Gusak, Lerner, 1980) has shown their closeness to an optimal, minimising ΔS . An optimal process is characterised by the tendendcy to decrease in the Hamiltonian* and real eigenvalues of the operator; its imaginary components vanish at the instant of equalisation of real $\lambda_i^t(\tau)$ (13). The mechanism of such levelling, and the existence of optimal controls (15), (16), were revealed experimentally in reconstructive automatic control processes. The condition of a minimum proper functional, as a macroscopic criterion of optimalisation, makes it possible to realise an optimal prediction of controlled non-equilibrium processes in a macrosystem which determines the conditions for their self-organisation and ordering. The problem is solved on the basis of an optimal space-time model, with consolidation of states, synthesised from the non-equilibrium entropy invariance condition (18) at the extremals (3). Because of the association with variational principles, the model properties are described by a system of local invariants which make it possible to speak of two types of optimal processes: dissipative processes, with discrete times τ_i (when imaginary components of the kinetic operator vanish in succession); vibrational processes, with a frequency $\omega_i^o \sim \beta_i^t \sim (\tau_i^{\prime})^{-1}; \hat{H}(\tau_i^{\prime}) = j\beta_i^t; \beta_i^t = Im\, \lambda_i^t$ and discrete times $\tau_i^{\prime}$ (determined by the disappearance of the real part of (14). In the intervals $\tau_i + \tau_i^{\prime}$ there exist dynamic solutions, which, when optimal processes stop, transform to stochastic oscillations of increasing frequency and constant dispersion (Lerner, 1981). Optimal dynamic processes start with analogous oscillations of frequency which decrease as $\tau_i^{\prime}$ increases. Further on, the development of these processes is determined and does not depend on random events within discrete intervals. The combination of stochastic oscillations and dynamic processes makes up a cyclic process with the formation of the structure of a microsystem, which contains self-organisation and ordering** as one mechanism.

* The Hamiltonian (12) plays the role of a dissipation function; its value depends on the past history of motion and the dynamics of state enlargement, due to which $\hat{H}(t) = Sp\, A(t^* + \tau) \neq Sp A(t^* - \tau) = \hat{H}(-t)$ and the macromotions (3) are irreversible; the dissipative function within discrete intervals is reversible and is an analog of Landau damping.

**Self-organisation is connected with an increase of the dissipation function, whereas in ordering the dissipation function decreases.

The sequence of the use of optimal controls, including 'needle' controls, which corresponds to (18), is selected through the condition

$$\min \hat{H}(\tau_i) = \min |\lambda_i^t(\tau_i)| = \min \alpha_i^t(\tau_i) > 0;\ \hat{H}(\tau_i) = Sp\,\lambda_i^t(\tau_i);\ \frac{d\hat{H}}{dt}(\tau_i) = 0 \quad (19)$$

where
$$\lambda_i^t(\tau_i) = \frac{1}{2}\frac{\tau_{i+1} - \tau_i}{\tau_i \tau_{i+1}} = \lambda_i^t;\ \lambda_i^t \geq \frac{2\lambda_{i-1}^t \lambda_{i+1}^t}{\lambda_{i-1}^t + \lambda_{i+1}^t};\ \alpha_{i0}^t \geq \frac{2\alpha_{i-1,0}^t \alpha_{i+1,0}^t}{\alpha_{i-1,0}^t + \alpha_{i+1,0}^t} \quad (20)$$

$$\alpha_{i0}^t \tau = b = inv;\ \alpha_{i0}^t \tau_i = \alpha = inv;\ \alpha_i^t(\tau_i)\tau_i = \alpha' = inv;$$
$$\frac{Im\,\lambda_i^t}{Re\,\lambda_i^t} = \gamma = inv;\ \frac{\alpha_{i0}^t}{\alpha_{i+1,0}^t} = \gamma_i^\alpha = inv.$$

This condition determines an optimal relation among the initial eigenvalues and sets up the line of switching controls - triplets, with $(\frac{n}{2} + 1)$ discrete intervals. An optimal model has an ordered discrete set $\{\alpha_{i0}^t\}$ determined by the function of its initial dimension n (or dispersion) and the vibration coefficient γ. To the optimal process (19) there corresponds a range of interacting processes $\{x_i(t)\}$ under which the state vector coordinates satisfy the condition

$$x_{i-1}(t_0) = \dot{x}_i(t_0) = x_k(t_0);\ \frac{\partial^2 x_i}{\partial t^2}\left(\frac{\partial x_i}{\partial t}\right)^{-1}(t_0) = \frac{\partial x_k}{\partial t}(t_0) = \lambda_{k0}^t = \lambda_{i-1,0}^t;\ \lambda_{i0}^t = \frac{\partial x_i}{x_i \partial t}.$$

If, at the initial moment, the phase coordinates of an optimal model are associated by the relation $x_{i-1}(t_0) = \lambda_{i0}^t x_i(t_0)$, then from (15) to (21) it follows that at discrete moments there holds the relation

$$x_{i-1}(\tau_i) = -2\dot{x}_i(\tau_i) \quad (22)$$

with a relative error

$$\varepsilon_i = \frac{\delta x_i}{x_i}(\tau_i) = \exp \alpha' \exp(\gamma_i^\alpha \alpha') - 2(1 - \alpha' \alpha^{-1}), \quad (23)$$

which, at $\gamma = 0.5$, does not exceed $\varepsilon_i = 0.0959 - 0.1118$. That is, in a system where (18), (20) are maintained, there exists process selection (filtration); as a result, a discrete set of ranged $\{\alpha_{i0}^t\}$ and an ordered succession of their threshold combinations (triplets) are observed, the succession being determined by the system properties*. This testifies

* In such a process, along with an automatic formation of the controls (17), (22), there may also appear 'needle' controls (16) (at the instant of triplet, τ_k, formation) containing successive joining of pairs: $\alpha_i^t(\tau_k) \to \alpha_{i+1}^t(\tau_k) \to -\alpha_k^t(\tau_k + O(\tau_k))$ to the smaller $(\alpha_k^t(\tau_k + O(\tau_k)))$.

to the possibility of independent appearance and existence of the optimal processes under consideration. To illustrate this let us determine the geometric phase space structure, due to spatial state vector variations, which provides entropy invariance in a controlled, irreversible process. The geometry of the optimal model will be characterised by the space displacement of the normal (6) to the surface (18) defined by solutions (3) and (4) with singular points at the instant of switching optimal controls: this corresponds to ordering. The motion, of this normal is proved to be reflected by the vectors $\vec{l}_i = \sum_{j=1}^{3} l_j^i \vec{e}_j^i$, which start from the singular point $\{l_j^{oi}\}$ and form the cone surface

$$\sum \frac{\partial^2 \Delta S}{(\partial l_j^i)^2} (l_j^{oi} - l_j^i)^2 = 0 \quad i = 1,..,n; j = 1,2,3 \tag{24}$$

at discrete times the vectors $\vec{l}^i(\tau_i)$ are mutually orthogonal. At the instant of triple union of operators (13), when

$$\lambda_i^t(\tau_k) = \lambda_j^t(\tau_k) = \lambda_m(\tau_k + O(\tau_k)) \tag{25}$$

$$k = 1,3,5; \; j = 1,2,\ldots,n; \; n = \text{rank } A^t$$

is valid, the vectors, $\vec{l}^i(\tau_k), \vec{l}^j(\tau_k), \vec{l}^m(\tau_k)$, which coincide in direction with the corresponding eigenvectors, $A^t(\tau_k)$, have the same direction, and this changes the structure, (A^t, A^e), and decreases the system dimension.

The velocity invariance condition (5) satisfies the parametric equation for a spiral curve on the cone, in which, the projection of the radius-vector onto the cone base is

$$\rho_i = b \sin(\varphi_i \sin \beta_i^o) \qquad i = 1,..,n \tag{26}$$

where $b_i = \mu_i \varphi_i^{-1}(\mu_i)$ is the parameter of the curve, β_i^o is half of the angle near the cone vertex, $\varphi_i = \varphi_i(l)$ is angular displacement of the radius-vector ρ_i . Transforming from one cone (25) to another at discrete points corresponds to the extremum (26) at

$$\varphi_i(\mu_i) \sin \beta_i^o = \pm \frac{\pi}{2}; \; \rho_i(\mu_i) = b_i(\mu_i). \tag{27}$$

At discrete instants the parameters (φ_i, β_i^o) of the curve (26) on each of the cones of the optimal model are constant, and satisfy:

$$\varphi_i^o = k\pi \; ; \quad \sin\beta_i^o = (2k)^{-1} \; ; \quad k = 1, 2, \dots \quad \beta_i^o = \beta^o .$$

The relative angles of rotation of the local coordinate system (Ψ) at the times when it reaches the hyperbola (20) and, during motion along this hyperbola, (Ψ_o) are constant

$$\Psi = \frac{\pi}{4} - \beta^o ; \quad \beta^o = \arcsin[(2k)^{-1}] ;$$
$$\Psi_o = \operatorname{arctg}[\sin\beta^o \cos\beta^o (1 + \sin^2\beta^o)^{-1}] . \qquad (28)$$

The optimal geometry is characterised by the volume, V, and the cross-section S:

$$V = \bigcup V_k^i ; \quad S = \bigcup S_k^i ;$$
$$V_k^i = \pi b_i^3 (3 + q\beta_i^o)^{-1} \sin(\varphi_i \sin\beta_i^o) ; \qquad (29)$$
$$S_k^i = \pi \varrho_i^2 (1 + (\sin\beta_i^o)^{-1})$$

of the macro-structure and by the discrete, three-dimensional intervals (6). To condition (27) there corresponds the extremum of the relative strength of the macro-structure determined by the ratio of the weight in volume V_k^i to S_k^i. By analogy with crystal structures, and, using the concept of symmetry order (S_T), we determine S_T for a certain set of m_T triplets: $S_T = 2\pi/(\frac{\pi}{2} \mp \Psi)$; $S_T' = 2\pi/\Psi$ which leads to a possible odd symmetry order: $S_T(-\Psi) \cong 7$; $S_T(\Psi) \cong 5$; $S_T' \cong 9$.

The specificities of a fixed macro-structure are characterised by its local Hamiltonian which has the meaning of specific (differential) production

$$\check{h} = \frac{\partial \Delta S}{\partial V} = \frac{\hat{H}}{\dot{V}} = \frac{\operatorname{Sp} A^t}{\dot{V}} = \frac{1}{4} M (\operatorname{div}(a))(c(\tau)S)^{-1} ; \quad \dot{V} = cS \qquad (30)$$

where H, C = C(τ) are determined by (n, γ) which are chosen from the conditions

$$\min_{\gamma, n=n_*} \overset{\vee}{h} = \overset{\vee}{h}_o; \quad \overset{\vee}{h} = \frac{Sp\, \alpha^t_{io}}{cS}; \tag{31}$$

$$\min_{n, \gamma=\gamma^*} \left(1 - \frac{\hat{H}}{\hat{H}_o}\right) = \varepsilon^o_M; \quad \hat{H}_o = \frac{1}{2} Sp\, \dot{r}_i r_i^{-1}(\tau_i)$$

The package of applied programs for a SM-I digital mini-computer constructed on the basis of (3) to (31) makes it possible to synthesise the dynamics and geometry of an optimal model of given dimension. As n increases, the range of α^t_{10}, at which there appear macrosystems (admitting state enlargement), increases minimum discrete intervals are narrowed, maxima are widened. Optimal controls extract structures with equal $\alpha^t_i(\tau)(n, \gamma)$ and join them: (15) and (16), necessary for fulfilment of (21), can be formed by means of control, and memorising, stochastically equivalent states by their doubling, for example, at the moment of their division (subdivision). The conditions (25) determine $\frac{n}{2} - 1 = m_T$ points of new triplet-structure formation; geometrically they correspond to the union of three cones (24) at single points, where three physical quantities change simultaneously. This corresponds to mutual engagement of interacting processes, which is in line with geometric 'key-lock' engagement produced by controlling, and controlled, processes. At these instants negentropy, for the interacting operators, is produced

$$\Delta S^{\delta} = \int \frac{d\Delta S}{d\tau}\Big|_{\tau\to 0} d\tau; \quad \frac{d\Delta S}{d\tau} = -H + \frac{\partial H}{\partial \tau}\tau; \tag{32}$$

$$\frac{\partial H_i}{\partial \tau}\Big|_{\tau_k} = 2H_i^2(\tau_k)\tau_k; \quad \Delta S^{\delta} = -2n\int_{\tau_i} \alpha^t_i(\tau_i)\tau d\tau = -\alpha^2 n.$$

At the point of triple cooperation, under the action of needle controls, there appears a difference of forces (an analog of the Van der Waals forces)

$$\Delta f_{ij}(\mu_k) = \alpha^l_i(\mu_k)^4 (\pi\varepsilon_o)^{-1}(1 - \exp(-\alpha^l_k(\mu_o)\mu_k)); \quad \alpha^l_{ko} < 0 \tag{33}$$

(where ε_o characterises the properties of the medium) directed to cooperation of the structures (i, j, m) with equal eigenvalues (25). The physical specificity of optimal controls is a portion (quantum) character of extracted (injected) energy:

$$E_i(\tau) = \frac{1}{2}\lambda_i^t(\tau)\tilde{x}_i^2(\tau) = \frac{1}{2}\dot{\tilde{x}}_i^2(\tau - O(\tau)),$$

at the moment τ there holds the symmetry condition of the kinetic operator, (14), and the variational principles; the stochastic system can be characterised by macrovariables and thus become macroscopic. The needle controls (16) give rise to asymmetric dynamic, and geometric, structures which do not admit of an exact coincidence by way of spatial rotation (S_T). It is mainly the diffusion and chemical processes which change the number of state variables* that are responsible for the formation of these structures. Self-organisation and ordering can also be caused, along with optimal controls, by a change in internal properties due to peculiarities of the geometric structure of the phase space. At discrete instants, ρ_i (26), a corresponding metric tensor of the geometric space (connected with $r_i = r_i(\ell)$ and ρ_i by (26)) undergoes jump-like variation: $\Delta\ell_i(\tau) = \rho_i(\tau)\Delta\varphi_i(\tau) = \mu_i(\tau)$, and, when the spiral transforms to a new cone,

$$\rho_i(\tau) = 0; \ \Delta\ell_i = \mu_i + O(\mu_i). \tag{34}$$

This is a geometric analog of the application of needle controls and at the same time, the massive nature of the set, the transition probability to get into which is $P = \exp \Delta S = 1$, determines the condition of automatic fulfilment of the variational principle for the proper functional (Lerner, 1980). In such a space** the optimal process in question develops independently and automatically. The negentropy (32) compensates the entropy of the system (18) under the condition

* In the vicinity of locally unstable stationary states - at $L(x(\tau)) = L(v(\tau), \delta v(\tau))$; $\lambda_i^t(\tau) > 0$; $\lambda_i^t(\tau + O(\tau)) < 0$ the optimal system is structurally stable with the attracting set of states $x_i(\tau + O(\tau))$.

**From (8) to (10), (16) to (22) there follows the law of variation of the correlation function $r_i = r_i(\ell)$ corresponding to (34), at which optimal controls are formed independently, by means of one control applied to the process with the largest $\alpha_{10}^t = \alpha_{10}^{(m)}$.

$$-\alpha^2 n + \frac{n}{2} = 0; \quad \alpha = \ln 2; \quad \gamma = 1/2 . \tag{35}$$

If $\alpha < \ln 2; \gamma < \frac{1}{2}$, the optimal system generates negentropy outwards and can control other objects: at $\alpha > \ln 2; \gamma > 1/2$ this negentropy is insufficient for the production (generation) of proper controls and, therefore, it is necessary to apply control from without. The condition of self-control is (35). The maximum negentropy generated by one needle control is $\alpha^2 \cong$ 0.59 (at $\gamma \to 0$) and is produced by one pari of interacting processes, when at the moment of their cooperation there arises the control (16). An excessive negentropy of a system consisting of n pairs of processes is equal to α^2 . If $n = \alpha^2(\alpha^2 - \frac{1}{2})^{-1}$ at $\alpha^2 = 0.59$, the minimum n = 6.6; the next integral of n which is a multiple of two is n = 8. A system of 8 pairs of processes is able to generate outwards (α^2 - 0.5) n = 0.72 negentropy units, i. e. not less than one control at $0.576 < \alpha^2 < 0.59$; $\alpha^2 = 0.583$. The opposite is also true: a system generates, outwards, at least one control and is capable of self-control at a minimum dimension of n = 8. A system of 8 pairs of interacting processes can be produced by a self-controlled system consisting of $m = n[(\alpha^2 - \frac{1}{2})2]^{-1}$ = 46 pairs of processes for $\alpha^2 = 0.583$, n = 8.

In a system formed by cooperative functioning of six, self-controlled subsystems, with n = 8 for each, three pairs of processes remain which are not united (one process at the last discrete interval for each of n = 8). In this case each of the subsystems generates one control for a subsequent subsystem, and the control generated by the sixth subsystem can be used for cooperation by only one of the three pairs indicated. Hence, out of 48 pairs of processes two pairs will not be controlled, which corresponds to a self-controlled system of n = 46 pairs of processes. Thus, we are led to the scheme of self-controlled processes: $(\{ n = 8 \to 1 v)6 - 2 = (m = 46)$ with a consecutive generation of one control $(1 v)$. In such a system processes control one another and automatically generate one another if, independently, there appears one subsystem with n = 8. To this end, the initial eigenvalues of such a subsystem must be ranged, and the optimal control should necessarily be applied to one of the pairs of processes. The ranging of eigenvalues is possible through the time function (by virtue of

(20), (21), as well as in an oscillation process at $\beta_i^t(\tau_i') = b/\tau_i'$ which corresponds to a purely Hamiltonian system. In the latter there may appear an eigenvalue levelling mechanism. For example, for the Lagrangian $\tilde{L} = \frac{1}{2}(x^2 + \dot{x}^2)$ generated by the processes $x_1(t) = x(t)$; $x_2(t) = \dot{x}(t)$, the Euler equation leads to the model of simplest interaction $\dot{x}_1 = x_2$; $\dot{x}_2 = x_1$, with the equation of phase trajectories

$\frac{dx_1}{dx_2} = \frac{x_2}{x_1}$; $x_1^2 = x_2^2 + C_t$, which at $C_t = 0$ provides levelling of the eigen-

values $\lambda_i^t = \frac{\dot{x}_1}{x_1} = \frac{\dot{x}_2}{x_2}\,\frac{x_2^2}{x_1^2} = \lambda_t$; $(v_1 = M[x_1^2] = M[x_2^2] = v_2)$ at any moment t.

If an external action, corresponding to the optimal control or to phase space geometry variation, is applied to such a system, according to (34), there arises an optimal process which is later maintained automatically. In a system of n = 8, an optimal process provides self-control and an extensive generation of negentropy with the appearance of 46 pairs of processes and a later generation of 8 pairs of processes.

A system of n = 46 contains $m_T = 20$ triplets corresponding, for example, to the number of coded amino acids; a subsystem of n = 8 forms during an interaction $m_T' = 4$ triplets, each of which can code one compound, AGTC. A macromodel, appearing from the conditions of matching of variational principles and randomeness, possesses a finite lifetime and the function of memorising the states $x_i(\tau)(n,\gamma)$ which, having been doubled (for example, by copying), form the optimal controls (15). Successive separation of the states indicated (at a velocity $\dot{x}_i(\tau_i + o(\tau_i)) \to 0$, proceeds on a spiral geometric structure which forms an information trace (IT) of coded triplets. An IT of controls requires a doubled spiral (26) to (28) which, in line with (22), is formed automatically. Applied to initial states with the same (n, γ) (an individual medium), an IT of controls is capable of reproducing (restoring) initial dynamic processes. In the framework of the model under consideration there arises the concept of adaptation as a capacity for conserving properties determined by the proper functional for random deviation of the parameters.

$$\Delta S(H(n,\gamma), t) = \Delta S_0 = inv. \tag{36}$$

The change of properties at the expense of new values of the operator and

Hamiltonian (mutation) induces the action of control only at discrete times and is, therefore, taken up by the macrosystem at these times as $\Delta\alpha_i^t(\tau_i)=\frac{1}{2}\Delta v_i v_i^{-1}(\tau_i)$. Fulfilment of (36) leads to preservations of local invariance and to relation of 'mutations' with the main parameters (n, γ)

$$\Delta\alpha_{i0}^t(n,\gamma)=\Delta\alpha_i^t(\tau_i)\frac{\alpha}{\alpha'}\,;\ \Delta\alpha_{i0}^t(n,\gamma)\geq\Delta\alpha_i^t(\tau_i)$$
$$\eta=\frac{\Delta H(\tau_i)}{H(\tau_i)}=\frac{\Delta H(t_o)}{H(t_o)}\,;\ H(\tau_i)=H(t_o)\alpha'/\alpha \qquad (37)$$

where the sensitivity of the system to the change of the properties of (η) can be expressed in the function of each α_{i0}^t , using, for example,

$$\sum_{i=1}^{n}\alpha_{i0}^t=\alpha_{10}^t(1+\gamma_{12}^\alpha+\gamma_{13}^\alpha)^{\frac{n}{2}-1}(1+\gamma_{12}^\alpha)\,;\ \gamma_{12}^\alpha=\frac{\alpha_{20}^t}{\alpha_{10}^t}\,;\ \gamma_{13}^\alpha=\frac{\alpha_{30}^t}{\alpha_{10}^t} \qquad (38)$$

We have arrived at the relations for the relative local increments

$$\frac{d\alpha_{10}^t}{\alpha_{10}^t}=\frac{d\alpha_{20}^t}{\alpha_{20}^t}-\frac{d\gamma_{12}^\alpha}{(\gamma_{12}^\alpha)^2}=\frac{d\alpha_{30}^t}{\alpha_{30}^t}-\frac{d\gamma_{13}^\alpha}{(\gamma_{13}^\alpha)^2}=\frac{d\alpha_{40}^t}{\alpha_{40}^t}-\frac{\gamma_{13}^\alpha d\gamma_{12}^\alpha+\gamma_{12}^\alpha d\gamma_{13}^\alpha}{(\gamma_{12}^\alpha\gamma_{13}^\alpha)^2}$$

from which, for all the admissible $\gamma\to(0\div 2,4)$, there follows

$$\frac{d\alpha_{10}^t}{\alpha_{10}^t}>\frac{d\alpha_{20}^t}{\alpha_{20}^t}>\ldots>\frac{d\alpha_{k0}^t}{\alpha_{k0}^t}\,;\ \frac{d\gamma_{12}^\alpha}{d\gamma}<0$$
$$\frac{d\gamma_{13}^\alpha}{d\gamma}<0\,;\ \gamma_{13}^\alpha d\gamma_{12}^\alpha+\gamma_{12}^\alpha d\gamma_{13}^\alpha<0 \qquad (39)$$

i. e. η is most strongly affected by the process with the maximum $|\alpha_{i0}^t|$ - the most rapidly-acting ($x_1(t)$). The relative action of a given α_i^t upon the properties of the system characterises the adaptation potential

$$\varphi_i^a=\frac{dH_i}{H_iH}\sim\frac{d\alpha_{oi}^t}{\alpha_{oi}^t H(t_o)}\,;\ \max\varphi_i^a=\max\frac{d\alpha_{i0}^t}{\alpha_{i0}^t\,\mathrm{Sp}\,\alpha_{i0}^t} \qquad (40)$$

whose maximum provides the optimal $x_1(t)$. The largest deviation of the parameters in a complex system, whereas in slower processes the parameters $|\alpha_i^t(\tau_i)|$ should be more stable. The maximum, $\eta=\eta^o$ characterises the limiting parameter deviations and the potential of the system as a whole, to maintain (36), i. e. the properties of the Hamiltonian. According to (37), mutations change (h, γ), whose optimal values may result from the action of random $\Delta\alpha_i^t(\tau_i)$ at minimum $(\Delta S, \tau_i)$, i. e. maximum $\alpha_i^t(\tau_i)^* = \Delta\alpha_i^t(\tau_i) + \alpha_i^t(\tau_i)$ corresponding to the maximum adaptation potential (40). The adaptation mechanism contains selection and memorising of the range in an optimal process) of states with properties (characterised by decreasing $H(\tau_i)$) improving because of the existence of discrete, consecutively expanding, filtrations intervals (20). This process is determined by cooperation between equiprobable states which form a space-time optimal filter, i. e. by the efficiency of the attraction forces (self-formation) (33), which increase with increase of $|\alpha_i^t(\tau_i)|$. Therefore, the most rapid-acting processes, concentrated at the beginning of optimal motion, are responsible for adaptation mechanisms and adaptation potential.

The Problems of Evolution

Non-Equilibrium Thermodynamics of Hypercycles

K. Ishida

Introduction

After the prebiotic chemical evolution, a self-organisation of primitive macromolecular species to replicicating 'individuals' may have occurred in a primordial soup. What is called 'hypercycle' can provide a concrete form for a mechanism of the self-organisation (Eigen, 1971; Eigen, Schuster, 1977, 1978).

The most simplified hypercycle, i. e., the elementary hypercycle is depicted in Fig. 49. Each of the macromolecular species organising the hypercycle carries the necessary information for its own synthesis. It is, therefore, named the information carrier (Eigen, 1971). The information, however, acquires its meaning only via the function which is encoded by it. The hypercyclic organisation is thus regarded as an integration of several information carriers into a functional order.

From the chemical point of view, the hypercycle is a non-linear reaction network in a closed loop. The non-linearity is due to the two-fold catalysis of information carriers I_i as follows: the replication of I_i is an autocatalytic reaction, and it is also catalysed by the precursor I_{i-1}, where I_0 is substituted for I_n.

The system considered here, other than information carriers I_i, contains energy-rich monomeric species required to synthesise I_i and energy-deficient species produced in the synthesis (Hofbauer et al., 1980). The former is poured into the system from its surroundings, whereas the latter is steadily removed. The system is metabolic and self-replicative, and so it manifests Darwinian behaviour at the molecular level.

Thermodynamics and Regulation of Biological and Processes

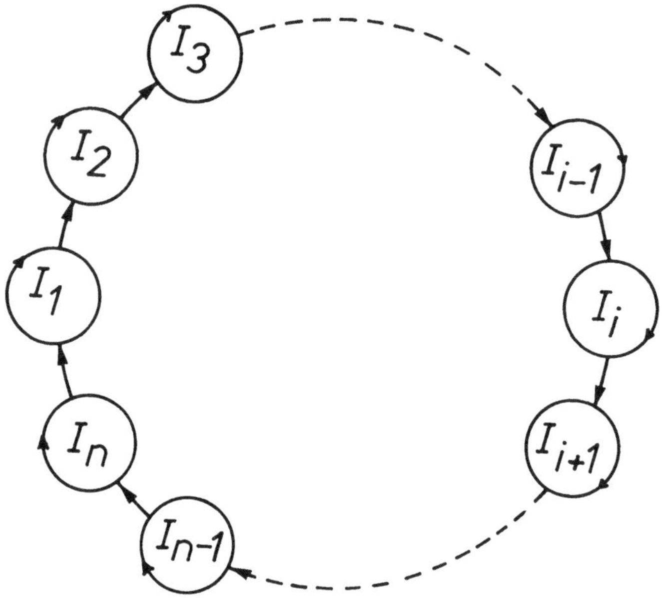

Fig. 49 N-membered hypercycle. The arrows on closed cycles (I_i) represent self-instructed replication of I_i and those pointing from (I_j) to (I_i) the formation of I_i catalysed by I_j , where $j = i - 1 + n\delta_{ij}$.

The system is assumed to operate under the following conditions:

1. It is an open system far away from thermodynamic equilibrium.
2. The transport of both monomeric species is regulated in such a way that their concentrations can be held constant. Accordingly, only the concentrations of I_i are the variables.
3. It is maintained at constant temperature.
4. It is in a mechanical equilibrium, and all the chemical species are univormly distributed over the whole space of constant volume V.
5. It is subject to the constraint referred to as constant overall organisation (Eigen, 1971),

$$\sum_{i=1}^{n} x_i = c \qquad (1)$$

where x denotes the concentration of I_i and c is a positive constant. This constraint brings about mutual competition among the information carriers.

A non-equilibrium thermodynamic analysis of such a system is as follows. The stability of the steady state of the hypercycle is studied on the basis of excess entropy production defined there (Ishida, 1981). The hypercycle accumulates information through its functional order. For an understanding of the first phase of biological self-organisation, this order seems to be much more important than spatial order (Eigen, 1971). An attempt to find the relation of the functional order to the excess entropy is made by use of the notion of information gain (Kullback, 1959).

We consider the system to consist of two distinct, short hypercycles, which are assumed to be in internal equilibration (Eigen, Schuster, 1978; Schuster et al., 1979, 1980). It is unstable. So they are in competition with each other and either of them is selected as a quasi-species (Eigen, 1976).

Finally we shall give a stochastic approach to selection through competition (Ishida, 1982). The fluctuation in the total number of information carriers organising each short hypercycle is anomalously enhanced near the unstable steady-state. Such an anomalous enhancement of fluctuation is the origin of the instability.

Rate equation and entropy production

The time evolution of an n-membered hypercycle is described by a set of non-linear differential equations (Eigen, Schuster, 1978)

$$\frac{dx_i}{dt} = k_i x_i x_j - \varphi/c \cdot x_i , \qquad (2)$$

where

$$\varphi = \sum_{i=1}^{n} k_i x_i x_j \qquad (3)$$

The term $k_i x_i x_j$ represents the self-replication of I_i catalysed by I_j, where j = i - 1 + n . At the same time it implies metabolism. That is, the rate constant k_i is a function of the concentrations of energy-rich monomeric species required to synthesise I_i. Under the condition (2) prescribed in the preceding section, however, these concentrations are involved in k_i as constants. As compared with metabolism and self-replication the prerequisite for evolution through natural selection, mutation is a secondary condition. We therefore are primarily interested in the metabolism and the self-replication free of error. The expression (3) for the dilution flux, φ (Hofbauer et al., 1980), comes from constraint (1) or, $\sum_{i=1}^{n} dx_i/dt = 0$.

We now introduce the dimensionless quantities y_i and a_i defined respectively by $y_i = x_i/c$ and by $a_i = k_i/k$, where k denotes the unit rate-constant with the same dimension as k_i. We can then write (2) as

$$\dot{y}_i = a_i y_i y_j - \psi y_i , \tag{4}$$

where

$$\psi = \sum_{i=1}^{n} a_i y_i y_j . \tag{5}$$

In equation (4) we have set $\dot{y}_i = dy_i/d\tau$, where the dimensionless time τ is given by $\tau = ckt$. And we have $\sum y_i = 1$ instead of (1). The steady-state solutions of (4) are given by

$$y_i = a_r / \sum_{i=1}^{n} a_i \tag{6}$$

where $r = i + 1 - n\delta_{in}$.

For the hypercycle organised in the system under the conditions (1) to (5), the deviation of entropy S from its equilibrium value S_e can be written in the form

$$S - S_e = -RV\sum\{x_i \ln \frac{x_i}{x_{ie}} - x_i + x_{ie}\} \leq 0 , \tag{7}$$

where R is the gas constant (Ishida, 1981). Of course, in the derivation of this equation, it has been assumed that the system is in local equilibrium.

For the system far from equilibrium, however, the non-equilibrium steady-state rather than the equilibrium state should be assumed as a reference state. The deviation of entropy S from its steady-state value $\bar{S}$, $\Delta S = (S - \bar{S})/RVc$, is expressed in terms of y_i as follows (Ishida, 1981);

$$\Delta S = -\sum_{i=1}^{n}(y_i - \bar{y}_i)\ln(y_i/y_{ie}) - \sum_{i=1}^{n}\{y_i \ln(y_i/\bar{y}_i) - y_i + \bar{y}_i\} . \tag{8}$$

The second sum on the right-hand side of this equation is of the same form as (7). Let A_i be the chemical affinity of I_i. It the represents the entropy change $\Delta S(\Delta A)$ due to the differences of the chemical affinities A_i from their steady-state values $\bar{A}_i$ (Ishida, 1975). That is to say

$$\Delta S'(\Delta A) = -\sum \{ y_i \ln(y_i/\bar{y}_i) - y_i + \bar{y}_i \} < 0 \qquad (9)$$

which is negative definite except in the steady state. We shall utilise (9) as a convenient test function in Lyapunov stability theory (Nicolis, Prigogine, 1977).

We now define the force X_i driving the species I_i to a given equilibrium state by the equation

$$X_i = -\frac{\partial}{\partial y_i}\frac{S_e - S'}{RVc} = -\ln(y_i/y_{i,e}) \qquad (11)$$

This is the thermodynamic force in the Onsager sense (Prigogine, 1967). The flow J_i corresponding to this force is the reaction rate $\dot{y}_i$; $J_i = \dot{y}_i$. As the case stands now, however, we are interested in the deviations of J_i and X_i from their steady-state values, $\bar{J}_i$ and $\bar{X}_i$. Since $\bar{J}_i = 0$, we use $\Delta J_i = J_i - \bar{J}_i$ instead of J_i. Differentiating (8) with respect to τ, we obtain for the entropy production, $\Delta P = \Delta \dot{S}$ around the steady state where

$$\Delta P = \sum \Delta J_i \bar{X}_i + \sum \Delta J_i \Delta X_i \qquad (12)$$

and where $\Delta X = -\ln(X_i/\bar{X}_i)$. The second sum on the right-hand side of this equation is called the excess entropy production which assumes the steady state as a reference state (Ishida, 1981). Suppose ΔP changes in such a manner that only the forces X_i vary with τ in keeping the flows J_i constant. It then follows from (12) that

$$\frac{d_x(\Delta P)}{d\tau} = \sum_{i=1}^{n} \Delta J_i \frac{d(\Delta X_i)}{d\tau} = -\sum \dot{y}_i^2/y_i < 0 \qquad (13)$$

except at the steady state. The evolution criterion (13) is independent of $\sum \Delta J \bar{X}_i$ in (12). This fact suggests that the stability of the steady state can be studied via the excess entropy (9) and the corresponding excess entropy production is

$$\Delta P(\Delta A) = \sum \Delta \mathfrak{Z}_i \Delta X_i = -\sum \dot{y}_i \ln(y_i/\bar{y}_i), \qquad (14)$$

where $\Delta P(\Delta A) = d\Delta S(\Delta A)/d\tau$.

The numerical computation of $\Delta P(\Delta A)$ has been made for different classes of hypercycles (Ishida, 1981). For simplicity's sake, it has been that $a_i = 1$ for all i. This assumption has no influence on the stability behaviour of the dynamic system described by (4) (Schuster et al., 1980). The results are summarised below.

For two- and three-membered hypercycles, the $\Delta P(\Delta A)$'s are always positive definite for $0 \leq \tau < \infty$. According to Lyapunov's theorems, therefore, their steady states are asymptotically stable. For a four-membered hypercycle, however, the time behaviour of $\Delta P(\Delta A)$ is somewhat complicated: Initially $\Delta P(\Delta A)$ is positive, but after a short time it starts to oscillate about the τ-axis and is damped. The region of $\Delta P(\Delta A)$ 0, in area, exceeds that of $\Delta P(\Delta A) < 0$. From this it is concluded that, globally, $\Delta P(\Delta A)$ is positive definite;

$$\int_0^\infty \Delta P(\Delta A) d\tau = \sum \{ y_i(0) \ln(y_i(0)/\bar{y}_i) - y_i(0) + \bar{y}_i \} > 0, \qquad (15)$$

where $y_i(0)$ is the initial value of $y_i(\tau)$. It is then said that the steady state is globally stable.

Hypercycles with more than four members are similar in their dynamic behaviour; that is, the steady state is unstable, and in the concentration space (y_1, y_2, ... , y_n; $n \geq 5$) a representative point starting from the immediate neighbourhood of it develops into the generation of a stable limit cycle. Let us now denote by τ_a the time at which the repesentative point enters the limit cycle with period T. We then obtain

$$\int_0^\infty \Delta P(\Delta A) d\tau = \int_0^{\tau_a} \Delta P(\Delta A) d\tau + \sum \int_{\tau_a+(p-1)T}^{\tau_a+pT} \Delta P(\Delta A) d\tau. \qquad (16)$$

The sum of integrals on the right-hand side of this equation vanishes because of $y_i(\tau_a + (p-1)T) = y_i(\tau_a + pT)$. Let the initial value $y_i(0)$ be equal to $\bar{y}_i + \varepsilon_i$, where ε_i is sufficiently small on the assumption that $y_i(0)$ is very close to $\bar{y}_i$. It is then found that (16) reduces to

$$\int_0^\infty \Delta P(\Delta A)d\tau \rightarrow -\sum\{y_i(\tau_a)\ln(y_i(\tau_a)/\bar{y}_i) - y_i(\tau_a) + \bar{y}_i\} < 0 \qquad (17)$$

as $\varepsilon_i \rightarrow 0$. The steady state is said to be globally unstable. (9) can be rewritten in the form

$$-\Delta S(\Delta A) = \sum_{i=1}^{n} y_i \ln(y_i/\bar{y}_i) > 0 \qquad (18)$$

By definition y_i is the relative frequency for I_i to be found at time τ. The quantity $\sum y_i \ln(y_i/\bar{y}_i)$, known as Kullback's information (Kullback, 1959), can be interpreted as the gain in information which is necessary for the state change from the steady-state distribution $(\bar{y}_1, \ldots, \bar{y}_n)$ to any other one $(y_1, \ldots, y_n)$ (Rényi, 1970).

For hypercycles with $n \leq 4$, and by reference of (15), the gain of information regresses with time; in other words information is lost. On the contrary, for hypercycles with $n \geq 5$, (17) tells us that the gain of information increases with time. Let us remark in passing that it was shown by Schlögl (1971 a, b) that gain of information plays a fundamental role in the non-equilibrium thermodynamics of open systems.

As we have just seen, information increases to the extent that dynamic complexity, L (Schuster, Sigmund, 1980), increases. This property is a useful measure for the degree of organisation and may be classified according to the nature and number of attractors on the simplex $(y_1, \ldots, y_n)$. For example, $L = 0$ for a sink and $L = 1$ for a limit cycle.

Competition of short hypercycles

An hypercycle with $n \leq 4$ is called a short hypercycle (Schuster et al., 1979). For the sake of simplicity we consider the competition of a three-membered hypercycle H_1 against a four-membered one H_2. The reason for doing this is that we do not know how to apply the notion of internal equilibration (Eigen, Schuster, 1978; Schuster et al., 1979, 1980) to long hypercycles competing against each other. The time evolution of each short hypercycle H_ρ (ρ = 1, 2) is described by differential equations of the same type as (2), namely

$$\frac{dx_i^{(\rho)}}{dt} = k_i^{(\rho)} x_i^{(\rho)} x_j^{(\rho)} - \varphi/c\, x_i^{(\rho)} \qquad i = 1, 2, \ldots, n_\rho \tag{19}$$

where $j = i - 1 + n_\rho \delta_{i n_\rho}$ and $\sum_{\rho=1}^{2} \sum_{i=1}^{n} x_i^{(\rho)} = c$. - We assume here that H_1 and H_2, individually, are in internal equilibration: i. e.

$$k_1^{(1)} x_3^{(1)} = k_2^{(1)} x_1^{(1)} = k_3^{(1)} x_2^{(1)} \tag{20 a}$$

and

$$k_1^{(2)} x_4^{(2)} = k_2^{(2)} x_1^{(2)} = k_3^{(2)} x_2^{(2)} = k_4^{(2)} x_3^{(2)} . \tag{20 b}$$

Using the same dimensionless quantities that we have defined in the derivation of (4) and assuming the relation (20), we can derive from (19)

$$\dot{Y}_\rho = Y_\rho (Q_\rho Y_\rho - \Phi), \tag{21}$$

where $Y_\rho = \sum_{i=1}^{n} y_i^{(P)}$, $\sum_{\rho=1}^{2} Y_\rho = 1$, $1/Q_\rho = \sum 1/a_i^{(P)}$, $\Phi = \sum_{\Upsilon=1}^{2} Q_\rho Y_\rho^2$.
Suppose the steady-state solution of (21) exists:

$$\bar{Y}_\rho = Q/Q \tag{22}$$

where $1/Q = \sum_{P=1}^{2} 1/Q_\rho$. Excess entropy can then be written as

$$\Delta S(\Delta A) = -\sum_{\rho=1}^{2} \{ Y_\rho \ln(Y_\rho/\bar{Y}_\rho) - Y_\rho + \bar{y}_\rho \} < 0 \tag{23}$$

except in the steady state. Corresponding to this, excess entropy production takes the form

$$\Delta P(\Delta A) = -Q Y_1 Y_2 (Y_1/\bar{Y}_1 - Y_2/\bar{Y}_2) \ln\{(Y_1/\bar{Y}_1)/(Y_2/\bar{Y}_2)\} < 0, \tag{24}$$

which is of the same sign as $\Delta S(\Delta A)$. The steady state is, therefore, unstable. This makes us expect the existence of another attractor. The competition has its origin in the instability. The above results are true also for any number of short hypercycles (Ishida, 1981).

Which of two short hypercycles wins the competition depends upon the initial conditions imposed on them. If, for instance, $Y_1(0) > Q/Q_1$ and

$Y_2(0) < Q/Q_2$, H_1 survives and H_2 dies out, that is, $Y_1(\tau) \to 1$ and $Y_2(\tau) \to 0$ in the limit $\tau \to \infty$. If, however, $Y_1(0) < Q/Q_1$ and $Y_2(0) > Q/Q_2$, the situation is reversed.

We then consider the non-equilibrium thermodynamics of the selection through competition. To start with, we rewrite (21) as

$$\dot{Y}_\varrho = Y_\varrho (Q_\varrho Y_\varrho - \langle QY \rangle), \tag{25}$$

where the mean $\langle QY \rangle$ has been substituted for ϕ:

$$\phi = \sum_{\varrho=1}^{2} (Q_\varrho Y_\varrho) Y_\varrho / \sum Y_\varrho = \langle QY \rangle \tag{26}$$

Equation (25) is of the same form as Eigen's equation for quasi-species (Eigen, 1976). Under the initial conditions $Y_1(0) > \bar{Y}_1$ and $Y_2(0) < \bar{Y}_2$ the inequality $Q_1 Y_1(\tau) > Q_2 Y_2(\tau)$ holds for $0 \leq \tau < \infty$. When $Q_1 Y_1(\tau) > \langle QY \rangle > Q_2 Y_2(\tau)$, H_1 is selected; thus $\Delta S(\Delta A)$ and $\Delta P(\Delta A)$ are given by

$$\Delta S(\Delta A) = -(Y_1 \ln Y_1 - Y_1 + 1) < 0 \tag{27}$$

and

$$\Delta P(\Delta A) = -Y_1 (Q_1 Y_1 - \langle QY \rangle) \ln Y_1 > 0 \tag{28}$$

respectively. These inequalities are the thermodynamic conditions for the state ($Y_1 = 1$, $Y_2 = 0$) to be asymptotically stable. At the same time it follows that $\langle QY \rangle$ tends to the maximum Q_1 as $\tau \to \infty$;

$$\lim_{\tau \to \infty} \langle QY \rangle = Q_1 .$$

Self-amplification of fluctuations

The instability, in general, appears in such a manner that fluctuations are self-amplificative near the instability point and their order of magnitude becomes macroscopic. The problem of the instability occurring in the competition of H_1 and H_2, should therefore be reexamined in the light of stochastic theory (Ishida, 1982).

For the stochastic variables we choose the ratios of the numbers of information carriers organising H_ϱ ($\varrho = 1, 2$) to the total number of information carriers: Let us denote by Z_ϱ the realised values which the stochastic variables assume at time τ and by $\Pi(Z_1, Z_2; \tau)$ the probability distribution. The time evolution of $\Pi(Z_1, Z_2; \tau)$ is governed by the non-linear Fokker-Planck equation,

$$\frac{\partial \Pi}{\partial \tau} = \sum_{\varrho=1}^{2} \left[-\frac{\partial}{\partial Z_\varrho} \{ (Q_\varrho Z_\varrho^2 - Z_\varrho \Phi) \Pi \} + \frac{1}{2N} \frac{\partial^2}{\partial Z_\varrho^2} \{ (Q_\varrho Z_\varrho^2 + Z_\varrho \Phi) \Pi \} \right]. \quad (30)$$

In this equation N is the total number of information carriers and the 'averaged' dilution flux is determined later on (refer to (33)).

We separate Z_ϱ into the most probable (or deterministic) paths $Y_\varrho(\tau)$ and the fluctuating parts η_ϱ, that is to say

$$Z_\varrho = Y_\varrho(\tau) + \eta_\varrho . \quad (31)$$

We assume here that the initial values $Y_\varrho(0)$ are given by $Y_\varrho(0) = \bar{Y}_\varrho + \delta_\varrho$, where the δ_ϱ 's are the deviations from the steady-state values $\bar{Y}_\varrho$ and $|\delta_\varrho| \ll 1$. When the smallness parameter $N^{-1/2}$ is small enough compared with the deviations $|N \delta_\varrho|$, the fluctuations which become anomalous owing to the self-amplification near the instability point are said to be in the extensive region (Suzuki, 1976; 1977, 1981). The fluctuating parts are Gaussian in this region. Then, following van Kampen (1976), we can write (31) in the form

$$Z_\varrho = Y_\varrho(\tau) + N^{-1/2} \xi_\varrho \quad (32)$$

Applying the system-size expansion to (30), we obtain the evolution equations as follows: For the most probable paths,

$$\dot{Y}_\varrho = Q_\varrho Y_\varrho^2 - Y_\varrho \Phi \tag{33}$$

in complete agreement with (21), and for the fluctuating parts.

$$\frac{d\sigma_{11}}{d\tau} = 2(2Q_1Y_1 - \Phi - Y_1\frac{\partial\Phi}{\partial Y_1})\sigma_{11} - 2Y_1\frac{\partial\Phi}{\partial Y_2}\sigma_{12} + (Q_1Y_1^2 + Y_1\Phi), \tag{34 a}$$

$$\begin{aligned}\frac{d\sigma_{12}}{d\tau} = &- Y_2\frac{\partial\Phi}{\partial Y_1}\sigma_{11} + \{(2Q_1Y_1 - \Phi - Y_1\frac{\partial\Phi}{\partial Y_1}) + \\ &+ (2Q_2Y_2 - \Phi - Y_2\frac{\partial\Phi}{\partial Y_2})\}\sigma_{12} - Y_1\frac{\partial\Phi}{\partial Y_2}\sigma_{22}\end{aligned} \tag{34 b}$$

and

$$\frac{d\sigma_{22}}{d\tau} = - 2Y_2\frac{\partial\Phi}{\partial Y_1}\sigma_{12} + 2(2Q_2Y_2 - \Phi - Y_2\frac{\partial\Phi}{\partial Y})\sigma_{22} + Q_2Y_2^2 + Y_2\Phi \tag{34 c}$$

where the variances $\sigma_{\varrho\varrho}$ are defined by $\sigma_{\varrho\varrho} = \langle(\xi_\varrho - \langle\xi_\varrho\rangle)^2\rangle$ and the covariance, σ_{12}, is $\sigma_{12} = \langle(\xi_1 - \langle\xi_1\rangle)(\xi_2 - \langle\xi_2\rangle)\rangle$. Supposing $Y_1(0) > Q/Q_1$ and $Y_2(0) < Q/Q_2$ as before, let us apply the anomalous fluctuation theorem (Suzuki, 1976, 1977, 1981) to (34). This theorem is available for the asymptotic estimation of the anomalous fluctuations observed in the extensive region ($N^{-1/2} \ll N\delta$). For example, it can be shown that $\sigma_{11}(\tau)$ takes the asymptotic form (Ishida, 1982):

$$\sigma_{11}(\tau) \simeq \{\sigma_{11}(0) + \sigma_{11}'\}\left\{\frac{Q_1Y_1^2(\tau) - Y_1(\tau)\Phi(\tau)}{Q\delta}\right\}^2 \tag{35}$$

for large time τ, which is of the order of magnitude τ_1 given by

$$\tau_1 = \frac{1}{Q}\ln\frac{1}{\delta}\,. \tag{36}$$

In (35), the constant σ_{11}' and the deviation δ are given by

$$\sigma_{11}' = \frac{Q}{Q_1}\left\{\frac{Q}{Q_2} - (1 - \frac{Q}{Q_2})\sigma_{11}(0) - 2\frac{Q}{Q_2}\sigma_{12}(0) + (1 - \frac{Q}{Q_2})\sigma_{22}(0)\right\}$$

and

$\delta_1 = - \delta_2 = \delta > 0$ respectively.

As expected, the relation $\sigma_{11}(\tau) \simeq \sigma_{22}(\tau) \simeq -\sigma_{12}(\tau)$ holds to a fairly good approximation, for large time τ of the order of magnitude τ_1.

The coefficient of variation in terms of the total number of information carriers organising H_1, $v_1(\tau)$, is given by

$$v_1(\tau) = \sqrt{\sigma_{11}(\tau)} / (\sqrt{N}\, Y_1(\tau)) \tag{37}$$

From this it follows that $\sqrt{\sigma_{11}(\tau)/N}$ is the fluctuation corresponding to $Y_1(\tau)$. An example of $v_1(\tau)$, together with the curves for $Y_1(\tau)$ and $Y_2(\tau)$, is shown in Fig. 50. Such behaviour of $v_1(\tau)$ is characteristic of the self-amplification of the fluctuation that will lead to instability.

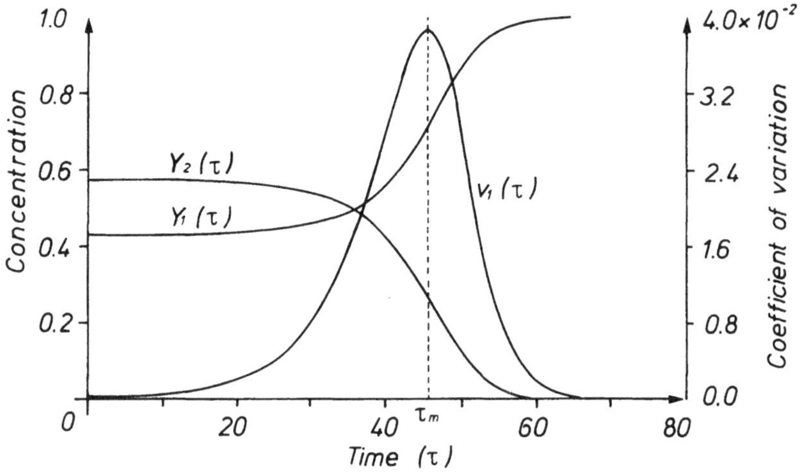

Fig. 50 Evolution of most probable paths $Y_1(\tau)$ and $Y_2(\tau)$ and time dependence of the coefficient of variation $v_1(\tau)$. On the assumption that $a_i^{\varrho} = 1$ for all ϱ and i, the initial steady state $(\bar{Y}_1, \bar{Y}_2)$ is given by (3/7, 4/7). The initial conditions are as follows: $Y_1(0) = 0.429$ and $Y_2(0) = 0.571$; $\sigma_{11}(0) = 2 \times 10^{-3}$, $\sigma_{12}(0) = -5 \times 10^{-4}$ and $\sigma_{22}(0) = 1 \times 10^{-3}$. Symbol τ_m denotes the time at which $v_1(\tau)$ takes the maximum value; then $Y_1(\tau_m) = 0.714$ and $v_1(\tau_m) = 3.86 \times 10^{-2}$ for $\tau_m = 45.7$.

We conclude this section with a few remarks. The evolution of self-replicating macromolecular species is due to the interrelatedness of the following three factors: gain of information (negative excess entropy,

$-\Delta S(\Delta A)$), instability and dynamic complexity. These are in close association with one another according to the circular scheme,

gain of information (negative
excess entropy, $-\Delta S(\Delta A)$)
↙↗ ↖↘
instability ⟷ dynamic complexity

We have pointed out, in particular, that the self-amplification of fluctuations in the competition of two short hypercycles brings about the instability to cause selection. We have yet to study how to relate the fluctuations to mutation. The self-amplification of fluctuations, however, seems to precede mutation.

The initial conditions for the selection through competition have been dealt with as though to be determined. Of course, the question arises as to which of two restrictions ($Y_1(0) > Q/Q_1$, $Y_2(0) < Q/Q_2$ or their inverse $Y_1(0) < Q/Q_1$, $Y_2(0) > Q/Q_2$) is preferred for selection. We suppose the preference is subject to chance, but the question remains untouched.

New Information in Evolution

N. M. Chernavskaya, D. S. Chernavsky

Biological information is usually understood as information written in DNA as a nucleotide sequence and realised, then, in protein biosynthesis.

Biological evolution is accompanied by the appearance, accumulation, and selection of valuable information. The character of these variations is different at different stages of evolution; two types of processes should be distinguished.

The first one takes place when a developed organism adapts itself most efficiently to permanent (within a certain time interval) conditions. The organism is assumed to possess, already, all necessary functions which now only have to be developed. This process has repeatedly been discussed; it can proceed due to point mutations and recombinations which may either improve or worsen survival. Owing to selection, positive, genetic changes are fixed and negative changes are eliminated. Such a process is evidently the basis of evolution. It may proceed rather rapidly and now stimulates no discussion. Notice, that the total amount of biological information in microevolution does not increase as a rule: it may even decrease (Chernavsky, Korogodin, 1982), the information value increasing.

The second, more interesting process, of the appearance of new information, takes place when there appears an essentially new and valuable function. For this reason the advantages of the organism strongly increase - this is like a 'big jump' in evolution. In effect, a 'big jump' is the very process of the appearance of living organisms, the translation apparatus, and a unique code.

Thermodynamics and Regulation of Biological and Processes

The problem of the origin of new information are important also at later stages of evolution, when the translation apparatus is completely formed. We mean, here, the appearance of enzyme-proteins with new functions, which precedes species divergence and the formation of new enzyme systems at such stages of evolution as, for example, the appearance of photosynthesising organisms, and later, organisms with respiration. The photosynthetic apparatus gave a colossal evolutional advantage by promoting the use of a new energy source (solar light) and new substrate materials (carbonic acid and water).

To this end, it was necessary to construct numerous enzymes, a chain of proteins with new functions. In a thorough analysis one can notice that not all the enzymes are actually new; most of them are forgotten, 'old' enzymes. So, in the Kelvin cycle a number of enzymes have functions close to those in the glycolytic chain; which is evolutionally more ancient. Nevertheless, a few ($\sim$ 3-5) functions (and, accordingly, enzymes) are actually new. Note, that a separate enzyme cannot perform a new function and, therefore, does not provide evolutionary advantages (in the evolutionary plan information on such enzymes can be considered as neutral). And only when enzymes have already appeared, it suddenly turns out that a complex of them can play a new unexpected role in the vital activity of organisms Just such 'big jumps' may be regarded as the basis of those forms of increase of the general organisational level which are usually referred to as aromorphosis.

From what has been said it is clear that the formation of a new protein (and, accordingly, information in DNA) is an act leading to spontaneous appearance of new, vast and valuable information. The probability of such an act is very small*, the more so as the probability of the appearance of a new enzyme system is small. The smallness of such a probability is the basis of the statements about the necessity to describe, consistently, the biological evolution, and the necessity for intervention of external forces etc. (Lewin, 1980; Monod, 1974; Crick, 1968).

* We speak of probabilities of the order of 10^{-50} - 10^{-100}; which can be considered negligibly small.

Below we consider this problem in more detail and discuss the hypothesis within which this problem may be solved.

Estimates of the amount of information and the probability of spontaneous appearance

Consider the appearance of new enzymes at a moment when the translation apparatus already exists and, therefore, Crick's dogma must be fulfilled. This means that, first, the information in DNA must arise, coding the primary sequence of amino acids in a protein which provides for the appearance of a new function.

Let us estimate this information. First we obtain the maximum estimate; this corresponds to the assumption that to provide functions it is necessary to set up the whole primary sequence, i. e. all the amino acid residues in the protein must be fixed. The characteristic dimensions of enzymes proteins are such that their sequence contains about 200 amino acid residues. Accordingly, the number of versions in such a sequence is $N = 20^{200}$. All possible permutations are taken into account. Hence, the information is equal to

$$\mathfrak{J}_{max.} = \log_2 N = 200 \times \log_2 20 = 867 \text{ bits}$$

The probability of spontaneous formation of such information is

$$W = N^{-I} = 2^{-\mathfrak{J}\,max} = 20^{-200} = 10^{-260}$$

The quantity $\mathfrak{J}_{max}$ is seen to be overestimated. Indeed, it is known that to perform a function it is not necessary to fix all the amino acid residues contained in the primary sequence. Many can be replaced by others without loss in function of a given enzyme-protein. In this case the quantitative characteristics somewhat change; namely, the Michaelis constant, the coupling constant, the turnover factor, etc., as is the case, for example, in isozymes.

Let us now present the minimum estimate. It is based on the assumption that, to perform a function, it is sufficient to fix the position of the amino acid residues present in the enzyme 'active centre'; in other words, the position of those amino acid residues that take part in chemical reactions catalysed by a given enzyme. The number of such amino acid residues is as a rule small, about $m \approx 5 - 10$. The global structure is not considered to play an essential role and, therefore, the position of the rest of the amino acid residues can be quite arbitrary. Then, we obtain:

$$\mathfrak{I}_{min} = \log_2 20^m = m\ 3,4 = 21 - 42 \text{ bits}$$

Accordingly, the probability of a spontaneous appearance of such information is equal to:

$$W_{max} = 2^{-\mathfrak{I}_{min}} = 10^{-7} + 10^{-13}$$

These quantities of necessary, valuable information are underestimated and, correspondingly, the quantity W_{max} is overestimated.

In reality, the sequence, and the properties, are important in function; the global structure provides the necessary mobility of amino acid residues which enter the active centre, the possibility for conformational transitions, etc.. These requirements impose limitations on the primary sequence of amino acid residues in the whole molecule. In other words, to provide the necessary properties of the molecule, information about the primary sequence is also necessary.

It is, as yet, difficult to estimate a realistic value for this information. In the paper by Csanyi (1980) it is accepted that to deduce the properties of the whole, it is sufficient to fix approximately a fourth of all the amino acid residues present. Thus, here, we have

$$\mathfrak{I}_{real} = \log_2 20^{50} = 50 \times 4,3 \quad 200 \text{ bits}$$

and the corresponding probability is given by

$$W_{real} = 2^{\mathfrak{I}_{real}} \approx 10^{-65}.$$

Comparing $\mathfrak{I}_{real}$ and $\mathfrak{I}_{min}$, we see that the information necessary to set up the whole construction exceeds the information required for the active centre of the enzymes. Note, however, that the value of $\mathfrak{I}_{real}$ is not generally accepted at present. So, for example, in papers by Ptitsyn, Finkelstein (1980) the opinion is expressed that the information necessary to specify the properties of the whole is much less, and is of the same order as or less than $\mathfrak{I}_{min}$. To put it differently, this means that practically any protein composed of a random sequence of amino acid residues already has the properties required in the whole molecule. From this it is clear that the estimates of $\mathfrak{I}_{real}$ are not exact. Nevertheless, we assume in what follows that $\mathfrak{I}_{real} \approx 200$ bits.

The quantity W is the probability of new information in appearing from one action. The probability of such a process during the period of evolution (the lifetime of the biosphere) is of interest and is equal to

$$P = \nu W / (1 + \nu W)$$

where ν is the number of attempts to form new information in the entire biosphere over a given time.

Estimate this number. It is equal to:

$$\nu = T \cdot V / \tau v$$

where V - is the volume of the sphere of existence of organisms

v - is the volume per one organism

T - is the characteristic evolution time (it is of the order of the Earth's lifetime):

$T \approx 10^9$ years $\sim 10^9 \cdot 360$ days $\sim 0.36\ 10^{12}$ days

τ - is the characteristic time of one attempt, i. e. the time over which information on a new protein due to mutation can appear; it is much more than the time of reproduction, therefore, we assume $\tau \geq \tau_{repr.} \approx 24$ hours.

The volume, V, of the sphere of existence of organisms is the volume of lakes, lagoons, off-shore waters, etc. and has been estimated in a number of papers (Quastler, 1964; Monod, 1974, and others), as of the order of 5 to 10 x 10^{10} cm³; the volume per cell (ϑ) is larger than the volume of the cell itself ($\vartheta \gtrsim \vartheta_{cell} \approx 10^{-6}$ cm³). With this in mind, we obtain $\vartheta \lesssim 10^{30}$. This quantity is, of course, overestimated since it corresponds to the assumption that the accessible volume is crowded with cells.

Using the value ϑ , one can estimate the probability P; for the formation of maximum information, P_{min}:

$$P_{min} \simeq \vartheta \cdot W_{min} \approx 10^{30} \times 10^{-200} = 10^{-170}$$

In a realistic case:

$$P_{real} = \vartheta \cdot W_{real} = 10^{30} \times 10^{-65} = 10^{-35}$$

Finally, in the case of minimum information, P_{max}

$$P_{max} = \frac{\vartheta \cdot W_{max}}{1 + \vartheta W_{max}} = \frac{10^{30} \cdot 10^{-13}}{1 + 10^{30} \cdot 10^{-13}} \approx 1 .$$

From this it is clear that in the first and second cases the spontaneous appearance of information necessary for a new function is, essentially, improbable. In the last example, on the contrary, the probability of the event is, approximately, to unity. From these estimates it also follows that, inaccuracy in the determination of ϑ (as defined above) is of almost no importance since the quantity ϑ , although large, is still insufficient to compensate for the smallness of W_{real}.

Thus, problems in the explanation of the rate of evolution arise if the necessary information (both about the active centre and on the global construction) is assumed to be sufficiently large. Below we consider possible ways of solving this problem.

Formulation of the hypothesis

The hypothesis which, in our opinion, solves the problem may be formulated as follows. Information about the construction of the whole protein performing a new function is large (of the order of $\mathfrak{J}_{real}$). However, a considerable part of this is already contained in the DNA regions that code the sequences of proteins which perform other, 'old' functions. Let us explain the essence of the hypothesis in more detail. Note that this hypothesis has a direct relation to the protein-machine concept (Khurgin et al., 1967; Chernavskiy, 1978) and the fundamental principle of this mechanism is as follows:

The whole protein structure is like a machine and consists of comparatively rigid blocks - levers jointed through mobile hinges. The rigid blocks are regions of the α-helix and contain about 30 amino acid residues. In addition, in the whole there exist energy reservoirs (springs) which are softer regions (β-structures and non-helix). The block structure provides for the conformational mobility of enzyme proteins which is important for the performance of the catalytic functions.

The character of mobility is, of course, individual, i. e. depends on the position of proteins and relates to the function performed. In other words, enzyme proteins are constructions of blocks and hinges. This concept is confirmed by experimental data, both on the structure, and, on the specificity of function (Chernavskiy, 1978; Ptitsyn, Finkelstein, 1980).

It is of importance that the details of protein-machines are quite standard- i. e. the number of types of blocks (q) which differ in length, rigidity, etc. is not large (about 10), that means: $q \sim 10$. On the contrary, in different enzymes there often exist blocks similar in physical characteristics. The number of blocks in each enzyme ($\nu \approx$ 5 to 7) can be either identical or different.

In view of this, in a new enzyme protein, one can use information on the protein structure which was already present in the 'old' enzyme. Essentially new information must be contained only in a new combination of proteins which exist already. The new information is, of course, $\mathfrak{J}_{min}$.

Mutations and recombinations which affect the reconstruction of blocks, or, more precisely, the DNA regions coding the blocks, are quite possible.

As a result, the probability of constructing a new protein from the blocks already contained in the 'old' protein essentially increases. What has been said can be explained through an example of a toy designer. The 'old' toy consists of details (levers, bolts, etc.) from which different constructions can be made: a bridge, a car, a house, etc.. These things are very different both in function and in construction, but most have common details. Therefore, any of the objects can be made by use of the details of other.

Let us present some estimates. Let there exist $q \approx 10$ to 20 different element-blocks, of which one can build an enzyme protein which performs a given function and consists of ν blocks (let ν be the order of 5). Let there exist a storage of N elements jointed in a sequence, containing n elements of i-th type ($i = 1, 2 \ldots q$; $\sum n_i = N$).

Let there exist a mutation or recombination mechanism which may arrange the element-blocks in any sequence.

Now estimate the probability of a certain sequence of N elements containing ν_i elements of the i-th type ($i = 1, 2, \ldots q$; $\sum \nu_i = \nu$).

We, thus, assume $N \gg \nu$ and $n_i \gg \nu$. This means that the information in the old protein is repeated in DNA many times and has, already, been changed.

Consider any region of the succession N, equal to ν. The probability of realisation in this region of a given succession is equal to $P_1 = q^{-\nu} \approx 10^{-5}$. The probability of the appearance of a required succession at any site in the chain, N, is equal to $P_1 \; (N - \nu) q^{-\nu} \approx 10^{-5}$.

The probability of a new construction (P_1 and P_1) is although small, much larger than the critical probability (i. e. $P_1 \gg 10^{-30}$; this means that such a process is quite possible in evolution.

The probability of the appearance of N new successions (N new blocks) is equal to

$$p^{(m)} = \frac{N!}{(v!)^m (N - vm)!} \cdot q^{-mv} \approx (N/v)^{vm} \cdot q^{-vm} .$$

The probability $P^{(m)}$ is seen to depend essentially on the number N, i. e. on reserve (stored) information. At $N \sim qv$ and the above values of m and v, the probability $P^{(m)}$ becomes equal to unity.

This shows the association of the hypothetical designer with the concept of accumulation of neutral information. The point is that prior to a 'big jump', i. e. before the appearance of a new integral enzymal system capable of providing new functions, the storage of information repetitions in DNA does not lead to evolutionary advantage, but is not, though, 'harmful'.

According to Kimura (Kimura, 1968; Kimura, Ohta, 1971), such information is considered neutral. It becomes, however, very necessary for the formation of new functions of the organism which make possible its existence and further development under qualitatively new conditions. To put it differently, accumulation of neutral information is a necessary premise for 'big jumps'.

Application of the hypothesis to a particular case

We shall illustrate the above statement by a concrete example. There is reason to suppose that, during evolution, the first photosynthesising protein was bacteriorhodopsin (BR). At the present time this protein has been well studied both in it's structural and functional aspects. We shall only refer to a number of reviews (Stoeckenius, Bogomolni, 1979; Ottolenghi, 1980). An efficient study of BR is possible since this protein is contained in halobacteria membranes. The main function of BR is to transform the energy of absorbed light into chemiosmotic hydrogen ion energy. In other words, in BR the protons are pumped through the membrane against a pH gradient and membrane electric field forces, i. e. BR functions as a proton pump.

Bacteriorhodopsin is built into the cell membrane of *Halobacterium halobium* and repumps protons from the internal medium (cells) into the external medium. There is reason to believe that for one light quantum absorbed, two protons are transported (Bogomolni et al., 1980; Kuschmitz et al., 1980; Chernavskiy et al., 1982).

The following is known about BR structure; that it is a protein with a molecular weight of 26.000 dalt.; in its primary succession there are 247 amino acid residues with a prevalence of hydrophobic residues (Ovchinnikov et al., 1978; Gerber et al., 1979). A light-absorbing co-factor in BR is retinal, chemically related to an apoenzyme, opsine, with the formation of a Schiff's base with N-lysine. The whole is represented in the form of 7 α-helix proteins joined by loops (hinges); each block contains 28 - 36 amino acid residues. The blocks are so joined that the external surface consists of hydrophobic amino acid residues, whereas hydrophilic residues face inside the molecule. Figures 51 and 52 show the primary succession in BR and an Xray pattern.

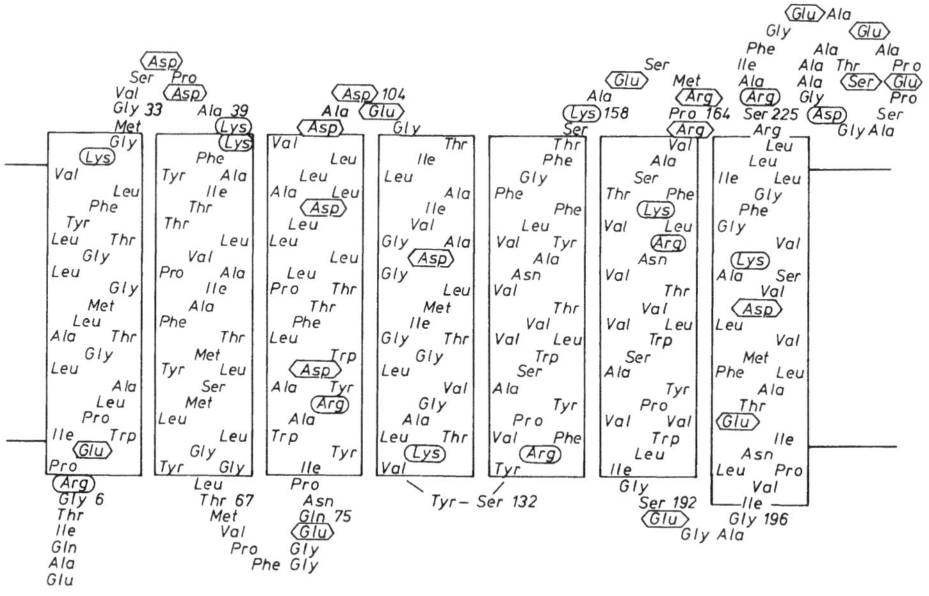

Fig. 51 Primary sequence of amino acid residues in bacteriorhodopsin (Ovchinnikov et al., 1978).

According to modern concepts, the BR mechanism consists of the following; light absorption induces in the retinal the electron density shift (Chernavskaya, Chernavskiy, 1981) and the formation of a strong local electric field. The latter is just the motive force which transports protons to the external medium against the pH gradient, and, against the electric forces of the external membrane potential. In the cyclic process the electron goes back to the initial state, and the newly formed proton vacancies are filled with protons from the internal medium. Proton motion from the retinal to the external medium and to the retinal from inside is realised by proton-exchange chains or channels which are chains of hydrophilic amino acid residues with certain pK values (Nagle, Morowitz, 1978; Nagle, Miele, 1980; Rastogi, Zundel, 1981; Chernavskiy et al., 1982). These chains must contain four different regions: two regions connect lysine (216) with the external and internal medium, and the two others connect the β -ionone ring with the external and internal medium.

Fig. 52 X-ray pattern of bacteriorhodopsin structure (Hayward, Stroud, 1981).

Let us discuss the number and the character of amino acid residues which compose the active centre of BR, i. e. participate in direct proton transport. They include, firstly, the amino acid residues that fix the retinal and there are a minimum of four of these: lysine (216) and three hydrophobic amino acid residues that fix the β -ionone ring. Secondly, they include the hydrophilic amino acid residues that fix the components of the proton-exchange chains or channels; there are too, a minimum of four of these, dependent on the regions of the proton exchange chain.

Thus, the quantity $\mathfrak{J}_{min}$ for BR is equal to

$$\mathfrak{J}_{min} = \log_2 20^8$$

since it is necessary to fix a minimum of eight amino acid residues.

It is more difficult to estimate the information necessary to fix the whole structure. If we assume, as has already been mentioned, the need for a 1/4 ratio in amino acid residues, then this means that one protein must contain information of the order of 34 bits. This information must guarantee, firstly, the structure of the block itself (the α -helix of 26-36 amino acid residues) and, secondly, the block's orientation with the hydrophilic amino acid residues facing the inside of the molecule, and the outwards facing hydrophobic residues. In other words, for the conditions indicated eight amino acid residues must be fixed in each block (approximately one per each coil of the α- helix in a column). The total, valuable information in a molecule is

$$\mathfrak{J}_{real.(BR)} \approx 34.7 \times \mathfrak{J}_{min} = 240 + 34 = 274 \text{ bits}$$

Thus, if the information about all the blocks and the information about the amino acid residues in the active centre is considered to have appeared randomly and at one instant, we obtain

$$W_{real/BR/} = 2^{-274} \approx 10^{-82}$$

$$P_{real/BR/} = 10^{-50} .$$

$W_{real/BR/}$ and $P_{real/BR/}$ are absurdly small quantities. The other quantities can be obtained from estimates on the basis of the hypothesistical design.

Firstly, pay attention to the fact that all the seven blocks are similar in the character and properties. Therefore, new, valuable information may be contained only in the construction of one block (column), the remaining six appearing as a result of six repetitions of the information about one block. In this case the new information is equal to:

$$\mathfrak{J}_{blocks} + \mathfrak{J}_{min} = 64 \text{ bits}$$

Correspondingly, the probability is:

$$W = 2^{-64} = 10^{-19}; \qquad P \approx \frac{\nu W}{1 + \nu W} \approx 1$$

This means that such an evolutionary process is quite probable. Now, assume the protein with the properties indicated to be already contained in one of the proteins functioning in the pre-photosynthetic period. Then the whole molecule could arise as a result of separation of a corresponding DNA succession and a seven-fold repetition. In this case, the new and valuable information is only equal to 34 bits. The corresponding probability is:

$$W = 2^{-34} = 10^{-10}$$

$$P = \frac{\nu \cdot W}{1 + \nu W} \approx 1 \ .$$

To put it differently, such a process may be considered quite probable.

Thus, we have shown the 'big jump' mechanism in evolution via a fixed example.

Our hypothesis is very close in essence to the considerations expressed by Gilbert (1978), Blake (1978) and Kretsinger (1972). The DNA successions containing information about functional blocks can be compared with exons discussed in the papers mentioned above.

The process of random permutation of blocks may be considered analogous to the process of entron elimination in transcription. More complicated comparison is also possible: for example, the situation where one exon codes several blocks connected by a hinge.

It should be noted, however, that in the papers mentioned above no attempt has been made to estimate the amount of new information during evolution and the probability of these processes.

In conclusion the authors express their sincere gratitude to V. I. Korogodin and O. B. Ptitsin for discussion and valuable comments.

State Spread and Evolution in Chemical Systems

L. A. Nikolaev

We assume, as is usually done, that the state of a system is known if some set of its parameter values is given. Then the process which results in acquisition by another, analogous system of the same parameter values will be referred to as state spread. If the original system transforms to a state different from the initial state, we shall call it state transfer. A very simple example of state transfer is the blow collision of two identical, elastic balls, one of which was moving and the other of which was at rest; when the direction of the blow coincides with the centre line there occurs a complete state transfer - the resting ball acquires the momentum and velocity of the moving ball, and the latter stops.

An example of state transfer in the microcosm is the inductive resonance which appears when the emission spectrum lines of one molecule are superimposed on the absorption spectrum of another molecule; the excited state is transferred when the distance between the molecules becomes of the order of ten molecular diameters. If molecules are identical, a sharp resonance interaction is possible also at large distances (Galanin, 1960; Shakhparonov, 1965). Resonance arises both between electron, and, between vibrational states of molecules (Shakhparonov, 1965). Both molecules form a single system, and the states where one of them is excited and the other is not are degenerate states - the eigenvalues of a corresponding energy operator for both wave functions of a system of two molecules are the same.

A macroscopic analog is the instance of two linked pendulums which exchange their energies.

In all the examples indicated one can single out, and describe, certain structures (a pendulum, an atom).

Thermodynamics and Regulation of Biological and Processes

In what sense can one understand similar phenomena in systems described in the language of thermodynamics, i. e. without using any structural considerations?

An essential difference of thermodynamic systems from those described above above is in the different nature of the state parameters. Thermodynamic parameters are, in fact, mean quantities and reflect statistical laws. If we restrict ourselves to describing a system in thermodynamic language, state transfer and spread processes would be, for example, the establishment of thermal equilibrium between a small body and a large heat reservoir (thermostat) whose temperature is acquired by the body.

A molecular-statistical analysis shows, however, that there is, strictly speaking, no state transfer at the molecular level. Transferred are only the averaged parameter values and the distribution function (if before the contact the small system is in a perturbed state).

In levelling the intensive parameters (temperature, pressure) the states are neither spread nor transferred. The reason is that, at the molecular level, elementary acts of molecular interaction (collisons) do not provide a complete state transfer, as is the case with state transfer to the resting ball.

It is just this incompleteness of state transfer that causes temperature levelling, a corresponding entropy increase and the irreversibility of this process. Otherwise molecules would just exchange states like weakly linked pendulums and there would be no energy dissipation at all.

Hence, if a state is described by statistically fixed parameters, its spread is possible only in the absence of state transfer at the molecular level; the principles of the possibility for spreading are evaluated by thermodynamic criteria, i. e. by the maximum entropy for an isolated system and by thermodynamic functions for closed systems.

The question acquires another meaning when we go from structureless thermodynamics to interaction of systems on whose components spatial limitations are imposed, i. e. to structural organisations.

Let us leave aside such self-organisation phenomena as, for example, crystallisation, where ordering is a consequence of the smallness of the term T.S in the thermodynamic function expressions, and consider interactions of non-equilibrium, dissipative organisations. Among them the most obvious examples of state transfer may be biological objects. Comparison of biological systems with simple physical systems shows, immediately, that in biological systems, the state spread implys identical variation of many parameters both at a molecular and, at a macroscopic, level. Indeed, to transfer a state in a system of linked pendulums, the conditions of similarity of their frequencies suffices, since the state is characterised only by the energy, or amplitude, value, whereas the molecular properties (composition) of pendulums are not taken into consideration at all. But, for example, in cell division a state is spread where both microscopic and macroscopic parameters are reproduced.

The structure of biological systems is such that equalisation of averaged values of macroscopic intensive parameters is not an appropriate mechanism of state spread.

In many cases this levelling leads to violations in structure organisation, for example, to elimination of specific density distribution, concentrations, charges, etc..

There exist some special conditions under which chemical potential equalisation is conjugated with self-organisation processes. Beginning with Turing (1952) and up to the present, self-organisation, and the appearance of dissipative structures, i. e. the appearance of inhomogeneities in an initially homogeneous medium have been under study (Nicolis, Prigogine, 1977). System rotation around the stationary state, entering a limit cycle, periodic oscillations of concentrations of intermediate compounds (for example, in Belousov-Zhabotinski reaction) depend, essentially, on kinetic constants. However, these constants reflect the chemical nature of the components of the process and the peculiar conditions of their interaction which, sometimes, provide state transfer. If these conditions are fulfilled, a conjugation of chemical potential levelling with state transfer is possible. So, a catalyst can lead the reaction along one of the thermodynamically possible paths, and, acting together with the matrix, and, using the energy of conjugate processes, can spread a state typical of the product molecule.

What are the conditions to be fulfilled for state transfer? To answer this question, one should pay attention to the character of atomic and molecular interactions with the medium without invoking averaging procedures.

Let us trace the evolution of inorganic but organised physical systems to biological ones and compare the nature of coding mechanisms at different organisational levels.

The simplest organisation is, of course, an atom. Free motion of its components is not quantised, whereas the nucleus-electron system is quantised, it can exist only in strictly defined states and can absorb or lose energy also only in definite portions: quanta. The magnitude of a quantum is connected with energy levels by the well-known Bohr frequency condition which follows from the fundamental Planck equation $\varepsilon = h\nu$. Only a certain frequency can cause an electron to pass over to a new energy level, an atom exists only in one of given discrete states. We call a discrete set of frequencies a code of the process of energy exchange between an atom and the medium. This is time coding and is a time based succession of values of the vectors E and H of an electromagnetic wave. At lower organisation levels spatial codes are of great importance too. In the interaction of complex molecules there are geometrical factors which occur and determine the possibility of energy and state transfer. Also the orbital symmetry factors limit the effects of external actions upon molecules. Discretely changing geometrical factors are of increasing importance in movement toward macrostructures. Conformational changes play an important role in enzymes, and in membrane systems they become dominant. The great variety of discrete spatial combinations makes it possible to code the channels of communication with the external medium at early states in biological evolution until a nervous mechanism develops. From that instant time codes once again dominate, but as trigger mechanism (Nikolaev, 1982).

The basis of a time code in the nervous mechanism is a sequence of electric impulses, i. e. a certain oscillation process. Oscillation processes possess almost unlimited possibilities for coded signal transfer. A signal can be coded by a given frequency. Remarkable in this respect is the analogy between neutron and atom: the more intense the excitation of radiation, the higher the frequency of the impulses generated by a neutron, and the greater the excitation of an atom from the ground state, the higher the frequency of the emitted radiation. A signal can also be coded by modulation, and finally the magnitude of the impulse, and the sequence of impulses and pauses is a convenient means of coding, too. This is simply the way coded information arrives at the eye, here light pulses are recoded into nervous pulses, and then there starts the subordinate stages of chemical coding in the organs that received different signals. Recoding of light pulses also requires inclusion of a chemical stage (cis-retinal - trans-retinal). A very important characteristic of coding, because of time variation in the oscillation process, is that the thermodynamic equivalence of work of the coding does not, in toto, depend on the meaning of the transferred information. All communications correspond to approximately equal and quite negligible energy expenditures and entropy increments. It is remarkable that a chemical link in the coding mechanism of the organ of sight also works with a negligible entropy (cis-to-trans form transformation) of the process.

Another characteristic feature of time coding by means of oscillations is the ability of a given code to subdue another. Pringl (in: Nikolaev, 1976) has shown that in a system of bounded, non-linear oscillators one of the oscillators subdues another so that the result depends on the side from which the frequency of a given oscillator approaches the frequency of the other. Approaching the stationary frequency from the right entails capturing this frequency by that of an approaching oscillator. If the varying frequency approaches the stationary frequency from the left, it itself is absorbed by the stationary frequency. From this it follows, first, that a system of bound oscillation systems will evolve, and, affected by random perturbations, can pass over to a more complex state (Goodwin, 1963) and, second, that rhythms in the external medium can subdue oscillation processes in the organism and, accordingly, the chemical processes controlled by them. Something of this kind is also observed in the case of spatial codes. The rigid, geometric structure of the molecules of some substrates in enzymatic reactions may cause conformational changes in the active region of an enzyme; the result is a catalytic transformation, after which the enzyme returns to the initial state.

Time codes are thus one of the important means of organism adaptation to external conditions. The possibility of transforming a chemical process into nervous excitation, i. e. into an oscillatory electric process explains why all forms of external actions are so often translated by an organism into the language of nervous impulses, for this is the most perfect and quick-acting coding system which operates with the least entropy loss. This is just why the simple chemical evolution of organisms was outlined long ago - development goes in the direction of the creation of macrostructures and of perfecting systems of connection and, primarily at the expense of time coding.

The weak dependence, between systems, of the coding mechanism of communication on chemical and thermodynamic limitations is a very important result of the evolutionary process and, at the same time, an expression of the general tendency of biological evolution. Thus, we obtain systems where there conjugate energy and mass fluxes with coding mechanisms, the latter having the control role: they maintain subsystems in a state which deviate from equilibrium and, regulate energy and mass fluxes.

The coding mechanism provides state transfer within a whole hierarchy of levels: state spread, or transfer, is realised on the enzyme-substrate level, on the DNA-RNA-protein level, where it leads to hereditary character transfer and, finally, at the level of nerve cell systems.

Higher coding systems, e. g. the brain, acquire, and create coded information without showing any tendency to transform into a state determined by some thermodynamic extremal conditions. At the same time, they are capable of transferring, to analogous systems, states which are characterised by a huge number of parameters and, finally, form the components of intellectual intercourse.

The self-organisation processes that lead to conjugation of fluxes and dissipative structures may become a first stage of evolution if, among the structures, there exist those whose further development depends on discrete time or space conditions not associated with thermodynamic prohibition. Discrete conditions (codes) change in the course of evolution towards an increasing liberation from chemical and thermodynamic limitations. Therefore, they become control mechanisms with respect to less free subsystems; in the course of perfection of higher coding mechanisms they become capable of transferring more and more complicated states to similar systems. The closer the systems, the more complete the state transfer, and vice versa; this is just what hinders men and animals in their verbal communication (state transfer).

The possibility of state transfer is a very important factor which unites components of highly organised systems, and, at higher organisational levels, it plays approximately the same role as chemical bonds at the atomic level.

Stochastic Models of Evolutionary Processes

R. Feistel, W. Ebeling

Introduction

Evolutionary processes are possibly the most complicated processes we observe in nature. Evolution may be defined as an infinite serics of self-organisation processes (Ebeling, Feistel, 1982). It appears appropriate starting from the established theory of self-organisation (Romanovsky et al., 1975; Zotin, 1976, 1978; Nicolis, Prigogine, 1977; Haken, 1978; Ivanizky et al., 1978; Volkenstein, 1981; Klimontovich, 1982) to develop a theory of evolutionary processes and several attempts have been made (see e. g. Zuckerkandl, 1976; Eigen, Schuster, 1977; Conrad, 1979; Schnol, 1979; Bremermann, 1979, 1980; Schuster, Sigmund, 1980; Eigen, 1981; Heinrich, Sonntag, 1981; Ebeling, Feistel, 1982).

However it should be stressed that all such approaches are to be considered merely as beginnings. There is still much to do before we may say that the physical basis of evolutionary processes is well understood. The approach presented here is devoted to the development of a new, formal model of evolutionary processes: the phenotype space model. In this model the evolution of species is considered as a stochastic process in many-dimensional space where the coordinates denote the phenotypic properties. Over this space a complicated value function with many minima and maxima - like stochastic potentials in disordered semiconductors - is defined which forms an adaptive landscape (Conrad, 1979; Feistel, Ebeling, 1982): Evolution is considered as hillclimbing in this rich-structured landscape, which itself depends on the distribution, and therefore is time-dependent due to internal and external (changes in environment) processes.

Thermodynamics and Regulation of Biological and Processes

The discrete model

Dynamics in population biology is usually expressed in a set of ordinary differential equations (see e. g. Poluektov, 1974; Svireshev, Logofet, 1978)

$$dx_i/dt = x_i w_i(x_1 \ldots x_n) \tag{1}$$

where $x_i \geq 0$ is the population (occupation) number of the species i and w_i its self-reproduction rate expressing all interspecies like trophic structures, competition, mortality, fertility, altruism etc.. Classical examples are Fisher-Eigen equations (Fisher, 1930; Eigen, 1971)

$$w_i = E_i - \langle E \rangle ; \ \langle E \rangle = \sum E_i x_i / \sum x_i \tag{2}$$

and Lotka-Volterra equations (Svireshev, Logofet, 1978)

$$w_i = a_i - \sum b_{ij} x_j . \tag{3}$$

The coefficients E_i, a_i, b_{ij} depend on the environmental conditions (assumed to be time-independent) and on the individual phenotypic properties

$$\underline{q}^{(i)} = \{q_1^{(i)}, q_2^{(i)}, \ldots, q_n^{(i)}\} \in Q$$

expressed as a property vector q. All real vectors, q, span the phenotype space, Q, which is a real Euclidian vector space. Due to the central dogma of molecular biology and the discrete nature of genetic information we assume that the set Q* of $q^{(i)}$ is a discrete countable set, which is not necessarily finite. The limited material ressources require that only a finite subset of Q* (Fig. 53) may be occupied ($x_1 > 0$), and thus most of the elements of Q* are empty ($x_i = 0$) (i. e. the corresponding species are already 'ruled out' or not yet 'invented'). Therefore the occupying species form 'point clouds' within the phenotype space.

The set Q* of possible species is restricted to certain compact areas of Q; the majority of the arbitrary combinations of real numbers $q_k^{(i)}$ contradicts physical laws (the relation of size and mass cannot exceed the density of water essentially, the metabolic efficiency cannot violate the Second Law etc.).

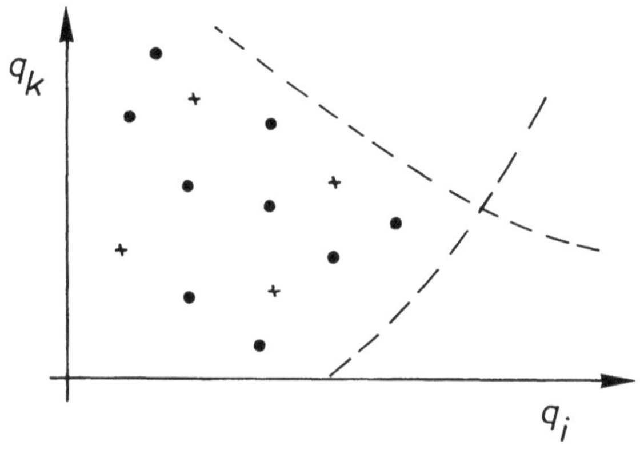

Fig. 53 Occupation of the phenotype space.
+ occupied states
· not yet realised species
- physical limitations

The point set Q* may be viewed as the vertices of a graph; then species interaction takes the form of directed, valued edges of the graph. An example is the community matrix $\{b_{ij}\}$ in (3): Its antisymmetric part corresponds to the food web (i. e. the transfer of free energy through the ecosystem) while its symmetric part reflects the collective dissipation (Ebeling, Feistel, 1982) of energy.

Within this framework evolution can be understood as changing the occupation, i. e. new points, Q*, enter by mutations and old ones leave by means of selection. Mathematically this can be expressed by a transition matrix A_{ij} giving the rate of passage from $q^{(j)}$ to $q^{(i)}$:

$$dx_i/dt = x_i W_i + \sum_j A_{ij} x_j \; ; \quad \sum A_{ij} = 0 \tag{4}$$

Again A_{ij} stands for a binary relation which can be represented by arrows $j \rightarrow i$ between points in the phenotype space. Very little is known about this graph in general, and as a first attempt we can consider it as a random network. It is then possible to calculate the probabilities for the appearance of cyclic or acyclic graphs, the size of connected subgraphs etc. (Ebeling et al., 1981; Feistel, 1979; Sonntag et al., 1981.).

For a given structure of the graph associated with A_{ij} it is possible to solve equation (4) qualitatively in the case of Fisher-Eigen dynamics, because the stationary solution of

$$dx_i/dt = x_i(E_i - \langle E \rangle) + \sum_j A_{ij} x_j \tag{5}$$

is an eigenvalue problem

$$\sum_j E_{ij} x_j^o = \langle E \rangle x_i^o \tag{6}$$

of a non-negative matrix $E_{ij} = E_i \delta_{ij} + A_{ij}$, where the eigenvectors are the occupation distributions and the eigenvalue $\langle E \rangle$ the effective reproduction rate of the population. Fortunately, the spectral properties of non-negative matrices are well understood (Gantmacher, 1971), and they can be derived immediately from the reducibility structure of the corresponding directed graph (Feistel, Ebeling, 1978; Ebeling, Feistel, 1982).

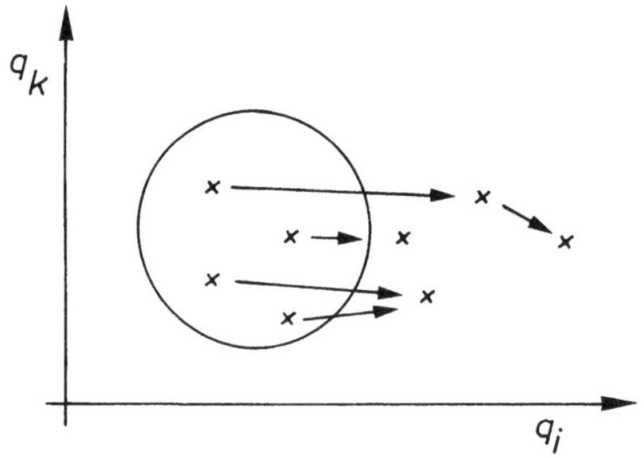

The continuous model

It is common in natural selection, as well as in artificial selection (cultivation), to think of phenotypes as continously variable items. Consideration of the continuous model brings many advantages but also complications. The vector

$$\underline{x} = (x_1, x_2, \dots) = (x(q^{(1)}), x(q^{(2)}), \dots)$$

now becomes a function of the phenotype 'coordinates' q:

$$(\partial/\partial t)\, x(q,t) = x(q,t)\, w(q; \{x\}). \tag{7}$$

The replication rate is now a function of q, but also a function of x(q). In both equations (2) and (3) this functional is linear (Feistel, Ebeling, 1982)

Fisher-Eigen $$W = E(q) - \frac{1}{C}\int dq' E(q') x(q',t) \tag{8}$$

Lotka-Volterra $$W = a(q) - \int dq' b(q,q') x(q',t). \tag{9}$$

The integrals are to extend over the whole phenotype space. Mutations can be expressed by a diffusion process in the phenotype space

$$\frac{\partial}{\partial t} x(q,t) = xW(q;\{x\}) + \frac{\partial}{\partial q} D(q) \frac{\partial}{\partial q} x(q,t) \tag{10}$$

where the diffusion tensor D(q) is related to the transition matrix A by

$$D(q) = \int dq'(q-q') \circ (q-q') A(q,q'). \tag{11}$$

Consider first the case of pure selection, i. e. D(q) = 0. By S we denote the set of species present initially

$$x(q, t=0) > 0 \quad \text{if} \quad q \in S \subset Q. \tag{12}$$

The stationary solution of (7) is

$$x^\circ(q)\, W(q;\{x^\circ\}) = 0 \tag{13}$$

i. e.

$$x^\circ(q) > 0 \text{ and } W(q;\{x^\circ\}) = 0 \text{ if } q \in V \subseteq S \tag{14}$$

for a certain set, V, of surviving species and, because of the required stability

$$x^\circ(q) = 0 \text{ and } W(q;\{x^\circ\}) \leq 0 \text{ if } q \in S - V. \tag{15}$$

If W is a linear function, equations (14) and (15) constitute a linear complementarity problem. Its solution is unique in the case of Fisher-Eigen dynamics, and for Lotka-Volterra systems if b(q, q') is positive and real.

This theorem can be deduced from the existence of Lyapunov functionals (dynamic evolution principles)

$$L_{FE} = \langle E \rangle = \int_S dq\, E(q)x(q) \Big/ \int_S dq\, x(q) \tag{16}$$
$$d/dt\, L_{FE} \geq 0$$

in the case of Fisher-Eigen and

$$L_{LV} = -\int_S dq \{ x(q,t) - x^{\circ}(q) \ln x(q,t)$$
$$d/dt\, L_{LV} \geq 0 \tag{17}$$

in the case of Lotka-Volterra.

Moreover, in structurally stable situations, (14) cannot have compact solutions, V, of the same dimension as the space Q, in other words, every set of surviving species V forms a subset of value zero in Q.

Mutations can enter in several ways. A first possibility is to extend the set V of survivors stochastically by some of its mutants, δV, and to start selection as described above again. This procedure can be repeated *ad infinitum* to model evolution:

$$\begin{array}{ll} S_1 & \\ \downarrow & \text{selection (deterministic)} \\ V_1 & \\ \downarrow & \text{mutation (stochastic)} \\ S_2 = V_1 + \delta V_1 & \\ \downarrow & \text{selection (deterministic)} \\ V_2 & \end{array}$$

This picture corresponds to the kinetic view of evolution - an unlimited series of self-organisation processes and instabilities. The question what pathways $V_1 \rightarrow V_2 \rightarrow V_3 \rightarrow \ldots$ through the phenotype space exist and whether they maximise a quantity 'fitness' is subject to the qualitative theory of selection (Ebeling, Feistel, 1982). The other way to look at mutations is the diffusion approximation (10). Consider the case of Fisher-Eigen equations in more detail.

First transform to new variables y (q, t)

$$x(q,t) = y(q,t) \exp\{-\int_0^t dt' \langle E(t') \rangle\} \tag{18}$$

This leads to the linear equation (Feistel, Ebeling, 1982)

$$\frac{\partial}{\partial t} y(q,t) = D \frac{\partial^2}{\partial q^2} y(q,t) + E(q)\, y(q,t) \tag{19}$$

(with D(q) = D supposed), which is equivalent to the Schrödinger equation with a potential - E(q). Simple examples for E(q) are:

(I) the linear case; $E(q) = E_o + E_1 q$, corresponding to a homogeneous external field in quantum mechanics and the 'shifting form of selection' in evolution theory (Timofeeff-Ressovsky et al., 1975)

(II) the quadratic case; $E(q) = E_o - E_2 q^2$, corresponding to the harmonic oscillator in quantum mechanics and to the 'stabilising form of selection' in evolution theory.

In practice little is known about possible shapes of the fitness landscape E(q); one route to a theory is the study of disordered potentials.

From solid-state physics it is known that approximate methods hold if E(q) varies much faster than y(q) in space or vice versa, i. e. if the Fourier spectra of E(q) and y(q) are separated. In the case of rare mutations the distribution, y(q), is sharply peaked, and Q-space can be decomposed into 'drainage areas' of the local maxima of E(q) and equation (9) solved with the quadratic, E, in each area:

$$\begin{aligned} y(q,t) &= \sum y_i(t) \exp\{-(q-q_i)^2/2\sigma_i^2\} \\ E(q) &\approx E_o^{(i)} - E_2^{(i)}(q-q_i)^2 \text{ in area } i. \end{aligned} \tag{20}$$

The result is a quantum tunneling of the population from one local maximum of E(q) to the next higher one (substitution of (20) into (19) yields

$$y_i(t) = x_i(0) \exp\{(E_0^{(i)} - D/\sigma_i^2)t\}$$
$$\sigma_i^4 = D/E_2^{(i)} . \tag{21}$$

And for x use of (18) produces

$$x(q,t) = \sum x_i(t) \exp\{-(q-q_i)^2/2\sigma_i^2\}$$
$$x_i(t) = y_i(t) \sum_j x_j(0) / \sum_j y_j(t) . \tag{22}$$

Discrete species in a continuous phenotype space

Discrete models, as discussed briefly in the second section, can be constructed on the basis of the continuous picture developed in section 3; we shall see that the idea of the continuous phenotype space upgrades the usual discrete schemes by additional statements.

From equation (10), then

$$\frac{\partial}{\partial q} x(q,t) = x(q,t) w(q;\{x\}) + \frac{\partial}{\partial q} D(q) \frac{\partial}{\partial q} x(q,t) . \tag{23}$$

The distribution x (q, t) is assumed to be localised at several sites $q^1, \ldots q^N$ in the phenotype space, the diffusion component of (23) is expressed by the existence of a narrow spread σ_i of phenotypic variation around its bary-centre q^i; further immediate influences are neglected. The bandwidth of variation should be small compared with the distances between the species

$$(\sigma_i^2 + \sigma_j^2) \ll (q^i - q^j)^2 \quad \forall\, i \neq j . \tag{24}$$

This allows calculation of the evolutionary motion of the centres $q^i = \langle q \rangle_i$ averaged over the vicinity of q^i: i. e.

$$q^i = \int_i dq\, q\, x(q,t) / \int_i dq\, x(q,t) \tag{25}$$

The time derivative gives Fisher's law

$$\frac{d}{dt} q^i = \langle qW \rangle_i - \langle q \rangle_i \langle W \rangle_i \approx \sigma_i^2 \frac{\partial W}{\partial q} \Big|_{q=q^i} . \tag{26}$$

The smallness of σ_i^2 makes the process (26) much slower than the levelling of the occupation by (23); applying Tikhonov's theorem to solve (23) without diffusion for a set of stationary points, q^i,

$$x^o(q) = \sum_i x_i \delta(q - q^i) \tag{27}$$

i. e.

$$W(q; \{x^o\}) = 0 \text{ if } q = q^i \quad i = 1, 2, \ldots, N . \tag{28}$$

Now in the whole space, Q, a landscape

$$p(q) \equiv p(q; q^1 \ldots q^N) \equiv W(q; \{x^o\}) \tag{29}$$

is formed by putting the solution (27) into the functional W. This landscape, p(q), is a scalar function over the phenotype space which depends on the parameters $q^1 \ldots q^N$, i. e. the location of the actual set of species in Q.

To make the construction of this adaptive landscape transparent we treat, briefly our examples. In the Fisher-Eigen case a species may be located at q^1. The functional

$$W(q; \{x\}) = E(q) - \frac{1}{C} \int E(q') x(q') dq' \tag{30}$$

must be zero at q^1 for the stationary case:

$$\begin{aligned} &W(q; \{x^o\}) = 0 \text{ at } q^1 ; \\ &x^o(q) = x_1 \delta(q - q^1) . \end{aligned} \tag{31}$$

The solution is $x_1 = C$. Substitution into (30) gives the landscape

$$p(q; q^1) = E(q) - E(q^1) . \tag{32}$$

In this special case the landscape is rigid; it can be shifted but not deformed by q^1.

The Lotka-Volterra functional is

$$W(q;\{x\}) = a(q) - \int dq' \, b(qq') \, x(q') . \tag{33}$$

Let the species be located at $q^1 \ldots q^N$, then the functional must vanish at these points for the stationary case:

$$\begin{aligned} &W(q;\{x^o\}) = 0 \quad \text{at } q = q^1, q = q^2, \ldots \\ &\text{with } x^o(q) = \sum_{i=1}^{N} x_i \, \delta(q - q^i) . \end{aligned} \tag{34}$$

The solution is achieved by solution of the linear system

$$\begin{aligned} &a_i = \sum b_{ij} x_j ; \quad x_i = \sum b_{ij}^{-1} a_j \\ &a_i = a(q^i) ; \quad b_{ij} = b(q^i, q^j) \end{aligned} \tag{35}$$

which is assumed to be positive, $x_1 > 0$. Replacing x(q) in (33) by this solution gives the landscape

$$p(q; q^1 .. q^N) = a(q) - \sum b(q, q^i) \, b_{ij}^{-1} a_j . \tag{36}$$

Obviously this function can change its shape in a complicated way if the parameters q^i vary.

Some general properties of p(q) are quite independent of the nature of the functional W.

I. p(q) is a local potential for the evolutionary process. The Fisher equation (26) now reads

$$\dot{q}^i = \sigma_i^2 \cdot p'(q^i) \tag{37}$$

so that every species changes its properties along the gradient direction of p(q). That is, evolution climbes uphill.

II. Occupied loci are zeros of p(q).
From (28) and (29) it follows that

$$p(q^i) = p(q^i; q^1 \ldots q^N) = 0 \quad \text{for } i = 1, .., N \tag{38}$$

That is evolution climbes uphill like an ice-breaker: nevertheless it remains on the zero level.

III. p(q) is 'elastic' with respect to phenotypic change within a single species.

We can arbitrarily fix some species at their positions and move other ones within certain limits, then (38) remains valid.

This is expressed mathematically by

$$dp(q^i) = 0 = \sum_j \frac{\partial}{\partial q^j} p(q^i) dq^j + \frac{\partial}{\partial q} p(q^i) dq \qquad (39)$$

or, because the dq_j are arbitrary

$$\frac{\partial}{\partial q^j} p(q^i) = -\frac{\partial}{\partial q} p(q^i) \delta_{ij} . \qquad (40)$$

If we change q^j, then p(q) remains unchanged at q^j. This elasticity is limited by the positive nature of the quasi-stationary solution $x^o(q)$.

IV. The landscape is always sinking at occupied loci. Let the $q^i(t)$ move in a manner corresponding to the differential equations (37). Putting this time behaviour into p we have $p(q;\, q^1(t), \ldots, q^N(t) = p(q;\, t)$ and with equations (37) and (40) it follows that

$$\frac{\partial}{\partial t} p(q^i) = \sum_j \frac{\partial}{\partial q^j} p(q^i) \dot{q}^j = -\sigma_i^2 \left\{ \frac{\partial}{\partial q} p(q^i) \right\}^2 \leq 0 . \qquad (41)$$

V. There is a general evolutionary principle of a Pfaffian type.

Consider the differential expression

$$\delta L = -\sum \frac{\partial}{\partial q^i} p(q^i) dq^i . \qquad (42)$$

By use of (37) and (40) we have

$$\delta L = \sum \sigma_i^2 \left[\frac{\partial}{\partial q} p(q^i) \right]^2 dt \geq 0 . \qquad (43)$$

This expression is, in general, a total differential dL, only if

$$\left(\frac{\partial^2}{\partial q^i \partial q^j} p(q) \right)_{q=q^i} = \left(\frac{\partial^2}{\partial q^i \partial q^j} p(q) \right)_{q=q^j} \qquad (44)$$

holds, where equation (43) expresses the maximisation of a function $L(q^1 \ldots q^N)$. Equation (43) determines the half-space the evolution is directed toward (Fig. 54). In this respect equation (43) has the same structure as the famous Glansdorff-Prigogine principle $d_x P \leq 0$.

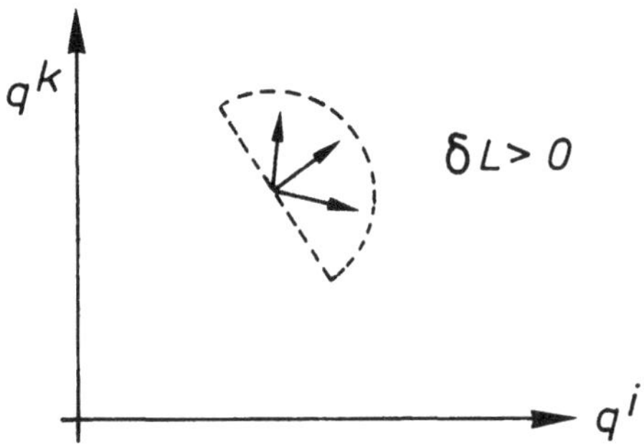

Fig. 54 The half-space of evolution given by an evolutionary principle of the Glansdorff-Prigogine type (Pfaffian form principle).

Statements IV and V may be illustrated in the case of Fisher-Eigen dynamics. As already mentioned, the landscape is

$$p(q;q^1) = E(q) - E(q^1). \tag{45}$$

The Fisher equation (37) is then

$$d/dt\, q^1 = \sigma^2 \partial E/\partial q \,|_{q=q^1}. \tag{46}$$

The change of p at q^1 is

$$\begin{aligned} \frac{\partial}{\partial t} p(q^1) &= \frac{\partial}{\partial q^1} p(q,q^1)\,|_{q=q^1}\, \dot{q}^1 \\ &= -\frac{\partial E}{\partial q^1}\, \sigma^2\, \frac{\partial E}{\partial q^1} \leq 0 \end{aligned} \tag{47}$$

and finally, the Pfaffian form (42) reads

$$\begin{aligned} \delta L &= -\frac{\partial}{\partial q^1} p(q;q^1)\,|_{q=q^1}\, dq^1 \\ &= \partial E/\partial q\,|_{q=q^1}\, dq^1. \end{aligned} \tag{48}$$

In this case we find a total differential $\delta L = dL$, which leads to

$$L(q^1) = E(q^1) \tag{49}$$

i. e. the replication rate becomes maximised by evolution.

Qualitative image of vanishing species or species differentiation

The Fisher equations in the form (37) represent a system of non-linear ordinary first order differential equations, independent of whether the functional W(q; { x }) is linear or not. They describe the long-time behaviour of a multi-species population with respect to their change in phenotype properties as described by Darwin's principle.

There is possible the full spectrum of qualitatively different behaviour as deduced from non-linear dynamics: i. e. multiple steady states, limit cycles, or even chaotic behaviour is possible. But one feature is completely different from dynamic systems and this is that moving points can vanish (the species dies out) or can divide (the species differentiate). The method of motion in an adaptive landscape can give a simple qualitative picture of how these catastrophes can happen.

First we consider the case that only one property is responsible for the qualitative change (one critical mode). We restrict the high dimensional phenotype space along this coordinate and plot the landscape p(q) versus q(Fig. 55, 56).

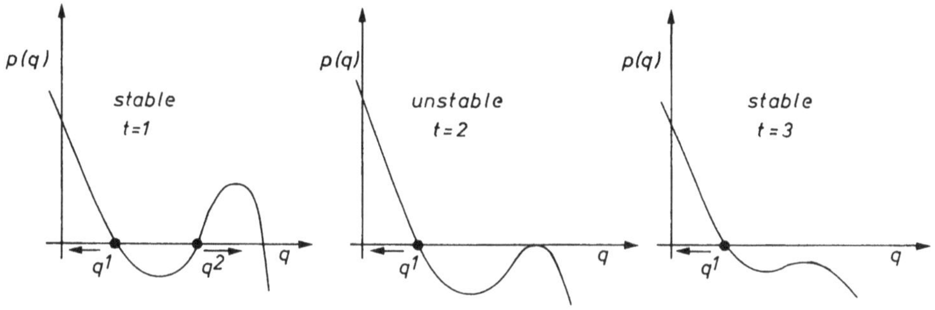

Fig. 55 Model for the disappearance of species.

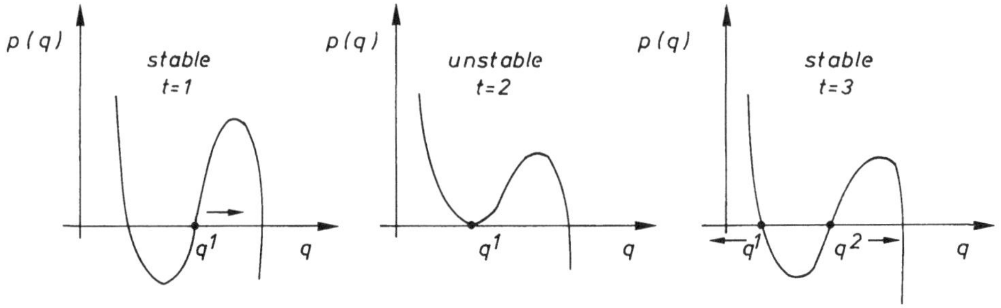

Fig. 56 Model for the (one-property) differentiation of species.

If two properties are responsible, a species may differentiate by passing through a saddle point (Fig. 57).

Here, the p(q) = 0 line is drawn in the two-dimensional Q-space as well as the points representing the species and their direction of motion.

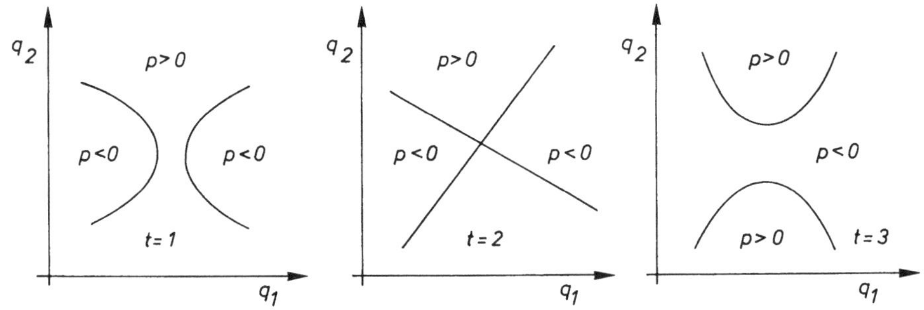

Fig. 57 Model for (two-property) differentiation of species.

Bioenergetic Trends of Evolutionary Progress of Organisms

A. I. Zotin

Paleonthological data definitely show that in the course of evolution there appear more and more complex and perfect forms of animals and plants. Therefore, there are no doubts as to the existence of evolutionary progress in the living nature. At the same time it is very difficult to present comparable concepts for quantitative characteristics of the evolutionary progress, although this phenomenon has been considered in detail by many authors (Severtsov, 1934; Yablokov, 1968; Timofeev-Ressovsky et al., 1969; Schmalhausen, 1969; Gilyarov, 1976; Zavadsky, 1971; Shnol, 1979, and others). More satisfactory, is the situation for the characteristics of bioenergetic perfection. Severtsov (1934) assumed that, in the course of progressive evolution, the energy of life activity in animals increased. Specification of this assumption in Ivlev's works (1959, 1963) led Dolnik (1968) to the idea that progressive evolution in animals is related to increasing respiratory intensity in organisms.

Bioenergetic progress and evolution

As is known (see e. g. Brody, 1945; Kleiber, 1961; Slonim, 1961; Prosser, 1973), energy metabolism in organisms depends strongly on their size. Therefore to compare standard and basal metabolism in organisms of differing size data on the relationship between respiration and body mass are usually employed; the respiration intensity of some conventional organism weighing 1 g is calculated, using parameter a from the equation

$$\dot{q}_{O_2} = aW^{-b} \qquad (1)$$

where $\dot{q}$ is the respiration intensity of organisms, W their mass, a and b constants.

Thermodynamics and Regulation of Biological and Processes

It has been shown (Dolnik, 1968, 1978, Konoplev et al., 1975; Zotin, Konoplev, 1978; Zotin, Krivolutsky, 1982) that the value of the parameter a increases with increase in complexity in organisation of the animals (Table 6). This means that respiration intensity increases during the course of evolution.

Table 6. Values of parameter for various animal groups (poikilothermal at 20 °C, reptilia at 30 °C, homeothermal in the thermoneutral zone) (Zotin, Krivolutsky, 1982)

Type or class	a mW	Type or class	a mW
Protozoa	0.098	Teleositei	1.66
Porifera	0.140	Amphibia	1.45
Coelenterata	0.192	Reptilia	2.46
Oligochaeta	0.512	Mammalia	18.0
Crustacea	0.727	Primates	36.5
Mollusca	0.895	Aves	
Insecta	3.15	Nonpasseriformes	26.1
Amphioxi	1.59	Passeriformes	43.2
Cyclostomata	1.65		

Dolnik (1968) believed that parameter a was equal in all the representatives of one animal class. However, it was shown that the value of a differs markedly in various orders of mammals, birds and reptilia (Zar, 1968; Konoplev et al., 1975; Bennet, Dawson, 1976). Table 7, e. g. presents data for various classes of mammals from which it can be seen that different orders within a class can be characterised by this parameter.

As seen from Tables 6 and 7, respiratory intensity increases markedly from protozoa to mammals and birds in the animal kingdom and from monotremata to primates in the class of mammals. This points to obvious energetic trends in the volutionary progress of these organisms.

Table 7 Value of parameter a for different orders of mammals (Konoplev et al., 1975)

Orders	a mW	Orders	a mW
Monotremata	6.7	Rodentia	24.2
Marsupialia	13.9	Carnivora	29.2
Edentata	14.2	Perissodactyla	33.6
Chiroptera	18.3	Artiodactyla	33.6
Insectovora	20.8	Primates (Pongidae)	36.5
Lagomorpha	23.0		

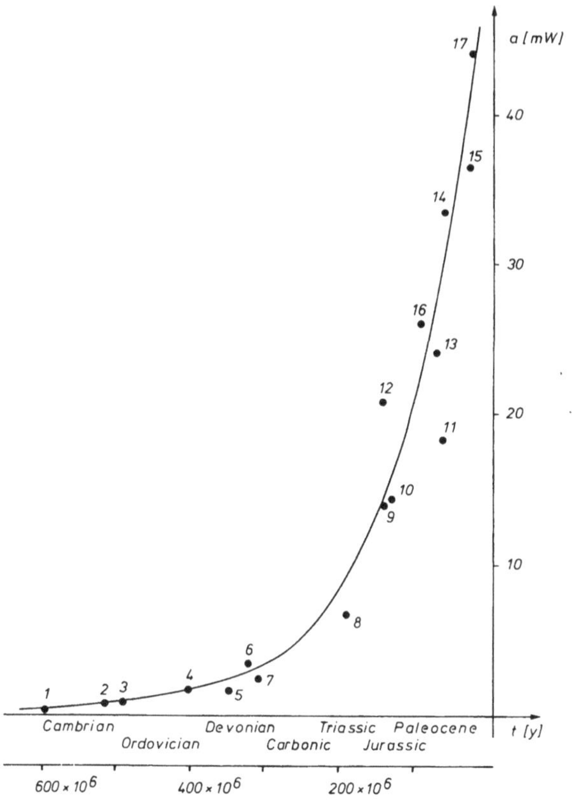

Fig. 58 Change of constant a in the course of evolution of animals (paleonthological data taken from Drushchits (1974) and Naumov and Kartashov (1979): 1-Coelenterata, 2-Crustacea, 3-Mollusca, 4-Teleostei, 5-Amphibia, t-Insecta, 7-Reptilia, 8-15-Mammalia (8-Monotremata, 9-Marsupialia, 10-Edentata, 11-Chiroptera, 12-Insectovora, 13-Rodentia, 14-Perissodactyla, 15-Primates), 16,17-Aves (16-Nonpasseriformes, 17-Passeriformes)

More interesting data were obtained on comparison of the respiratory intensity of each group of animals and the time when this group appeared in paleonthological annals (Zotin, 1979; Zotin, Krivolutsky, 1982). Fig. 58 shows this comparison which implys that in the course of evolution animals with higher levels of respiratory intensity appear in succession. The biological significance of this process consists in an increase in the level of basal metabolism, and consequently, in an increase in the total power in animals (Lomov, 1975), the value of maximal metabolism (Dontsova, Zotin, 1982), the rates of biochemical reactions, i. e. the total pattern called by Shnol (1979) kinetic perfection.

These are the facts that reveal bioenergetic trends in the evolutionary progress. Certainly, considerable work is required to account for the curve in Fig. 58 because of its importance in many problems of evolution, systematics and paleonthology. Nevertheless these data support the view that progressive evolution is associated with an enhancement of the bioenergetic state of the organism. From the data in Fig. 58 this progress can be quantitatively expressed in mW per each million of years:

- during the Cambrian period the value was about 0.005;
- in the Ordovician period 0.011;
- in the Devonian period 0.014;
- in the Carboniferous period 0.024;
- in the Triassic period 0.076;
- in the Jurassic period 0.099;
- in the Cretaceous period 0.192;
- in the Paleocene period 0.269;

i. e. the rate of progress increased.

Physical sense of progressive evolution

The biological significance of bioenergetic trends in evolutionary progress is evident: a higher level of respiratory intensity confers considerable advantages in the struggle for life, in colonising new territories, during adaptation to changing environmental conditions etc.. What, however, is the physical meaning of continuous improvement in an organism's energetic state over the course of evolution?

Many suggestions can be advanced. One of them has been developed in our studies (Zotin, 1979; Zotin, Krivolutsky, 1982). It is as follows. In terms of thermodynamics, the equilibrium state is the most probable state of the system. If the system deviates from equilibrium, the probability of that state decreases and the level of order increases. The further from equilibrium is the thermodynamic system, the more ordered will it be in the steady state. Life on Earth was initiated from 'primordial soup' in a state near equilibrium (Oparin, 1964). The first living systems, therefore, were close to equilibrium and consequently the probability of their state was great, and the order small. During the course of evolution organisms grew further and further away from their initial, close to equilibrium state, and their orderliness increased. This appears to be the physical sense of progressive evolution in organisms: ever greater distance from a state of equilibrium, i. e. from the initial environment in which the living systems originated. This conclusion is proved below on the basis of some thermodynamic concepts.

It is accepted, in the thermodynamics of irreversible processes, that the further from equilibrium is the thermodynamic system, the higher is the value of specific dissipation function:

$$\Psi = \frac{T}{V} \frac{d_iS}{dt} \qquad (2)$$

where T is the absolute temperature, V the volume, d_iS/dt the rate of entropy production within the system.

In the equilibrium state, the specific dissipation function of the system is zero. As the system moves away from equilibrium the value increases. This is valid only for systems close to an equilibrium or stationary state. In non-linear regions the dissipation function is much more complex (Glansdorff, Prigogine, 1971). We have tried to avoid these difficulties by introducing the concept of an external dissipation function (Zotin, 1974; Zotin, Zotina, 1977; Zotina, Zotin, 1982):

$$\Psi_d = \dot{q} \simeq \dot{q}_{O_2} \tag{3}$$

where $\dot{q}$ is the heat procution intensity, $\dot{q}_{O_2}$ the respiration intensity. Unfortunately, by setting the Ψ_d -function equal to respiration (3), which is necessary to use the data on respiration intensity (Table 6, 7, Fig. 58), brings us back into the field of linear non-equilibrium processes. Consequently, all the arguments presented below are within the validity limits of the thermodynamics of irreversible processes.

It has been shown (Zotin, Zotina, 1977; Zotina, Zotin, 1982) that the function is in inverse proportion to the state probability of the system:

$$\Psi_d = \alpha T p_0 / p \tag{4}$$

where p is the probability of the state of the system, p_0 the probability of its state at equilibrium, α constant.

Proceeding from (3) and (4) the probability of the state of the animal group under study can be estimated by means of the concept of relative probability (Zotin et al., 1975). For this purpose, proceeding from (4), we write

$$\Psi_d^{(1)} / \Psi_d^{(2)} = T_1 p_2 / T_2 p_1 . \tag{5}$$

The symbols with subscript 1 refer to the simplest organisms (microorganisms) and with subscript 2 to any other animal. Applying (1) and (3) we obtain:

$$\frac{p_2}{p_1} = \frac{a_1 T_2 W_2^{b_2}}{a_2 T_1 W_1^{b_1}} . \tag{6}$$

Assuming that the probability of a state, P_1, of a common account with mass 1 g at 20 °C is 1, and, taking into account that according to Hemmingsen's (1960) observations a = 0.10 mW, b = 0.25, we obtain the result that the relative probability of the state of any organism of 1 g mass is

$$p_a = 3.41 \cdot 10^{-4} \frac{T}{a} . \qquad (7)$$

For clarity it is more convenient to use a measure reciprocal in meaning to relative probability $C_r = 1/p_a$, the orderliness criterion (Konoplev et al., 1975; Zotin et al., 1975). It follows from (7) that the orderliness criterion is

$$C_r = 2.93 \cdot 10^{3} \frac{a}{T} . \qquad (8)$$

Consequently relative orderliness in the organism, and the distance from the equilibrium state is related to parameter a in equation (1). Thus, the suggestion as to the physical meaning of bioenergetic trends during evolution, i. e. continuous withdrawal of the organism from the equilibrium state during evolution receives some thermodynamic support.

Methodological problems

Three methodological problems arise on assessment of the facts and considerations noted above:

1. the seeming discrepancy between bioenergetic trends in evolution and basic thermodynamic concepts: the systems themselves, and with increasing rate, deviate from the most probable, equilibrium state, and this contradicts the Second Law of thermodynamics and the causality law;
2. the possibility of applying the thermodynamics of linear, irreversible processes to such a complex phenomenon as the evolution of organisms;
3. the relationship between macro- and microevolution, i. e. the specificity of the mechanism of progressive evolution.

The discrepancies associated with the first problem are often perplexing and sometimes result in rather extreme theories (Trinscher, 1967). However, at present bioenergetic trends in evolution can be accounted for without the need for extravagant conclusions. The fact is that bioenergetics progression involves only some of the organisms inhabiting the biosphere; indeed comprising only a small percentage of the total living mass of the biosphere. In a large thermodynamic system such as the biosphere thermodynamically unlikely, processes are possible and, they proceed in one part of the system at the expense of coupled processes taking part in the rest of the system. Consequently, one can believe that, evolution in the biosphere as a whole, proceeds, in accordance with thermodynamic principles, towards the equilibrium state, but that some groups evolve in the opposite direction. A solar energy gradient exists and natural selection stimulates organisms to use this energy more completely. But this is possible only if the organism came into being with a high energetic metabolism. Thus, two factors influence the evolution, i. e. thermodynamic principles urging the process towards the equilibrium state and natural selection forcing the organism to look for new ways of utilising, more completely, the material and energetic resources of the Earth.

The application of the thermodynamics of linear irreversible processes in description of the evolution of organisms is applicable only to those systems which are near equilibrium or stationary states. Therefore its usage in solution of problems of evolution can only be justified if it can be shown that organism evolution is realised through a number of stationary states. As in the course of ontogenesis, and in the adult state, organisms are in a stationary state, such proof is quite possible.

Finally there is no doubt that microevolutionary processes underlie macroevolution, and evolutionary progress. However, that bioenergetic progress is realised by means of specific, for macroevolution, mechanisms is not forbidden. Some of these mechanisms have been discussed in earlier publications (Zotin, Krivolutsky, 1982).

Ontogenesis

Continuous Phase Transitions and Morphogenesis

J. L. Rius, B. C. Goodwin

Introduction

Living systems though relatively stable are in some respects far from thermal equilibrium. Order and functioning are achieved by maintaining a flux of energy and matter through the systems. Biological processes are characterised by their great complexity and order, and one of the most striking features of life is the transition from disorder to order. In physics, a number of phenomena which are well understood clearly show a transition from disorder to order. This ordering can occur with respect to spatial positioning, but also other quantities may undergo an ordering process ('phase transition'), for instance, in superfluid helium, and in superconductors in which a subtle type of long-range order is found due to the collective (cooperative) behaviour of a great number of their constituents. Biological systems also may be expected to exhibit some order of a non-trivial nature. Here, living, ordered states are maintained by a flux of energy through them and certain functions can be achieved only when energy flux is increased.

Biological development is characterised by a sequence of transformations which are initiated at fertilisation. These transitions occur in a continuous manner, the organism changing form and structure in a progressive way, with a gradual increase in order and complexity. A convenient starting point in the study of morphogenesis is the basic pattern of early development. The phenomena which follow fertilisation is characterised by a period during which the egg divides and subdivides to produce an increasing number of cells. This is the period of cleavage or segmentation. In this phase, which is common to all multicellular animals, the single zygote is partitioned into a finite number of smaller cells, or blastomeres. The way

Thermodynamics and Regulation of Biological and Processes

in which the egg is subdivided into the daughter blastomeres is usually very regular.

Fertilisation elicits from the egg a highly regulated but simple cell cycle. Once induced, cleavage goes on by means of a series of mitotic cycles, characterised by the fact that no growth periods separate them. The main processes of division are not only related in time, but also in space. The direction of successive division planes in the cleaving egg are further related, so that in every species a real cleavage pattern is realised.

The cleavage process in embryogenesis follows a very regular spatial and temporal organisation. In order to capture the general features of this order, an analytical description which takes the form of a field theory for developing organisms is used. Considering our developmental field to be expressed mainly on the cortex, the successive stages in the cleavage pattern can be seen as a series of continuous (second order) transitions as metabolism proceeds, where the principle of minimisation of our 'surface free energy' gives the successive stable states that the egg acquires. The cleavage process is then presented as a series of transformations to successively higher energy characteristic states of the morphogenetic field as metabolism goes on. This sequence of states is described by characteristic surface harmonics which define the cleavage planes as nodal lines, particular solutions being selected by the constraints imposed by boundary conditions and the binary division process.

An internal cleavage field involving the mitotic apparatus is seen to depend on the surface cleavage field or cortical field. Relevant biological variables are spatially organised by the internal cleavage field and contribute to it. These factors are incorporated into a unified treatment of the cleavage process involving not only spatial variables but also their temporal behaviour with the metabolic activity supplying the necessary energy flux.

The suggestion that biological systems belong to those systems in which the supply of energy imposes a characteristic order was put forward some time ago by Fröhlich (1968, 1969). He indicates that we must consider biological systems in terms of their activities which require a high degree of organisation. Such organisation may be of a complex nature but it does require consideration of the system as a whole. In microscopic terms it thus requires the introduction of collective properties (Fröhlich, 1980). The possibility of cooperative phenomena in biological processes has been discussed in a series of papers by Fröhlich, who proposes a theory of long-range coherence in biological systems based on the high dielectric polarisability of biological material (Fröhlich, 1968, 1969, 1970, 1972, 1973, 1975, 1977, 1978, 1980).

Continuous Phase Transitions

Events of early embryonic development exhibit a remarkable dynamic order which is expressed in precise spatial and temporal organisation. The co-operation of many individual parts of the embryonic system leads to a sequence of defined spatial and temporal structures which undergo a change of behaviour at the onset of each cleavage cycle. These phenomena, through the supply of metabolic energy, emerge out of phases of lesser 'structure'. Due to the supply of energy the old structure may become unstable and a new structure arises. As metabolism proceeds the new structure can again become unstable being replaced by another and so on for each cleavage stage. A field description is adequate to represent this sequantial emergence of order, and the concept or order parameter, as pointed out by Haken (1978), might play an important role in the understanding and adequate treatment of biological processes.

In the case of early development each new cleavage state grows continuously and homogeneously from the old as metabolism proceeds. Such transitions are said to be second order or continuous. Thus a second order transition is continuous in the sense that the state of the system changes continuously, but it acquires a qualitatively new property at the transition point (Landau, Lifshitz, 1969). Landau's theory of second-order phase transitions provides a relatively simple view of phenomena near the critical point, and it gives a qualitatively correct view of the kind of things which do happen. All second order phase transitions have the same qualitative features.

The most fundamental idea which helps elucidate the behaviour near a critical point is the concept that this transition is describable by an order parameter (Landau, Lifshitz, 1969). This parameter u is a numerical measure of the amount and kind of ordering which is built in the neighbourhood of the critical point. According to Landau, to describe the change in our system from a less ordered phase to a more ordered phase when it passes through the phase transition point, we can define a quantity u, called the order parameter in such a way that it takes non-zero (positive or negative) values in the more ordered phase (less symmetrical phase) and is zero in the less ordered phase (more symmetrical phase).

The symmetry of our system is changed (namely, increased) only when u becomes exactly zero; any non-zero value of the order parameter, however small, brings about a lowering of the symmetry. A passage through a phase transition point of second order has a continuous change of u to zero.

The appearance of order can be viewed as the breaking of a symmetry. In group theoretical language, the ordered (or lower symmetry) phase is then viewed as a subgroup of the parent (or higher symmetry) phase, with the reduced set of symmetry operations being represented by the order parameter (Fleury, 1981). In general the order parameter need not be a simple, one-component scalar quantity; it may be quite complicated depending on the complexity of the two phases it connects. The complexity of the order parameter is related to the number of components n, needed to specify it completely.

According to Landau's theory of second order phase transitions (Landau, Lifshitz, 1969) the energy difference between the ordered and disordered phases is expressible solely in terms of the macroscopic order parameter u. Landau's theory is based on the idea that a phase transition could be characterised by some kind of order parameter, and on a simple postulated form for the dependence of a potential function on the order parameter. The basic idea of Landau's theory is that a phase transition can be regarded as going from an ordered to a disordered phase. If u is our order parameter then as the variable σ decreases u decreases becoming zero at the critical value σ_c . Landau's theory is concerned with the region near the critical value σ_c in which u is small. The stable phase is the one in which the potential function F is a minimum.

The continuity of change of state in a second order phase transition is expressed mathematically by the fact that the quantity u takes arbitrary small values near the transition point. Considering the neighbourhood of this point we expand F in powers of u

$$F = F_n + au + bu^2 + cu^3 + du^4 + \dots \qquad (1)$$

where a, b, c, ... are analytic functions of $\sigma - \sigma_c$, and as we shall see, in order that the transition point itself should be a stable state (i. e. in order that F should have a minimum at that point when u = 0) the third order term must vanish and the fourth order term must be positive-definite.

The stable state of our system is determined from its potential function, F, analogous to the free energy, and the equilibrium phase corresponds to a minimum in F

$$\partial F / \partial u = 0 . \qquad (2)$$

As we have seen, in the normal phase, we must have a minimum at u = 0 which implies a = 0 for $\sigma < \sigma_c$ and since a can be expanded as a power series in $\sigma - \sigma_c$, this means we must have a = 0 for all σ . Due to symmetry considerations in a number of cases the odd order terms become zero. We are thus left with

$$F = F_n + bu^2 + du^4 + \dots \qquad (3)$$

(2) for the value u_o, of us at the minimum gives

$$bu_o + du_o^3 = 0 \qquad (4)$$

with solutions $u_o = 0$ and $u_o^2 = -b/d$. Now we want $u_o = 0$ to be the only solution for $\sigma < \sigma_c$, whereas for $\sigma > \sigma_c$ we must have a solution with $u_o \neq 0$. We can achieve this if we take the σ dependence so that -b/d is positive for $\sigma > \sigma_c$ and negative for $\sigma < \sigma_c$. In addition, we must have d positive at all σ, since if d were negative F would decrease indefinitely for large values of u. We therefore want b to be negative for $\sigma > \sigma_c$ and positive for $\sigma < \sigma_c$. The simplest σ dependencies that give this are

$$b(\sigma) = B(\sigma_c - \sigma)$$

and (5)

$$d(\sigma) = d(\sigma_c) = d > 0 .$$

With b and d given by (5) the solution for u_o, from (4) is

$$\begin{aligned} u_o &= 0 & \sigma < \sigma_c \\ u_o &= B^{1/2}(\sigma - \sigma_c)^{1/2}/d^{1/2} & \sigma > \sigma_c \end{aligned} \qquad (6)$$

In addition, $u_o = 0$ remains a solution above σ_c, but as we shall see, it corresponds there to a maximum not a minimum. (6) gives a rapid parabolic increase in u_o as σ increases from σ_c, as shown in Fig. 59.

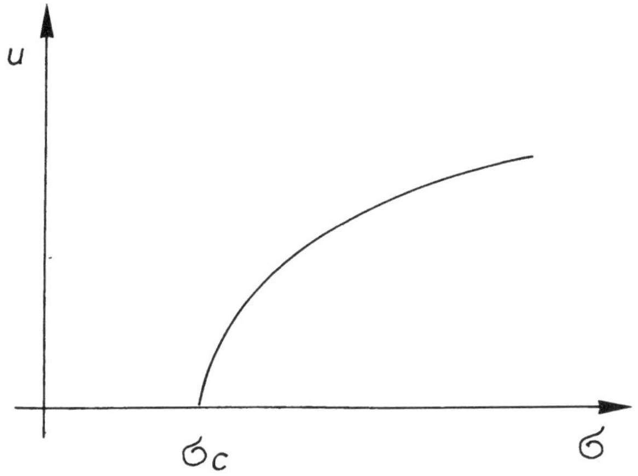

Fig. 59 Explanation in the text.

Fig. 60 shows the dependence of F on u for various values of σ. For σ well below σ_c, F is simply a parabolic-shaped curve, with a well defined minimum. As σ approaches σ_c and b(σ) decreases, the minimum becomes more shallow. When σ passes through σ_c, and b (σ) becomes negative, the minimum at u = 0 becomes a local maximum, and minima develop at $u = \pm u_o$. As σ increases further, the minima move out in u and slowly become deeper. F always increases for large enough u, as we ensured by taking d to be positive.

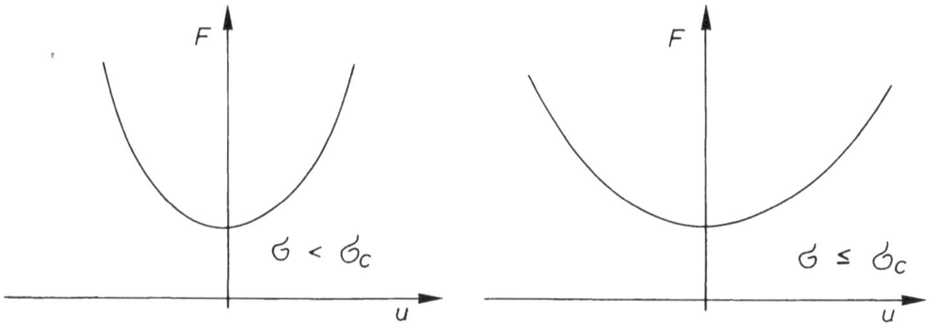

Fig. 60 a Explanation in the text.

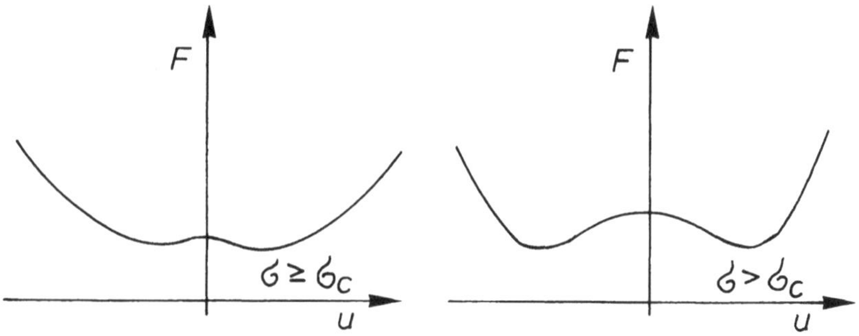

Fig. 60 b Explanation in the text

When the order parameter u is not constant in space, the 'free energy density', i. e. the 'free energy per unit volume' will depend not only on u but also on its derivatives with respect to the coordinates (in the first approximation, on the first derivatives). Accordingly f (F per unit volume) must be expanded in powers of u and of their gradients ∇ u near the transition point. If the 'free energy', i. e. the integral $\int$ f dv over the whole volume is to be a minimum for constant u, it is necessary that the first-order terms in the gradients in this expansion should vanish identically (the terms quadratic in the derivatives must be positive-definite).

Near the transition point, the function F(u) can be expanded in powers of the function u; since this function is slowly varying, we need only consider expansion of terms of the lowest order in its derivatives (and the lowest powers of these):

$$F = \int f dv = F_n + \int (bu^2 + du^4 + \alpha(\nabla u)^2) dv. \qquad (7)$$

The final term in this equation makes the 'free energy' larger when u(r) varies in space. Hence this term serves to damp out spatial variations in u(r).

The 'free energy density' is then given by

$$f = bu^2 + du^4 + \alpha(\nabla u)^2. \qquad (8)$$

Minimising the 'total free energy' $\int f\, dv$ leads to the Euler-Lagrange equation

$$\alpha \nabla^2 u + bu + du^3 = 0. \qquad (9)$$

Field description of cleavage

Soon after fertilisation the zygote becomes rapidly converted into a population of cells through a series of cell division. The period of cleavage is considered to extend from fertilisation to the time of formation of the multicellular embryo, i. e. blastula stage. Generally, there is little evidence of growth during this time and the shape of the embryo does not visibly change.

Embryonic development is typically polarised (Nieuwkoop, 1977). In the growing oocyte or the mature egg, dorsoventral polarity cannot yet be distinguished. Usually only a single primary, animal-vegetal polarity is evident from the very beginning. The primary dorso-ventral polarisation of the egg often becomes manifest only shortly before cleavage.

A cell consists of three basic components - nucleus, cytoplasm and cell membrane. Although the nucleus is a very complex structure, it does not seem to have an intrinsic polarised organisation: nuclear transplantation does not influence the polarity of the cell. It seems much more likely, therefore, that the polarisation of the cell depends upon that of the cytoplasm and/or the cell membrane. Although the cytoplasm shows a dynamic ultrastructure consisting of microtubules, membranous structures, etc., which may undergo drastic changes e. g. during cytokinesis, it seems at present most likely that polarising functions of the egg, such as its animal-vegetal and dorso-ventral polarities, reside primarily in the cortical layer. It must however be realised that there are intricate interrelations between the cortical layer and the rest of the cytoplasm; but these may be of a second nature only.

Polarity is essentially a global property since it may either encompass the entire egg (animal-vegetal polarity) or at least a considerable portion of it (dorso-ventral polarity). During cleavage, through which the unicellular egg cell is subdivided into a logarithmically increasing number of daughter cells, polarity remains a global property of the developing embryo. In other words, cleavage does not affect the polarity(ies) of the egg. During cleavage the original single and global cortical layer of the egg remains functional as a continuous layer in the pluricellular embryo by means of a network of interconnecting desmosomes. The latter link the cortical layer of the individual blastomeres together into a global system, despite the fact that it is cut into ever smaller fragments. Therefore, although during cleavage the cytoplasm of the egg becomes organised into an increasing number of individual units, each with its own nucleus, the cortical layer maintains its global character.

This situation probably continues as long as the cortical layer of the embryo maintains its individuality and no other structures are able to take over its unifying function. This is certainly the case during early development, and possibly even well into the postneurela stages. It seems, however, that the unifying function of the cortical layer begins to weaken as soon as mesodermal differentiation becomes manifest. In the mesodermal mantle the individual cells seem to be more loosely connected with each other; where in the mesodermal marginal zone of the blastula a continuous cortical layer existed previously, it seems to break down during or after invagination and endodermal tube formation.

The plasma membrane is a selectively permeable barrier, i. e. it is not equally permeable to all types of molecules, and it is this selectivity that allows the distribution of substances inside the cell to be totally different from that outside the cell. A system consisting of two compartments separated by a membrane that is permeable to some but not all charged particles present will, at equilibrium, have unequal concentrations of the permeable ions and an electrical potential difference across the membrane. A steady state (Gibbs-Donnan equilibrium) will be reached when the electrical potential gradients exactly counteract the net diffusional gradients including the osmotic gradients.

If one measures the concentrations of the two most important cations, Na^+ and K^+, their distribution across the cell membrane is not that expected from the Donnan electro-chemical equilibrium. Cells have the ability to develop extreme concentration gradients across their plasma membranes by means of active transport. The distribution of ions across the membrane is a steady-state condition, in which sodium ions are continually leaking in by diffusion and being pumped out by active transport. Similarly, the movement of postassium ions occurs in the reverse direction by the same type of processes.

Due to the presence of nondiffusible anions inside the cell, the resulting voltage across the membrane is negative inside with respect to outside. Typical values for living cells would be -50 to -90 mV, the negative sign by convention indicating negative inside with respect to the outside. The resting potential is affected by the ionic composition of the external medium.

Resting potentials have been measured in a wide variety of egg cells (Hagiwara, Jaffe, 1979), and an increase in membrane potential during early cleavage stages has been reported for several eggs. Tupper (et al., 1970) observed an increase in membrane potential for the echinoderm

Asterias (starfish). A similar increase in membrane potential was observed by Ito and Hori (1966) during cleavage in the newt *Triturus pyrrhogaster*. It was also shown that the resistance of the surfaces at the level of the junction between blastomeres, i. e. the resistance across the former cleavage planes are very small. Voltage attenuates very little from one blastomere to another. This close electrical coupling between embryonic cells opens the possibility that embryonic cells in some eggs communicate through low junctional resistance during their development.

Palmer and Slack (1970) measured the membrane potential during cleavage in eggs of the toad *Xenopus laevis*. The membrane potential increased from -6.5 to -57 mV at the mid-blastula stage. Current spread between cells apparently occurs less easily during mid-blastula stages than at earlier stages in development perhaps indicating an increase in junctional resistance during development.

The measurements of membrane potential in *Xenopus laevis* showed the same trends as that reported by Ito and Hori (1966) in *Triturus* embryos, namely, that this rises progressively during the stages prior to gastrulation. During the first cleavage no voltage drop was recorded across the intercellular junctions, due to the low junctional resistance, and the embryo may be considered electrically as a uniform polarised sphere. Also, isolated pairs of blastomeres from early amphibian embryos (*Xenopus*) were shown to be electronically coupled (Spray et al., 1979). This coupling between blastomeres provides a global property to the egg during early development. Due to uncoupling of specific cells or cell groups at specific times large differences in resting potential betwen different regions of an embryo can be developed, thus providing a means for differentiation (Blackshaw, Warner, 1976).

Experimental evidence and theoretical considerations support the view that the global properties of the egg resides in the cortical layer, which acts as a functional unit. Considering the developmental field to be expressed mainly in the cortex, we shall see that the remarkable spatial and temporal order of the successive stages in the cleavage process can be seen as a series of continuous transitions as metabolism proceeds, where the principle of minimisation of the 'surface free energy' gives the successive stable states that the egg acquires just before the onset of each new series of cell divisions and which determine the cleavage planes.

As cleavage proceeds the egg uses its metabolic energy in order to acquire each new cleavage state. Supply of energy causes a change of behaviour, comparable to a phase transition, imposing a characteristic order in our biological system as proposed by Fröhlich (1968; 1969). The cleavage process is then seen as a series of transformations to successively higher energy, characteristic states of the morphogenetic field as metabolism proceeds.

In some physical systems, as we decrease the supply of energy σ below a critical threshold σ_c , (8) for the 'free energy density' describes the passage through a phase transition point of second order in which the state of the system changes continuously, but it acquires a qualitatively new property at the transition point, such as the onset of superfluidity or superconductivity. In the cleavage process we have the situation in which supply of energy above a critical threshold gives a second order phase transition in which the new phase grows continuously from the old, but at the transition point of each new cleavage state it acquires a new characteristic defined by our order parameter u.

Considering the cortex (the ectoplasm) of the egg as a spherical shell, its cortical field is described on this thin surface layer of the organism. Our starting point in the field description of cleavage is the expression for the 'cortical free energy' as function of $u(\theta, \phi)$. In accordance with the general ideas of Landau's theory, this is found by expanding the 'surface free energy density' in powers of the small (near the transition) order parameter $u(\theta, \phi)$ and its derivatives. In spherical polar coordinates it is given by an expression of the form

$$\mathcal{E}(\theta,\phi) = A\left[\left(\frac{\partial u}{\partial \theta}\right)^2 + \frac{1}{\sin^2\theta}\left(\frac{\partial u}{\partial \phi}\right)^2 + \beta u^2 + \delta u^4\right] \qquad (10)$$

where A , β , δ , are constants incorporating the relevant physiological properties of the surface. Following the description of the cleavage process given by Goodwin and Trainor (1980) we suppose that the characteristic cleavage states correspond to a minimum of the 'total surface free energy' E

$$\delta E = \delta \int_0^{2\pi} \int_0^{\pi} \mathcal{E}(\Theta, \phi) \sin\Theta \, d\Theta \, d\phi = 0 \tag{11}$$

leading to the Euler-Lagrange equation

$$\frac{1}{\sin\Theta} \frac{\partial}{\partial\Theta} \left(\sin\Theta \frac{\partial u}{\partial\Theta} \right) + \frac{1}{\sin^2\Theta} \frac{\partial^2 u}{\partial\phi^2} - \alpha u - \gamma u^3 = 0 \tag{12}$$

and due to the smoothness property of developmental fields (Goodwin, 1977) we expect Landau's theory to be valid near the transition point where u(θ, ∅) takes arbitrary small values, therefore in the neighbourhood of this transition point (12) takes the form:

$$\frac{1}{\sin\Theta} \frac{\partial}{\partial\Theta} \left(\sin\Theta \frac{\partial u}{\partial\Theta} \right) + \frac{1}{\sin^2\Theta} \frac{\partial^2 u}{\partial\phi^2} - \alpha u = 0 . \tag{13}$$

which is the well known Helmholtz equation.

If we consider our order parameter u to be finite, single-valued and continuous over the sphere, the possible solutions are restricted to a characteristic (eigenfunction) set, viz. The spherical harmonics (real part taken)

$$u(\Theta, \phi) \rightarrow Y_\ell^m(\Theta, \phi) = \sqrt{2} N_{\ell m} P_\ell^m(\cos\Theta) \cos m\phi \tag{14}$$

where ℓ takes on the integral values 0, 1, 2, etc. and for given ℓ, the m values are integers ranging from $-\ell$ to $+\ell$. The parameter α is restricted to the corresponding characteristic values (eigenvalues) $\ell(\ell + 1)$. In (14), the P_ℓ^m are associated Legendre polynomials and the $N_{\ell m}$ are the normalisation constants given by

$$N_{\ell m} = \left[\frac{2\ell+1}{4\pi} \cdot \frac{(\ell - |m|)!}{(\ell + |m|)!} \right]^{1/2} \tag{15}$$

The 'surface free energy' corresponding to each characteristic cleavage system state is given by

$$E_\ell = A \left[\ell(\ell+1) + p \right] \tag{16}$$

However as Goodwin and Trainor (1980) show, not every characteristic state is realised in the cleavage process since the mitotic apparatus imposes a biological imperative that the number of cells is doubled in each cleavage stage.

After cleavage a new surface field will be present due to the new constraints imposed by the cleavage furrow. Each cleavage modifies the surface field and hence the position of subsequent cleavage planes. Thus, the constraints imposed by boundary conditions and the binary division process determine the whole pattern of cleavage. The (ℓ , m) pair selected at the pth synchronous division, where the harmonic function $Y_\ell^m(\Theta, \Phi)$ describes only the new planes occurring at the division, is given by the following constraints:

$$\ell = 2^{n-1} \begin{cases} m = \ell \text{ when p even} & (p = 2n) \\ m = 0 \text{ when p odd} & (p = 2n + 1) \\ \text{except for the case } p = 1 \ (n = 0) & \\ \text{where } \ell = 1,\ m = 1. & \end{cases}$$

Since the 'free energy' involved in cleavage is solely a function of ℓ as shown in (16), there is a degeneracy of the pattern for p = 2n and 2n + 1 for any n. This is resolved by selecting the maximum value of m as the solution which is expressed first for given ℓ , a selection rule which, as previously described (Goodwin, Trainor, 1980), may be interpreted biologically as the action of a weak field in the direction of the animal-vegetal axis which selects vertical over horizontal cleavages when there is a choice due to equal values of the 'free energy'. This field is evident in most holoblastically-cleaving eggs as a pigment and/or yolk gradient, revealing the animal-vegetal polarity. Thus the cleavage pattern is interpreted to arise from four factors:

- a global field constraint,
- 'free energy' minimisation,
- binary cell division,
- a weak polar field which resolves degeneracies. The following table illustrates the correspondence between cleavage states and harmonic functions for the first 9 cleavages:

Cleavage number (p)	Number of cells (2^p)	Number of cleavage planes (2^{n-1})	(ℓ, m) pair	$u(\bar{\theta}, \phi)$
1	2	1	(1,1)	$Y_1^1 = C_1 \sin\theta \sin\phi$
2	4	1	(1,1)	$Y_1^1 = C_2 \sin\theta \cos\phi$
3	8	1	(1,0)	$Y_1^0 = C_3 \cos\theta$
4	16	2	(2,2)	$Y_2^2 = C_4 \sin^2\theta \cos 2\phi$
5	32	2	(2,0)	$Y_2^0 = C_5(3\cos^2\theta - 1)$
6	64	4	(4,4)	$Y_4^4 = C_6(\sin^4\theta \cos 4\phi)$
7	128	4	(4,0)	$Y_4^0 = C_7(35\cos^4\theta - 30\cos^2\phi + 3)$
8	256	8	(8,8)	$Y_8^8 = C_8(\sin^8\theta \cos 8\phi)$
9	512	8	(8,0)	

Fertilisation induces a number of changes at the surface of the egg, and in some cases, it is evident that these changes have a profound influence upon the organisation and future development of the embryo. The action of the spermatozoon in part is to break the inital radial symmetry of the cortex, due to the original animal-vegetal polarity, producing a bilateral one. The main directional axes of the future embryo (dorsoventral; left-right) which reside in the cortical layer, are organised and determined by the site of sperm penetration. If we consider that our surface field $u(\theta, \phi)$, due to changes occurring in the cortex by sperm penetration, takes a zero value $|u| = 0$ at the point of sperm entry, then our first solution $Y_1^1(\theta,\phi)$ becomes stabilised. For subsequent cleavages, the modifications induced on the surface by the new furrows give new boundary conditions, the cortical field taking a maximum value $|u|_{max}$ in the regions where the daughter cells meet. The following diagram (Fig. 61) illustrates this correspondence for the sequence of states which occur during the typical cleavage pattern.

The internal cleavage field which involves the mitotic apparatus, depends on the surface cleavage field or cortical field $u(\theta, \phi)$. The internal field establishes the cleavage furrow which appears during telophase in the

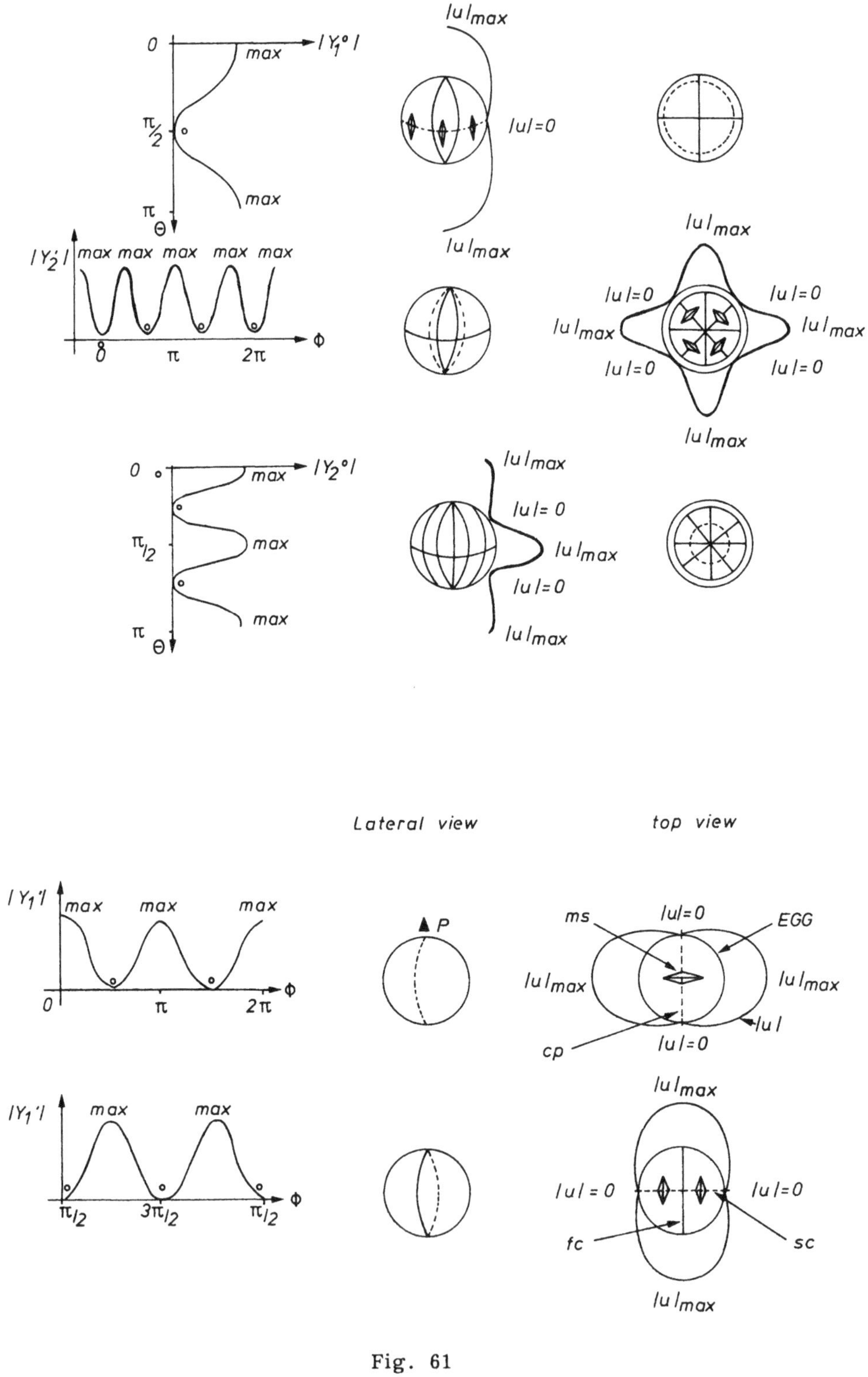
|Y₁⁰|
0
π/2
π
Θ
max
|u|max
|u|=0
|Y₂¹|
Φ
2π
|Y₂⁰|
Lateral view
top view
|Y₁¹|
P
ms
EGG
cp
|u|
3π/2
fc
sc

Fig. 61

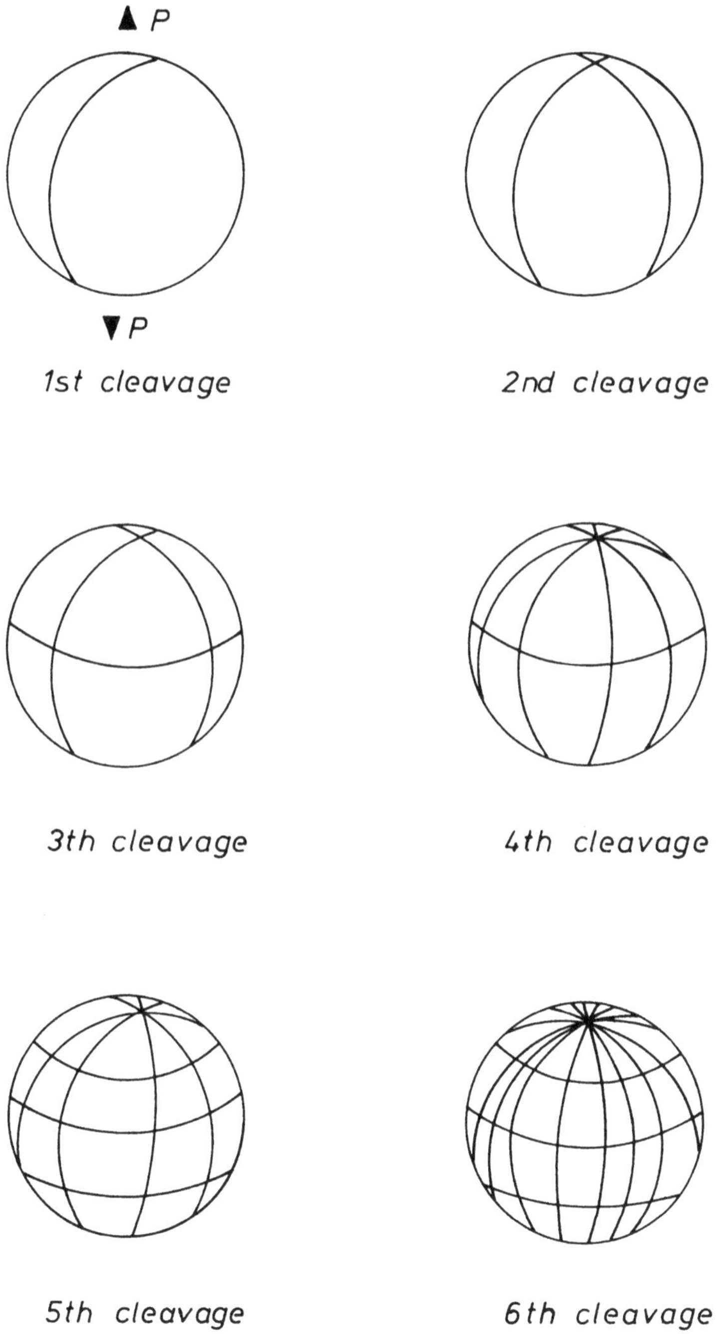
P
P
1st cleavage
2nd cleavage
3th cleavage
4th cleavage
5th cleavage
6th cleavage

Fig. 62

plane of the metaphase plate. Centrioles, asters, mitotic spindles, condense chromosomes, and other variables must be spatially organised by the internal cleavage field and contribute to it. These factors can be incorporated into a unified treatment of the cleavage process, involving not only spatial variables but also temporal behaviour, using Fröhlich's concepts of long-range coherence in biological systems.

Long-range coherence in biological systems

The participation of the vibrational modes of dipolar constituents in biological process is discussed by Fröhlich, who considers that coherent behaviour is possible if energy is supplied to the dipolar elements above a certain critical rate which causes the oscillating subunits to condense into the lowest energy mode in a form characteristic of Bose condensation. Fröhlich proposes that through metabolic energy supply coherent electric vibrations in the region of 10^{11} Hz can be excited.

Strong excitation of a polar mode tends to deform the system in which it acts and such deformations are counteracted by elastic restoring forces. A second proposal assumes that due to non-linear coupling of the elastic field with the polarisation field, an excited metastable state with very high electric polarisation exists.

An essential feature in the two cases is the basic importance of non-linear characteristics in conjunction with a supply of energy. The established new order is shown in one case to be dynamic in terms of coherent excitations of electric vibrations. This type of excitation could have very important biological consequences for it would lead to selective long-range interactions. The selectivity of the long-range forces depends on the frequency of the oscillating systems, which in turn depend on their structure.

Cell water contains considerable amounts of small ions which seriously affect the long-range interactions of charged groups. For, statically, all charges surrounded by cell water are screened within a very short distance through mobile ions. Dipoles or higher poles are not necessarily completely screened because short-range interactions may affect positive mobile ions in a different way from negative ones so that the screening in such cases need not be complete. Such screening does not hold, however, in the case of dynamic interactions arising from the fields of oscillating charges in the frequency regions considered.

Membranes of biological systems usually have a thickness of about 10 nm across which an electric potential difference of the order of 100 mV is maintained. An enormous electric field of the order of 10^7 V/m thus acts on the membrane which, therefore, is strongly polarised electrically. Oscillations with displacement perpendicular to the surface exist so that for the longest wavelength the membrane thickness equals half the wavelength. Assuming elastic properties corresponding to a sound velocity of 10^3 m/s this yields a frequency $\nu \sim v/\ell$ of 0.5×10^{11} Hz. The membrane usually contains a considerable number of proteins, dissolved in it, so that small sections of the membrane (e. g. between two proteins) might vibrate separately from the rest. Due to the high electric field present, the whole membrane system is strongly polarised. Hence excitation of vibration of a particular section of the membrane is connected with a vibrating electric dipole. Individual proteins within the membrane will also be strongly polarised. Their oscillations may also give rise to oscillating dipoles, probably with higher frequencies. Molecules outside the membrane may, in principle, also oscillate as many other sections of a cell, and a great variety of possible frequencies exist. Sections of giant molecules such as DNA or larger complexes might oscillate with lower or higher frequency.

Oscillating polar systems of this type interact with each other, and their resulting oscillations can usually be presented in terms of normal modes. Due to this interaction one would expect the appearance of bands of modes. Some of these modes will extend spatially over relatively large regions; others may be restricted to regions around certain molecules. Fröhlich pointed out that the lowest frequency of these modes can be excited very strongly and coherently through the supply of energy from metabolic processes. The existence of such coupled vibrations will have consequences which are far-reaching and relatively independent of the manner in which they are excited, provided they involve oscillating electric dipoles. The possibility of switching on or off the collective mode through metabolic processes might give rise to changes in the configurations of large regions such as required prior to cell division. The supply of metabolic energy, in turn, would have to be regulated so as to be available at the correct time. A whole temporal succession would thus arise.

In order to find an interpretation of the field descriptors for the cleavage process in terms of relevant variables, we note that during cleavage the original single and global cortical layer of the egg remains functional as a continuous layer in the pluricellular embryo, this coupling between blastomeres provides a global property to the egg during early development; and electrically we can consider the embryo as a uniformly polarised sphere. Due to the high electric field present the whole membrane system is strongly polarised and vibrations of particular sections of the membrane are connected with oscillating electric dipoles. Therefore, for the embryo, we have the possibility of coherent excitation of electric vibrations (polarisation waves) through supply of metabolic energy, as proposed by Fröhlich.

The descriptive field $u(\theta, \emptyset)$, associated with the membrane of the egg, i. e. the 'surface cleavage field' may be taken as an order parameter which measures the amplitude of the coherent polarisation wave generated by Fröhlich's phonon condensation. At the transition the coherent excitation of the lowest frequency polarisation mode of the membrane system will form a pattern which yields the minimum of a certain not yet defined free energy.

Fertilisation initiates the development of the egg which begins a highly regulated but simple cell cycle. Once induced, cleavage goes on by means of a series of mitotic cycles. The cell cycle in the early blastula period is a simple one, having only M and S phases (Graham, Morgan, 1966). Using Fröhlich's concepts Cooper (1979, 1980) has proposed a model for the eukariotic cell cycle, and from this we have that cleavage will proceed due to the coherent excitation of the lowest frequency polarisation mode of the system through supply of metabolic energy. After each cleavage the cells will enter another cycle. For these embryonic cells, the rate of supply of energy to the polar modes through metabolic processes, σ , should be larger than the critical value, σ_o , necessary for Fröhlich's phonon condensation, $\sigma \geq \sigma_o$. After the reproductive cycle of states, the embryonic cells upon passing out of the M or mitotic state, revert to the reproduction cycle again.

The smoothness property which characterises developmental fields, makes possible the validity of Landau's theory near the transition point where the order parameter u is very small. In the cleavage case the ordered phase is achieved by maintaining a flux of energy. If u(θ, ∅) is our order parameter which measures the amplitude of Fröhlich's coherent electric vibrations, then as the supply of metabolic energy to the cellular polarisation modes increases, a critical value σ_o will be reached at which u will increase continuously from zero (Fig. 63).

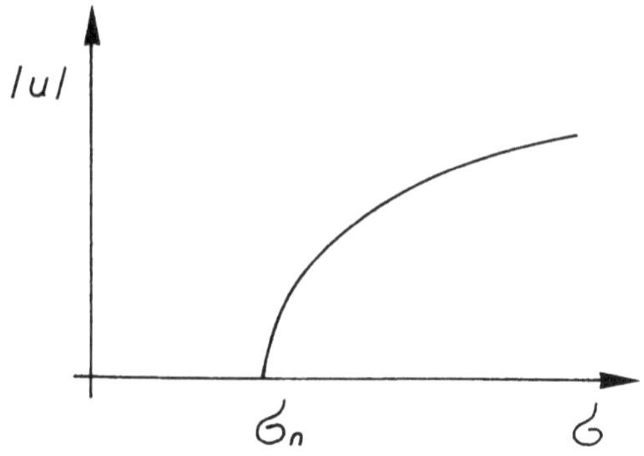

Fig. 63.

The corresponding phase transition is enforced by the supply of energy, and the lowest frequency polarisation mode gets very strongly (coherently) excited when the energy supply is larger than the critical level. Landau's theory is concerned with the region near the critical value σ_o in which the 'free energy' is a minimum.

One of the consequences of excitation of coherent electric vibrations is the establishment of long-range selective interaction between systems with nearly equal frequencies (Fröhlich, 1980). If the lowest frequency in a band of polarisation waves is very strongly excited $n_- \gg 1$, then in the case of near resonance $w_1 \simeq w_2$, the interaction between the systems has very long-range $I = -(R_0^3 / R^3 \varepsilon_-)E_-$, where E_- is the energy of excitation ($E_- = \hbar n_- w_1$). Since the magnitude of the interaction depends on the strength of the excitation, then a temporal sequence of events will follow depending on whether or not metabolic energy is available.

At certain periods during the development of the cell, due to metabolic activity, different strongly excited modes will appear. The long-range interaction between systems, being frequency selective, can lead to long-range recognition and attraction if w_- is excited. This might, for instance, hold for the attraction of centrioles to specific sites on the membrane, when the appropriate frequency on those sites, interacting selectively with the mitotic centres has been excited. Our cortical field $u(\theta, \phi)$, associated with the magnitude of the coherent polarisation wave, could be related to the orientation of the mitotic spindle; centrioles being attracted toward the regions of the membrane which are strongly excited $|u|_{max}$.

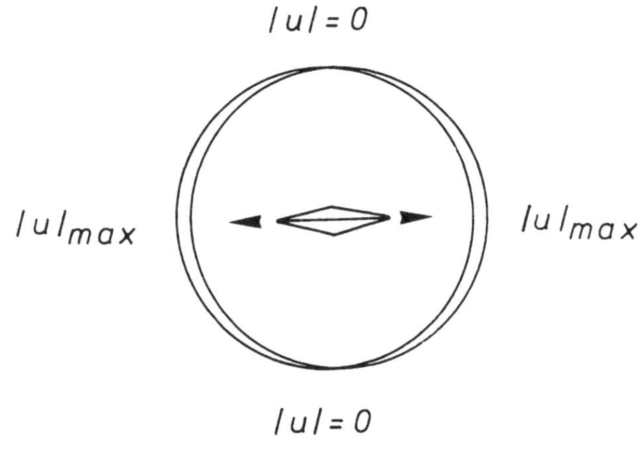

Fig. 64

As the cell progresses through the cell cycle the centriole pairs, associated with spindle assembly, separate and are oriented by the cortical field. By the end of prophase the two pairs of centrioles define the poles of the mitotic spindle. The separation of centriole pairs and their association with the spindle during cell division ensures that each daughter cell receives two centrioles, giving binary cell division.

Conclusions

The approach to morphogenesis described in this paper, wherein spatial patterns result from global fields defined by an order parameter which describes relevant molecular and cellular order, has been applied to one particular problem in embryogenesis but is clearly of much more general applicability. Long-range order is a basic feature of morphogenetic processes, and although coherent states of the type involved in continuous phase transitions are only one way of achieving such order, there is much circumstantial evidence to implicate them. Direct experimental evidence for the occurrence of coherent electromagnetic oscillations in the microwave range in dividing cells has been obtained by Webb (1980) and by Zimmermann (et al., 1981), and with the development of improved techniques for the observation of coherent states one may confidently expect an extension of such studies to embryonic processes. What these promises provide for biology is an important contribution to our understanding of certain categories of force involved in the systematic transformations of spatial order which cannot be accounted for by short-range molecular and cellular interactions of the type which have been so successfully studied in molecular biology.

Stability of Ontogenesis

A. I. Zotin, T. A. Alekseeva

The problem of stability arises not only on consideration of living systems, but on consideration of any other non-equilibrium system as well: only in the equilibrium state is the physical system absolutely stable. Non-equilibrium systems are unstable, and, to maintain their stability and consequently their reliability certain conditions must be fulfilled.

Special stability criteria are developed in thermodynamics, and, as they are of general pattern, they are valid for any system, including living systems. At the same time living processes have some specific features which makes impossible the direct application of the stability criteria developed in thermodynamics. This specificity is determined mainly by the fact that all animals and plants are organised systems, i. e. systems in which regulation and control are realised. In such systems the validity of thermodynamic criteria is not so evident, since they are concealed within other systems more specific to living systems. This does not mean that the thermodynamic criteria of stability do not hold for organised systems - they are simply transformed because of the regulatory and control processes proceeding in the system.

Stability criteria in thermodynamics

Thermodynamic stability criteria are closely related to extremal principles in thermodynamics, which, finally, depend upon the Second Law of Thermodynamics. The physical sense of these principles is, in practice, reduced to the statement: at a steady state of the thermodynamic system the rate of energy dissipation is minimal.

Thermodynamics and Regulation of Biological and Processes

Energy dissipation is accompanied by entropy production in the system. For an isolated system, this brings about a continuous rise of entropy which reaches the maximal value in the equilibrium state. After that, the entropy does not change further, and therefore, in the equilibrium state the rate of energy dissipation is minimal, and equals zero.

In open, non-equilibrium systems, changes in entropy proceed, not at the expense of entropy increase within the system, where irreversible processes proceed but, as a result of an entropy current from the environment to the system, and *vice versa*. Therefore, the rate of dissipation of energy ought to be compared only with the rate of entropy production inside the system; rather than with the total entropy change of the system.

Open systems may be divided, conventionally, into linear (near equilibrium), slightly non-linear (close to linear systems) and strongly non-linear (far from equilibrium). There are thermodynamic criteria of stability for the first two types only. So, for linear systems the rate of energy dissipation may be compared with the rate of entropy production within the system. In this case the stationary, steady state of the system is characterised by a minimal rate of entropy production, Prigogine's theorem (Prigogine, 1967). For non-linear systems, close to linear systems, a theory has been proposed (Glansdorff, Prigogine, 1954, 1964; Glansdorff, 1960; Glansdorff, Prigogine, 1971), according to which of the two sources of energy dissipation within the system (related to currents and to forces) that dissipation which depends on forces is minimised.

Another approach to slightly non-linear systems is concerned with consideration of the fate of dissipation of the heat produced as a result of irreversible processes (Zotin, 1974; Zotin et al., 1975; Zotina, Zotin, 1982). In this case the dissipation function

$$\Psi = \frac{T}{V}\frac{d_iS}{dt} \tag{1}$$

is divided into two parts:

$$\Psi = \Psi_u + \Psi_d \tag{2}$$

where Ψ is the dissipative function; T the absolute temperature, V

the volume of the system; d_iS/dt the rate of entropy production in the system; Ψ_u the bound dissipation function, Ψ_d the external dissipation function ($\Psi_d = \dot{q}$, where $\dot{q}$ is the intensity of heat production in the system).

It has been shown that, in a stationary state of a slightly non-linear system the external dissipation function takes the lowest value (Zotina, Zotin, 1982):

$$\Psi_d = min. \qquad (3)$$

The problem of stability of strongly non-linear systems is more complicated. We have no stability criteria here as yet. It seems, however, evident that such criteria ought to be somehow or other related to the minimal rate of energy dissipation in the system.

Stability of non-equilibrium processes

Stability in equilibrium or quasi-equilibrium processes is evident, since in these cases the systems pass through a number of equilibrium (quasi-equilibrium) states, i. e. through a number of stable states. The pattern in steady, non-equilibrium processes is not so evident.

As discussed by Prigogine (1967), when there are two time measures, the evolution of an open system toward equilibrium is subdivided into two types of changes: manifestation of the stationary state and, evolution of the entire system towards equilibrium via a number of stationary states. In this case the principle of 'fastest fall' (Presnov, 1978) is applicable to the evolutionary processes of the system. This principle is a criterion of stability in a non-equilibrium process proceeding in a linear or slightly non-linear system. It should be noted that, in this case, the stability of the process can be characterised by the integral thermodynamic criterion: a non-equilibrium process is stable when the total, specific energy dissipation in the system is minimal.

For slightly non-linear systems this means that

$$\int_0^t \Psi_d \, dt = \min \tag{4}$$

i. e. the variational principle holds

$$\delta \int_0^t \Psi_d \, dt = 0 . \tag{5}$$

Steady state of living systems

There is stored energy in living systems which provides for some definite level of energy metabolism in the organism. Simultaneously, the system of energy reservoirs requires special mechanisms for their replenishment. However, these additional internal energy sources do not determine the specific features of living systems. Regulation and control are the most characteristic features and they largely determine all the processes proceeding in living systems, including energy metabolism. Consequently, in addition to ordinary thermodynamic flows and forces, which determine the evolution of an open system and its stationary state, there exists a special system of parameters in living systems which describes the processes of regulation and control. By use of internal sources of energy it is possible to create stability on the system when it is not in a stationary state. In biology, such a steady state is called homeostasis. A great number of investigations have dealt with the determination of homeostasis and homeostatic mechanisms of individual physiological functions (Cannon, 1932; Brody, 1945; Grodins, 1963; Rosen, 1967; Ivanov, 1972; Malenkov, 1976; Gorizontov, 1981; Eroshenko, 1982; Shidlovsky, 1982 and others). In general by homeostasis is meant the property of a living system to maintain the stability of internal parameters, when the environmental parameters change, within some limits. Consequently, unlike the stationary state of simple physico-chemical systems, homeostasis is not completely determined by the external parameters of the system. The second specific feature is that the rate of energy dissipation at homeostasis always exceeds that in the stationary state of the same system which would evolve, if there were no control and regulation. Finally, as was mentioned by Balmukhanov (et al., 1970), biological homeostasis, in cell division, involves a number of stationary states, which form a closed cycle. A cyclic pattern to homeostasis no doubt arises in other cases too, Thus the energetics of the organism as a whole has a circadian rhythm associated with the active and sleep periods of the organism. In spite of the above the stability of homeostasis can be defined by some minimal value in the rate of energy dissipation. When the system is stimulated the rate of dissipation increases or decreases, but, after the stimulation ceases the system returns to the minimum value of the rate of energy dissipation. Since homeostasis in organisms is changed in cycles over 24 hours, it can be satisfactorily characterised by some average value of heat production intensity; which corresponds to the standard metabolism of the organism.

The notion of homeostasis is more complicated than that of the stationary state, since it is related to specific, regulatory and control processes. Therefore, when some value of heat production intensity in the homeostasic state is established not only thermodynamic but also purely biological regulation occurs. The latter is determined by the forces of evolution, primarily by natural selection. This concept is realised in different expressions of the optimisation principle in biology (see Rosen, 1967; Khanin et al., 1978). It appears that a combination of evolutionary principles of optimisation and thermodynamic regulation determines stability in non-equilibrium systems and permits the development of a quantitative theory of homeostasis in the living systems.

Stability of ontogenesis

A concept similar to homeostasis has been advanced, in developmental biology, to characterise stability and reliability during ontogenesis of organisms, called homeorhesis (Waddington, 1968). According to Waddington, homeorhesis is a steady, stabilised torrent of changes over the course of development of the organism. Any deviation from this steady process under the effect of external or internal factors initiates processes in the system such that they return into channeled direction trajectory, homeorhesis, and this was christened creod by Waddington.

As noted above, at any moment the organism is in the state of homeostasis. Therefore it is possible to say that homeorhesis is a continuous line of hemeostasic states. Thus the approach of the organism to the final, stationary state over the course of ontogenesis will be realised via a number of stationary states (homeostases). Therefore, an attempt can be made to apply the principle of minimal energy dissipation to identify the characteristics of homeothesis. Since homeostasis is characterised by minimal heat production intensity in the system, homeorhesis, by analogy with the stationary process, is characterised by minimal, total intensity of heat production in the organism over the ontogenetic period under consideration, i. e. by the variational principle (5).

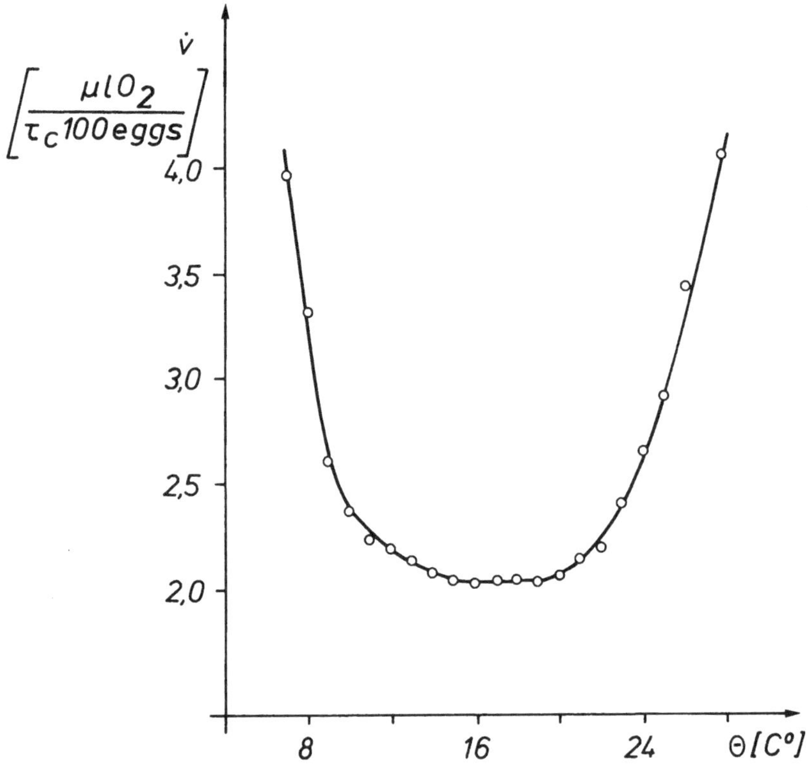

Fig. 65 Oxygen consumption during one division (τ_c) by loach eggs at different temperatures (Zotin, Ozernyuk, 1966).

Fig. 65 shows experimental data on total oxygen consumption over one ontogenetic period, namely one division of loach eggs within the temperature range 8 °C - 28 °C (Zotin, Ozernyuk, 1966). 15 °C to 20 °C is the optimal temperature range for loach embryonic development. As can be seen in Fig. 65, the minimum oxygen consumption per division, i. e. the minimum heat production intensity is observed within this temperature range. Similar results were obtained on using data of respiration intensity in animals in other periods of ontogenesis. Janda and Kocian (see Keister, Buck, 1964) for example, investigated, in detail, respiration and development over time of *Tenebrio molitor* pupae (meal worm) at different temperatures. Calculation of total oxygen consumption for the pupal period at different temperatures showed (Fig. 66) that minimum oxygen consumption was observed within the range 15 °C to 30 °C. Similar regularity was observed for other animals. Calculations by Crescitelli (1935) of total oxygen consumption by *Galleria melonella* pupae at different temperatures showed miminum oxygen consumption at 30 °C. At lower and higher temperatures oxygen consumption rapidly increased (Fig. 67).

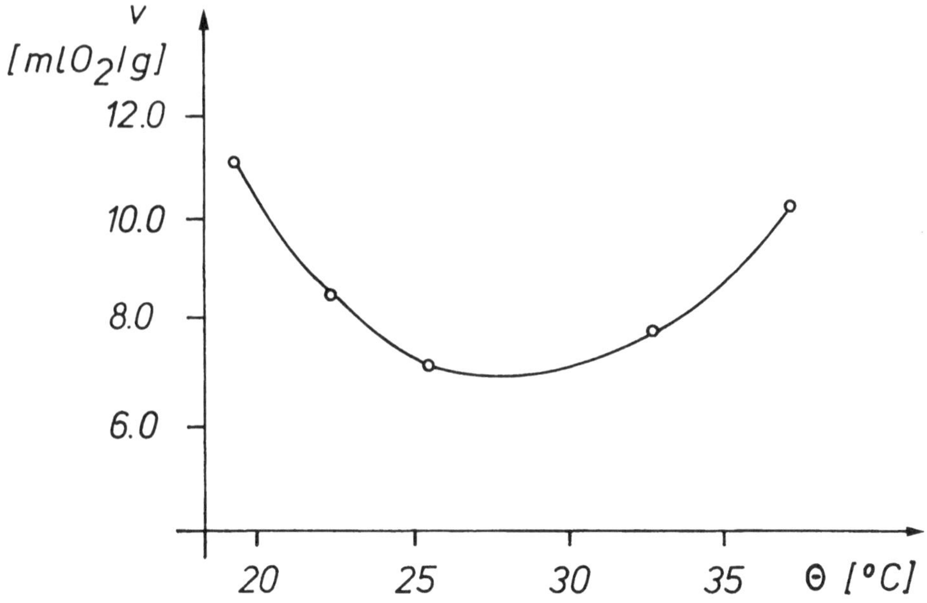

Fig. 66 Oxygen consumption during pupae development of *Tenebrio molitor* at different temperatures (according to Yanda and Kosian; Keister, Buck, 1964).

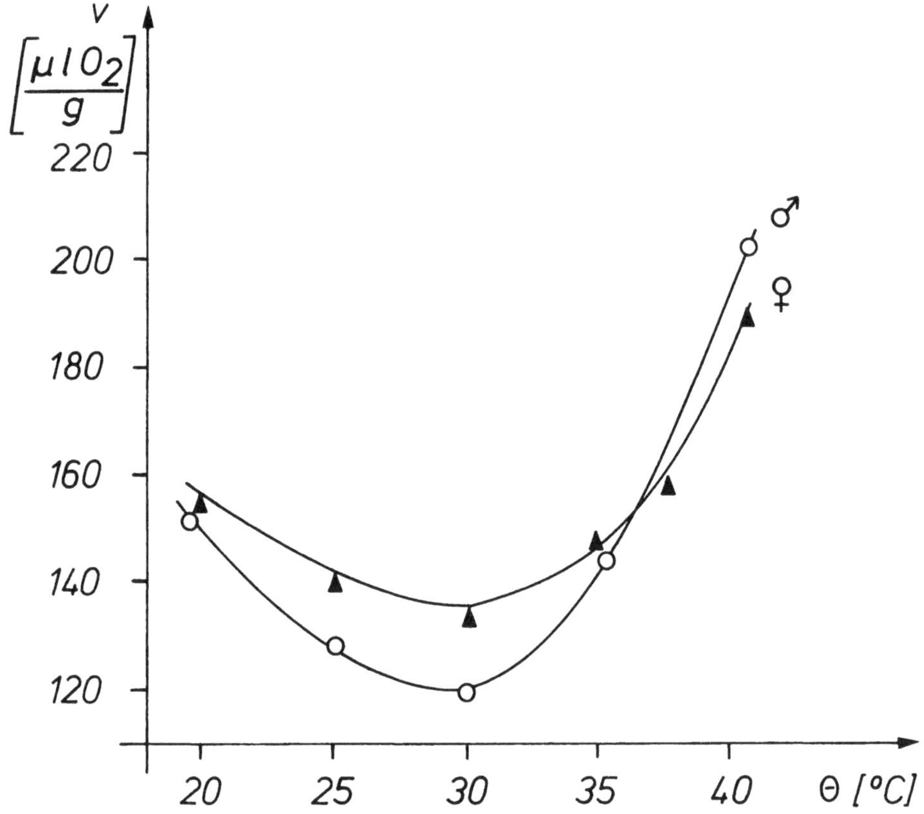

Fig. 67 Oxygen consumption during the development of bee moth pupae at different temperatures (according to Crescitelli, 1935).

Thus, homeorhesis, or creod, is characterised by the minimum total energy expenditure for the particular developmental period. Deviation from creod trajectory brings about excessive and unjustified energy waste. This must affect the results of the complete ontogenesis: excessive energy dissipation at each developmental stage of the organism must finally result in reduction of the ontogenetic period, i. e. in a decrease in the life span of the organism. These facts allow us to formulate a purely biological problem concerned with criteria for the duration of life and biological age.

The Rubner constant as a criterion of the life span

Recently a number of ways to increase the specific life span of animals have been proposed (see Dubina, Razimovich, 1975; Frolkis, Muradyan, 1982) and it is possible, experimentally, to increase the life span of some animals by 30 % to 50 %. The question, however, arises as to whether the calendar time is an adequate criterion for the efficiency of some effect on life span. For example, a change in environmental temperature abruptly increases or decreases the life span of both poikilothermal and homeothermal animals. It is evident that specific life span does not depend on environmental conditions, and a shift of time-of-year with change in environmental temperature means only that this criterion is not valid for life span. A special criterion must be developed which includes properties related to the active pattern over the life time. In gerontology, this problem is designated as that of biological age (Dubina, Razumovich, 1975; Voytenko, 1982).

The famous German physiologist Rubner (1908) was one of the first to propose an adequate criterion of complete animal ontogenesis. He believed that specific life span could be characterised by the amount of energy used by the organism throughout its life - from birth to death. On this basis Bauer (1931, 1935) proposed designation of the product of respiratory intensity over an animal's life time as the Rubner constant. In the 30's Bauer's disciples Shcherbakov and Winberg published a series of studies in which the Rubner constant was reported for different species of Drosophila, and the effects of some environmental factors on this constant analysed (Shcherbakov, 1933, 1935, a, b, 1937; Winberg, 1935; Winberg, Shcherbakov, 1937). This work was not developed any further, although more and more evidence has been accumulated on the inverse relationship between respiratory intensity and the life span of mammals. Sacher (1976) has even discussed the inverse relationship between life span and the rate of entropy production. This approach is too simplified. It appears that, at present, the concepts of the Rubner constant ought to be related to homeorhesis, and that thereby this value would acquire some new biological and thermodynamic meaning.

There is, however, a substantial difference between the Rubner constant and a thermodynamic understanding of homeorhesis. According to the definition, the Rubner constant determines the total life span of the animal under optimal conditions, i. e. it corresponds to creod of complete ontogenesis. In contrast, homeorhesis and creod can characterise any period of ontogenesis; creod corresponding to the minimum, total oxygen consumption (minimum total heat production intensity) at any state of individual development. The Rubner constant refers to complete ontogenesis, and, in order to calculate it one must know the time of death. This time cannot be determined by morphological indications. Therefore, in calculating the Rubner constant we use calendar time, rather than ontogenetic stage. Naturally, under optimal life conditions calendar life time will be the longest, therefore the Rubner constant, in contrast to the creods of separate, developmental stages, acquires the maximum value under optimal conditions.

These views can be proved by experimental evidence on the respiratory intensity and life span of Srosophila at various temperatures and of different organism densities. Shcherbakov (1937) measured total CO_2 production throughout the life of adult individuals at different temperatures and the maximum value of the Rubner constant for Drosophila was observed at 20 °C (Fig. 68). At lower, and higher, temperatures the Rubner constant significantly decreased. Similar results were obtained by Muradyan and Timchenko (1982) who reported that the optimum in total CO_2 production was at 20 °C to 25 °C.

Shcherbakov (1935 b) also studied the effect of organism density in *Drosophila melanogaster* in test tubes on their respiratory intensity. The optimal condition for Drosophila in these experiments was found to be 20 to 40 individuals per test tube. Calculation of the Rubner constant according to Shcherbakov (Fig. 67) show that the maximum value of the Rubner constant corresponds to the optimum maintenance conditions of Drosophila. The same conclusion was drawn by the author himself.

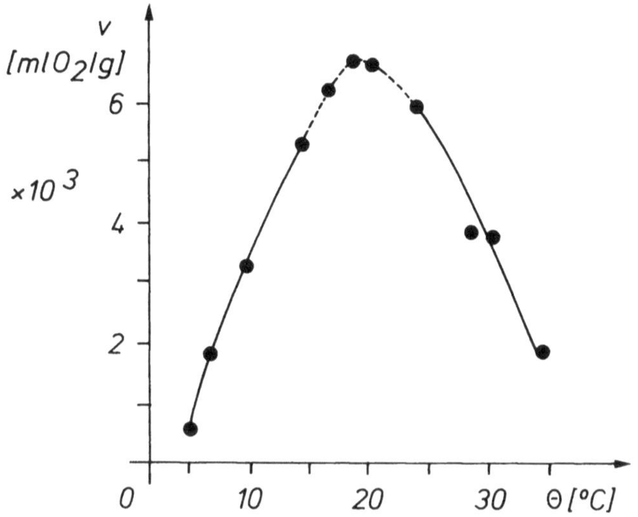

Fig. 68 Oxygen consumption during the whole life of *Drosophila* imago at different temperatures (Shcherbakov, 1937).

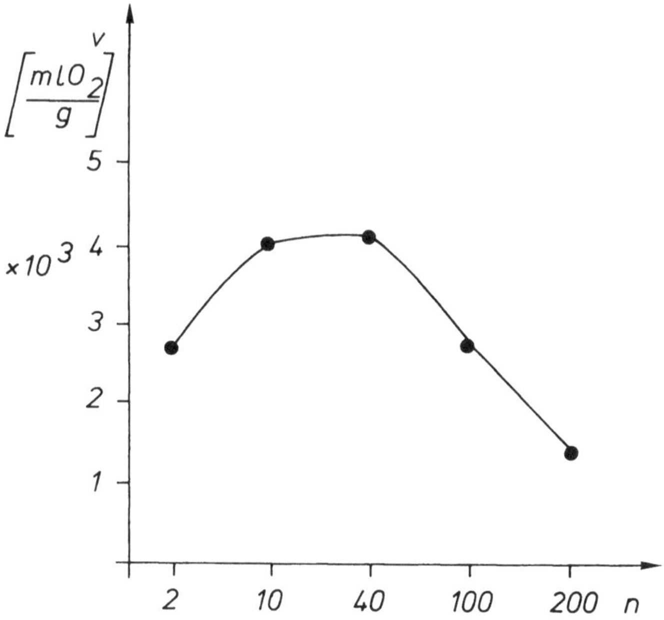

Fig. 69 Dependence of oxygen consumption during the whole life of *Drosophila* imago on number of flies in test-tube (Shcherbakov, 1935).

Thus the life span of animals can be characterised by the Rubner constant, its maximum value conforming to possible life time. This constant gives a more satisfactory estimation of life span as compared to calendar time since it takes into account the important parameter: the energy metabolism of the animal. Its advantage lies in the fact that it can be associated with a concept as common as homeorhesis, since, as is clear from the above, the Rubner constant is a special case of the creod concept; it corresponds to creod of complete ontogenesis.

Analysis of Homeorhesis in Ontogenetic, Population and Evolutionary Aspects

V. M. Zakharov

Homeorhesis, in a general form, is understood as a stabilised flow of trajectories in development of the group of individuals under consideration (Waddington, 1968). This characteristic which reflects specific features of individual development, may be also considered as a population character, since it can be estimated only in the group of individuals by analysis of the scatter of developmental trajectories of a number of individuals.

It is possible to estimate the level of homeorhesis from the width of the trajectories for the group of individuals analysed. A relatively narrow flow corresponds to high stability of individual development. At the same time, there can be distinguished the main trajectory, the creod (Waddington, 1957) around which the developmental trajectories of separate individuals are located. Creod is changed as a result of certain genetic peculiarities and definite environmental changes. This means that the differences between some trajectories are due, not only to change deviations from the definite creod, but, in the extreme, they result from switching development over to another creod (Fig. 70).

When the width of the trajectory flow at one of the developmental stages has been estimated, the level of homeorhesis reflects the ability to develop similar phenotypic character under definite environmental conditions, notwithstanding some genotypic differences between the individuals analysed. The phenotypic uniformity of the group under analysis corresponds to a high level of stability, while its diversity points to disturbed developmental stability.

Thermodynamics and Regulation of Biological and Processes

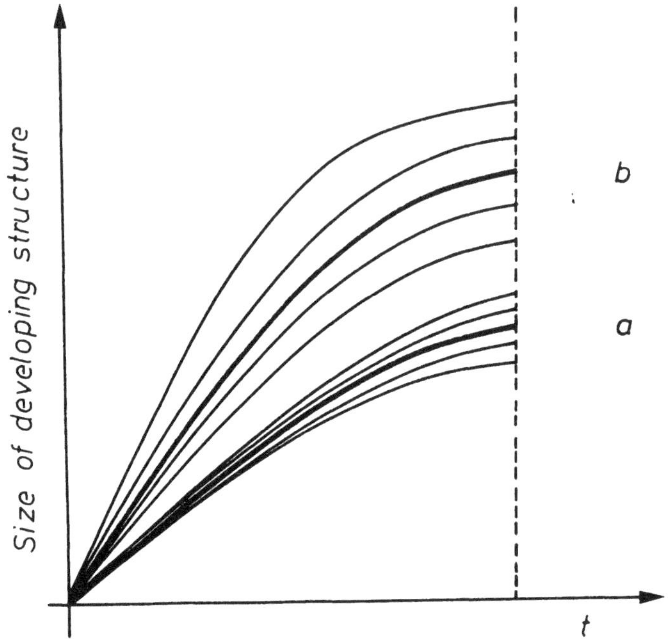

Fig. 70 Scheme of changes in the stability of development. A: narrow flow of trajectories corresponding to high stability level; B: broad flow of trajectories indicating disturbed development. Thin lines correspond to trajectories of development of individuals, the thick line corresponds to creod, the dotted line shows width of trajectory flow.

In estimating the level of homeorhesis the phenotypic peculiarities of developing individuals, which can be considered as consequence of disturbed development, are of primary importance. Although the task in a general outline is quite clear, its practical realisation is methodologically complicated. It is often difficult to discriminate, exactly, which component of phenotype diversity in developing individuals is caused by developmental disturbance.

One possible approach to the problem may be the quantitative estimation of chance variability in development.

Homeorhesis and chance variability in development

Differences between homologous structures of the same individual can be analysed, quite unambiguously, as a manifestation of chance variability in development (Astaurov, 1927, 1974; Reeve, Robertson, 1953; Reeve, 1960; Van Valen, 1962; Rokitsky, 1974). These are the structures located along the longitudinal, or transverse axis, of the body. Differences between homologous structures not related to definite structural and functional differentiation may, naturally, be attributed to the results of chance developmental disorders.

It has been shown that differences between homologous structures of an individual cannot be explained by genetic differences or varied developmental conditions, since their development proceeds on the basis of the same genotype under essentially identical conditions. In spite of this fact the mean value of such difference in the group of individuals analysed proves to be under genetic control and shows certain dependencies on environmental conditions (Mather, 1953; Thoday, 1958; De Marinis, 1959; Beardmore, 1960; Adams, Niswander, 1967; Brückner, 1976). The nature of the differences between homologous structures in the same individual shows that they can only be the result of chance disturbances of development whereas stability towards these disturbances is under the control of the genotype; and their frequency increases under normal developmental conditions. Due to the stochastic nature of these disturbances neither the genetically determined resistance, nor the dependence of their frequency on certain environmental factors can be revealed when individuals are analysed. Their influence can be detected only at the populational level. The fact of the existence of some genetic control, and dependence on the environment, does not conflict with the stochastic nature of the phenomenon under consideration.

Here a complete analogy with the notion of noise in the information theory seems to be possible. Noise is understood as uncertainty in the source of the information, however the total noise level can be decreased either by improving the transmitting system as a whole, or by strict maintenance of optimal environmental conditions. Furthermore an analogy can be drawn here with spontaneous genetic variability. Although its level can be changed substantially under the effects of definite environmental factors, it remains random and spontaneous by nature (Bresler, 1981).

The next important questions to answer are: 'To what level does chance variability in development rise?' and 'What role does it play in the observed changes in the level of total phenotypic variability?'. It is quite natural to expect that such chance variability is not only manifested in the difference between homologous structures, but is also widely distributed and makes great contribution to phenotypic variability observed in various structures. The difficulties in estimation of the value of chance variability are related to the complexity in discriminating this kind of variability. It has been shown, more than once, for various objects, that chance variability in development, determined from the differences between homologous structures of an individual, may attain an appreciable value which is comparable with the level of total phenotypic variability between the individuals (Rokitsky, 1974; Robertson, 1957). Moreover, it has been shown that, the values of total phenotypic variance and that of intra-individual differences frequently change concordantly. This fact indicates that the rise in total phenotypic variance value may be due to an increase in chance variability in development (Zakharov, 1981 a; Mather, 1953; Bader, 1965).

In addition to intra-individual differences, chance variability is also observed on analysis of the differences between monozygotic twins (Lundström, 1967; Astaurov, 1974). It seems that, to chance developmental variability, phenotypic variation in highly homozygous inbred lines increases, as compared to more heterozygous individuals (Rasmuson, 1951; Robertson, Reeve, 1952; King, 1961; Voroshilov, 1969). This increased variability cannot be explained by either genotypic diversity, which is much lower in inbred lines, nor by varied environmental conditions, which remain constant. Chance variability seems also to play an important role, for developing individuals, in an increase in phenotypic diversity under the effect os unusual environmental conditions (Tatarko, 1977; Osgood, 1978; Zakharov, 1982). Thus under any temperature regime, optimal or non-optimal, environmental diversity and phenotypic variation amongst individuals are similar. An increase in phenotypic variation under non-optimal conditions is observed in genetically heterozygous groups, as well as in highly homozygous inbred lines in which genotypic variation is significantly lower. The fact that phenotypic variation increases, in this case as a result of increased chance developmental variability, is also shown by an abrupt increase in differences between homologous structures in an individual which can be, unambigously, considered the consequence of chance developmental disturbances (Thoday, 1958; Siegel, Doyle, 1975 a; Baardmore, 1960; Zakharov, 1982). Thus, increased phenotypic diversity under normal conditions seems to be due to increased chance developmental variability as a result of a non-specific increase in the frequency of different chance disturbances over the course of development.

As indicated above, of importance in estimation of the level of homeorhesis are not those phenotypic differences between individuals which result from directed changes of creod under the effect of certain genetic or environmental peculiarities, but those which proceed because of some deviation in development from a definite creod. The only type of variability meeting these requirements is chance developmental variability since phenotypic differences, initiated during the process, cannot be considered as a direct consequence of definite genetic or environmental differences between the developing individuals. These must occur even when the genetic and environmental factors are identical (Astaurov, 1974).

Adequate estimates of the level of developmental stability can be obtained from the measure of chance variation and, from the analysis of real groups of individuals where, as a rule, there are both genotypic and environmental diversities. Genotypically controlled stability towards a developmental disturbance is determined by the balanced pattern of the co-adapted, complete gene complex. The level of this balance proves, in principal, to be similar, in spite of some genotypic differences between individuals taken in sequence or from the natural population. In addition, because of the stochastic nature of the disturbances detected on analysis of chance developmental variation, some genotypic or environmental differences between developing individuals cannot bring about directed phenotypic differences between individuals. Genotypically determined resistance to the disturbances, as well as their frequency under the given conditons, can be estimated only by populational analysis.

Thus chance developmental variability can apparently, be not only characterised as an index of developmental stability, but it must be considered as one of its main criteria.

In terms of these concepts, the level of intra individual differences between homologous structures; a measure of the non-genetic component of total phenotypic variance of quantitative character; the frequency of phenodeviations, may serve as criteria for estimation of the level of homeorhesis (Rasmuson, 1960; Zakharov, 1981 a). The first is often used as the main criterion. Tentative estimation of the level of two of the other criteria is possible by either examining the agreement between total phenotypic variability and the value of the first index, or, by using other indirect approaches (Zakharov, 1982).

Changes of stabilisation and channeling of development

As has been shown in a number of investigations, on the basis of balanced co-adapted gene complexes, a high level of developmental stability (narrow flow of developmental trajectories and low level of disturbances) is maintained under certain, common conditions for a given species. Destabilisation of development, manifested in a broader flow of developmental trajectories, is observed on disturbances of the balanced, co-adapted gene complex (when individuals genotypically very different from each other mate) and, under abnormal conditions too. This disturbance in developmental stability proves to be an unspecific reaction to different changes in the environment (Tebb, Thoday, 1954 a, b; Thoday, 1958; Beardmore, 1960; Siegel, Doyle, 1975; Brückner, 1976).

In addition to changes in the width of the flow of developmental pathways, the main developmental pathway, the creod, may also change. This is observed under certain genotypic or environmental changes. Different pathways of development can be discrete, but this appears to take place rather rarely and only in most important adaptive structures where it is initiated in early ontogenesis. More often creod changes smoothly with changes in a certain environmental factor, and only the extreme variants can be considered as discrete. In addition, irrespective of whether the creod is changed or not, under all conditions, except the relatively optimal conditions, destabilisation of development, i. e. broadening of the flow of developmental trajectories takes place (Tatarko, 1977; Zakharov, 1981 b, 1982; Osgood, 1978; Thoday, 1958).

Stabilisation of development, by its very nature, has non-specific characteristics, and reflects only the level of developmental disturbance, and this increases when developmental conditions deviate from the optimum. Channeling of development is, on the contrary, very specific and reflects directed developmental changes under certain environmental conditions.

Possible characteristics of optimal conditions for development according to the level of homeorhesis

Since homeorhesis is one of the universal characteristics of development, it is of substantial interest to discover, in general, its possible application in the estimation of optimal conditions for individual development.

Let us explore the possibility of such an approach by analysis of *Lacerta agilis*, as an example. The offspring obtained from females of a natural population were incubated at constant temperatures of 20 °C, 25 °C, 30 °C and 35 °C (Zakharov et al., 1982). The level of homeorhesis was determined through three parameters reflecting developmental disturbances in each experimental group of newborn lizards (Fig. 71). The highest level of homeorhesis accorded with the smallest values found for all three indices used, and was observed at 25 °C (Zakharov, 1982).

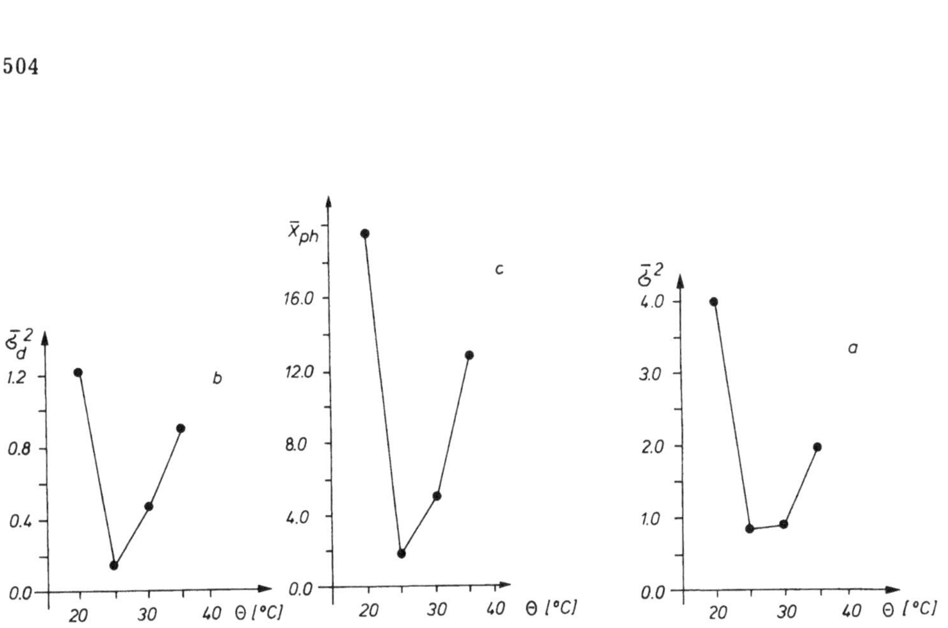

Fig. 71 Temperature relationship of changes in three parameters of phenotypic variation reflecting the level of developmental disturbance in *Lacerta agilis*. a: change in mean level of total phenotypic dispersion $\bar{\sigma}^2$ (13 characters of scutum formation); b: change of the mean level of fluctuating asymmetry $\bar{\sigma}^2_d$ of the same characters; c: change in the mean number of disturbances per individual $\bar{X}ph$ (of 150 alternative deviations from usual normal location of scuta).

Thus the optimal conditions determined by the level of homeorhesis on the one hand, can be considered as the optimum for individual development, in general, because of the general pattern of the criterion applied. And, on the other hand, the range of such conditions proves to be quite narrow, and this points to high sensitivity in this index.

Level of homeorhesis as a characteristics of the state of natural populations

Populational investigations of homeorhesis have shown that a high level of homeorhesis is observed in many populations of a species which, over a long period, exists under the given conditions at the expense of the formation of balanced co-adapted gene complexes (Zakharov, Zyuganov, 1980; Zakharov, 1981 c).

Disturbances in homeorhesis take place in populations which live under abnormal conditions for the given species, i. e. in extremes of the ecological range of the species, in populations introduced into a new environment and following industrial contamination of the environment (Soulè, 1967; Soulè, Baker, 1968; Valentine et al., 1973; Zakharov, 1981 c). Destabilisation of development is also observed in the margins of zones where sufficiently divergant forms exist which are adapted to different conditions, since here the co-adapted gene complexes seem to be disturbed (Zakharov, Zyuganov, 1980).

Comparison of experimental estimates of optimum conditions with data obtained on natural populations is of special interest. As shown for *Lacerta agilis*, the optimal range of conditions for developmental stability is rather narrow, being similar for many populations of a specie derived from different parts of the geographic range of existence. It is much narrower than the range of natural, living conditions of the species. In view of this it is natural to suggest that some populations of the species live under non-optimal conditions. Actually it appears that in certain populations developmental stability is disturbed in the same way as was observed during experiments under non-optimal conditions (Zakharov, 1981 c, 1982). It seems that such events may be found in nature because the phenotypic peculiarities revealed by this approach are not under the control of natural selection and are confined within wide limits. This does not affect the viability of individuals and only points to some disturbances of stable development which are under the influence of the relatively, non-optimal conditions.

In addition, specific changes of creod found by experiment can also give information about the direction of change in conditions from the optimum for real natural populations. Disturbances of developmental stability were found in two populations of *L. agilis* inhabiting the ecological periphery of their range. On analysis of specific characteristics in these populations changes were found in one of them (those from the Issyk-Kul lake) which are characteristic of temperature conditions below optimal and in the other (the Sevan lake) above optimal.

Thus analysis of homeorhesis in natural populations gives information about the state of the population, its change under different disturbing effects, and some information about the direction of deviation of real conditions from the optimum.

Changes of creod as a mechanism of evolution

Changes of creod, initially due to certain alterations in developmental conditions, and then, as a result of genotype change can be considered as a possible route for evolutionary transformation, and is known as the Baldwin effect (Waddington, 1957; Schmalhausen, 1968; Shishkin, 1981). The mechanism of such transformation can be presented in the following form. In some populations of a specie under normal conditions some phenotypic switching of development to another creod can take place, as compared to that found under optimal conditions for the given specie. When the population exists under these conditions for a sufficiently long time (a great number of generations) reconstruction of the genotype may occur on the basis of existing, and new, genotypic variability, under the force of natural selection. This reconstruction ensures successful development, the necessary phenotypic effect having being achieved, When the initial and final points of this evolutionary process are compared, a substitution of phenotypic changes for genotypic changes of creod is observed. During the process stabilisation, by the new creod, of development takes place, because during phenotypic creod changes induced by unusual environmental conditions, some destabilisation of development occurs, i. e. the width of the developmental trajectories around the new creod increases.

As an example the data mentioned above on lizards can be discussed. Finite changes of creod at temperatures above optimal were observed for a number of morphological structures in offsprings obtained from females of different populations (from different regions within the range of existence) of the widely distributed species *Lacerta agilis* incubated under experimental conditions. The number of femur pores increased, the number of postnasal and frenale scales decreased. Another more southern species, *Lacerta strigata*, has morphological features which seem to be similar to changes of creod observed in *L. agilis* on developing at higher temperatures. *L. strigata* is characterised by a greater number of femure pores, as compared to *L. agilis*, and a relatively small number of postnasal and frenale scales. Experimental analysis has shown that in *L. strigata* the same relationship is observed between creod changes and developmental conditions as in *L. agilis*, but at each temperature regime the measures of the structures analysed are significantly higher. This means that, evidently, *L. strigata* and *L. agilis* have similar relationships between the changes in some structures and environmental conditions. However they are realised via a genetic basis which could have been developed in the course of evolutionary transformations and which is supplied in the related species *(L. agilis)* only under the effect of a relatively high environmental temperature.

This trend in evolution seems to be more widespread than was believed earlier.

Prospects of homeorhesis investigations

The multi-factorial analysis of homeorhesis as a universal characteristic of ontogenesis, gives insight into the nature and role of both chance and directed change during development and may form one approach to the formulation of general, theoretical concepts of individual development as a whole.

Analysis of phenotypic diversity in terms of the concept of hemeorhesis can create a basis for the further development of the theory of phenotypic variability, and of the concept of chance variability in development in particular.

Because of the general pattern of homeorhesis and it's sensitivity to different changes in developmental conditions this characteristic can be used for creation of theoretical concepts of the optimum, and, for experimental determination of optimal conditions for individual development. The data obtained in the process are of interest not only for developmental biology, but for the study of natural populations too. It becomes possible to reveal those peculiarities which result from development under non-optimal conditons. Determination of the level of homeorhesis in natural populations permits one to obtain the characteristics of the population state and to discover its response to different environmental stresses.

In it's evolutionary aspect the study of homeorhesis permits one to analyse the mechanisms of evolutionary transformation and to evaluate the latitude allowed by natural selection.

Complete analysis of homeorhesis, one of the universal characteristics of individual development, requires a complex approach with the application of different methods. The main attention of this study was focused on characterising homeorhesis in terms of morphology. But other viewpoints yield some interesting results in the analysis of homeorhesis. It has proved possible to characterise homeorhesis in terms of energetics (Zotin, 1974) and through the duration of development to estimate its level (Detlaff, 1981). It is apparent, that a number of other physiological and biochemical criteria may, in future, be used for this purpose.

Thermodynamic Aspects of Plant Ontogenesis

W. Zelawski, W. Galinski

The structural complexity of living matter and its apparent orderliness are the basis of the earlier views that life is not a purely physical phenomenon and that ontogenesis of an organism does not conform to the Second Law of Thermodynamics. In fact, it was the essay written by Schrödinger (1951), which initiated discussions as to the physical nature of life, although the elements of such a discussion had appeared much earlier in the famous paradox of 'Maxwell's demon'. In the sixties, it became clear that the paradox itself, and the apparent irrelevancy of the Second Law to ontogenesis, had found sufficient justification in the direct relationship between the aspects of thermodynamics and information theory (Brillouin, 1956; Trinscher, 1967).

For the majority of physicists today, that life processes conform to the laws of thermodynamics is obvious. Many treat the phenomenon of life in a quite deterministic manner and some are even inclined to understand it as a long-term perhaps unavoidable, complex, tendency to approach thermodynamic equilibrium in the process of development of the universe (Baranowski, 1979, 1981). On the other hand, biologists' views of the physical nature of life are much more differentiated. Apart from some metaphysical attempts to solve the puzzle of life, the physical interpretation of information processes, which are important in creating biological structures (but do not occur in physical structures), appeared to be a very difficult task.

A detailed analysis of such problems as; the equivalence of thermodynamics and information entropy; the relationship between orderliness, structure and organisation of matter; the measurability of structural and organisational properties of living matter; achievement of a steady state by biological

Thermodynamics and Regulation of Biological and Processes

systems etc. indicates their permanent topicality. Particularly controversial seems to be attempts to apply hitherto existing concepts of thermodynamics to the development of autotrophic organisms. Here concern for assimilation of negentropy contained in food (Schrödinger, 1951), and accumulation of information from the environment (Trinscher, 1967), are hardly discernible. Nor is there described any trend towards a final steady state (Prigogine, Wiame, 1946; Zotin, 1974). In fact Górski (1966) has already pointed out that the environment cannot be a source of negentropy for plants but it must be an unknown 'organisation capacity', not quite defined by the author, that is involved in producing and maintaining the specific structure of living matter. Nontheless previous considerations have not yet resulted in an understanding of the thermodynamic aspects of plant ontogenesis, and, simultaneously, have created numerous doubts as to the concepts and terminology used. Although many people find it hard to believe, a thoughtless use of certain expressions or even slight extensions in terminological conventions that had been established earlier may create artificial problems, bringing about confusion and sterile discussion. The misuse, in biological thermodynamics, of such concepts as orderliness, structure, organisation and information are good illustrations of this. Either these concepts are not quite precisely defined - e. g. orderliness is often mistaken for organisation or structure - or the word is applied in a sense that is different from its common usage - e. g. information in the quantitative theory of information - or, last but not least, an expression in introduced because of similarity in mathematical formulae occurring in various fields, e. g. informational entropy. It is worth stressing that long ago several authors pointed out the improper use of the expressions mentioned above. Pierce (1961), for instance, questioned the equivalence of information and thermodynamic entropies, Denbigh (1979) raised the problem of interchangeable misuse of the notion: 'order', 'orderliness' and 'organisation' and Chernavski (1975) questioned the applicability of Prigogine's principle as a basis to describe the development of a biological system; the cautions expressed by these authors remained, however, unnoticed.

It is not easy to undertake, again, a problem so controversial as to seem insoluble. The right to participate in this discussion should derive from our knowledge of the physiology of plants, which are seldom the object of biophysical research, and our belief in the rightness of Einstein's thesis that only those who are not conscious of the insolubility of the problem have a chance to solve it.

Entropy, structure of matter, and thermodynamic order

In the 1850's Clausius definitively formulated the Second Principle of Thermodynamics using the concept of entropy - gr. εντροπια - to define the fraction of heat that cannot be converted into work in a cyclic process. The atomistic concept of the structure of matter was, in those days, strongly affected by classical theory and a natural consequence of this was an attempt to interpret the Second Principle within the scope of mechanics. As a result of such an approach Boltzmann formulated his famous H-theorem, in which he related the value of entropy with the number of possible states of the system. Since, in certain spontaneous, thermodynamic, conversions a loss of regular structure of the matter occurs, it has generally been taken for granted that the entropy increase means diminution in order, i. e. augmentation of chaos in the system*, quantitatively this would be characterised by an increase in the number of possible states of a system. A consequence of the assumption that an increase in entropy means a decrease in order, and also of the implicit assumption that the measure of order is reciprocal to the measure of chaos, was the interpretation of entropy, i. e. the logarithm of the number of possible states, as a measure of chaos, and order in general. Order as defined in the above sense, however, is not identical with the common use of the word (Jones, 1979; Tonnelat, 1979) and has nothing to do with regularities in the

*It is, however, not so obvious since there exist spontaneous thermodynamic conversions related to the emergence of regular structure, e. g. crystallisation of a super-cooled liquid.

structure of living matter that represents biological organisation rather than thermodynamic order.

The confusing usage of the concepts 'organisation' and 'order' has undoubtedly been due to the fascination of biophysicists by the complex, but still quite regular - apparently 'ordered' - structure of a living system, it appears, in fact, to be a state of very low probability as compared with the chaotic dispersion in the environment, of all the elements combining to produce this structure.

Unfortunately, in all attempts to describe a living organism from the thermodynamic standpoint, very important features distinguishing physical and biological structures, have been neglected. Most significant is the way in which they have emerged: structure in a physical sense is a result of a random realisation of a certain class of possible states whilst the spatial distribution of the elements, which combine to produce this structure, results solely from the condition of minimum energy; on the other hand, it would be difficult to speak of a purely casual combination of all elements of biological structure, since, here, not only energetic but also informational constraints, related to functional properties of organisms, are involved. These informational constraints might even be considered as one of the defining characteristics of the biological structure.

A subjective 'intuition of order' that biologists experience when they investigate biological structures, reinforced, by a judgement that the amount of order can be defined in terms of the measure resulting from the statistical definition of entropy, long affected the thermodynamic approach to biological systems. The tendency to identify order in a physical sense with the 'special order' found in biological systems is recognisable in the idea of 'aperiodic crystals' which was introduced by Schrödinger (1951) to define the singularity of the sequence of elementary units in biopolymers. As a measure of 'biological order' Schrödinger invented the concept of 'negentropy', i. e. entropy with a negative sign. To maintain the existence of a biological system, food, containing a great amount of negentropy, would have to be consumed, i. e. advantage must be taken of the resources of 'order' present in the environment. In the case of animals it is 'order' contained in the structures of nutritional compounds that could be taken into account but, in the case of photosynthesising plant organisms the source of entropy is much less defined. There were some suspicions that high energy solar radiation was the source of orderliness (Tribus, McIrvine, 1976). However, such a source of negentropy, even if quantitatively sufficient, could not create any biological structure which differed from a non-biological structure because of the existence of additional constraints of an informational character, as mentioned above.

The distribution of elements which combine to produce physical structures, such as occur in closed systems or in dissipative structures which sometimes appear in non-biological open systems, is spontaneous in character. On the other hand, the biological structure surely means something more than the creation of a certain configuration of atoms, particles, and ions, in space that is occupied by an organism. Postulating the measure of this structure to a function of the total number of possible states would require a purely statistical approach. Yet, the processes which occur during the development of a living organism do not correspond with changes in thermodynamic order, since the way such an organism is 'ordered' has no direct analogy with the orderliness of crystals or even of dissipative structures.

Information, biological structure, and organisation

Cellular and subcellular structures in biological systems are formed from elements integrated physically and in a sufficiently stable way to enable the existence of this state and the simultaneous formation of the spatial conditions for physiological process to function. Biological structure is organised according to an informational programme contained in the genome and modified by environmental effects. Because of the action of receptors, which are unique for each biological system, the informational signals undergo translation processes and become physiological stimuli which are amplified and then transformed into an action of the system which, as a rule, is not cmparable in magnitude with the received signal. Thus one can assume that there are no informational processes in nature other than those which occur in biological systems or in the technical devices elaborated by these systems.

The consciousness that any biological system is formed under the influence of constraints, which result mainly from the genotype programme, has long inspired some authors to search for the interrelation between the quantity of information contained in the coded signals of this programme and the degree of 'orderliness' in the system, expressed by negentropy. As the result of Brillouin's (1956) attempts to generalise the work of Szilard (1929) and Shannon (1949), some authors thought, erroneously, that the quantity of information, expressed in binary units, could even be recalculated into the change in thermodynamic entropy (expressed in $J \cdot K^{-1}$).

It follows from Pierce's (1961) and some other author's considerations that both entropies, informational and thermodynamic, have nothing to do with each other, except the name and the similarity of mathematical formulae. The similarity results from using the same functional of the mean logarithm of probability to calculate the values of both entropies, whereas the discrepancy is a consequence of non-comparability and, in particular, the incommensurability of the physical quantities on whose probability distributions the above functional is defined. Hence, the search for equivalence of both entropies seems only to be a sad mistake. Furthermore, the triumphant announcement of the 'expulsion of Maxwell's demon' (Brillouin, 1956), based on the hypothesis that the received information signals cause a diminution in the entropy of the system, also seems to be a misunderstanding. For an effective segregation of particles the demon not only must receive a signal regarding the approaching object but he also must possess knowledge - what he should do with the given particle. It is this knowledge and not the quantity of information ifself, i. e. informational entropy, which makes possible an action that prevents the natural, spontaneous course of events to take place. In this case, the problem of realisation of an *a priori* assigned programme (in which the reception of information signals is only one of a number of elements and cannot be 'recalculated' into the entropy decrease) is involved. Hence, everything indicates that the problem of Maxwell's demon also exceeds the frames of present thermodynamics.

Three unfortunate mistakes occurred simultaneously in undertaking efforts to describe living organisms in terms of thermodynamics; firstly - the emergence of biological structure has been identified with the formation of physical structure; secondly - the concept of thermodynamic entropy was mistakenly used as an equivalent for the concept of informational entropy; and thirdly - the fact was ignored that the final shape of a biological structure depends not only upon genetic but also upon environmental information (a phenotypic variation in plants is a particularly good example of this).

The quantitative theory of information, invented for a very definite purpose of describing transfer of signals in telecommunication, appeared to be an in insufficient tool to describe the structure and organisation of biological systems. Numerous authors intuitively felt a lack of a qualitative theory of information and pointed out the non-comparability (inequivalence) of physical and biological structures. Various authors attempted to overcome the consequent difficulties by creating new, auxiliary concepts such as: aperiodic crystal (Schrödinger, 1951), structural entropy (Górski, 1966), ordering work (Trinscher, 1975), quality of information (Blumenfeld, 1974), value of information (Volkenstein, 1978; Volkenstein, Chernavsky, 1978); however, the heart of the matter lies in the lack of a definition, expressed in physical terms, of such concepts as: information, biological structure and organisation.

The following terminological conventions are suggested: Information is the impact of the environment upon the system such that it causes non-trivial courses in spontaneous processes taking place within this system; the non-trivial course of a spontaneous process is defined by its probability, *a priori*, being not maximal. From the above definition it follows that, to enable such a reaction in a system, it is necessary to translate the external impact into the internal stimulus and, on the other hand, in order to make the course of spontaneous processes non-trivial it is necessary to amplify and transform this stimulus using an additional source of energy.

A biological system structure differs from a purely physical structure mainly because it not only appears to be a consequence of purely energetic actions of external forces but also a result of the informational constraint mentioned above. Thus, the biological structure reflects an interaction between the informational and energetic activities of the system and expresses limitations imposed by the information upon spontaneous thermodynamic changes.

It follows from the above statements that the formation of a biological structure, i. e. its organisation, can also be defined as a non-trivial - - since imposed by the informational constraints - way of realisation of spontaneous processes; it enables both the separation of the system from its environment and the maintenance of this system in a non-equilibrium state. Organisation does not mean order in the sense of entropy decrease but only a prevention of the naturally expanding chaos that would inevitably advance if the forces resulting from the informational activity were not involved.

From the above definitions a direct relationship between the notions of information, biological structure, and organisation can be deduced. Likewise, as for information itself, all the notions discussed above, specifically characterise living organisms.

Stationary and non-equilibrium in ontogenesis of plants

The biological system, being an open system in a non-equilibrium state, is characterised by an exchange of matter with its environment, inevitably related to a continuous energy flow. Among the authors investigating the problems of biological thermodynamics there is, however, no uniformity of views as to the degree of departure of biological systems from an equilibrium state. Some of them are of the opinion that these systems remain in a linear, non-equilibrium range (Prigogine, Wiame, 1946; Prigogine, 1955), while others point out that a non-linear range is also conceivable (Zotin, 1974). And there are a few who consider even non-linear, non-equilibrium thermodynamics to be an insufficient tool to describe biological systems constructively because these systems function on the basis of processes which are often cyclic in character.

The idea that the biological system as a whole tends to achieve a state of minimum rate of entropy production, as is the case in non-biological, non-equilibrium, linear, open systems, has been expressed in papers by Prigogine and Wiame (1946) and Prigogine (1955, 1972). However, if one tries to compare the changes in the rate of heat production, which take place during ontogenesis in certain animals, with the decrease of the rate of entropy production, postulated by Prigogine's principle, serious interpretational difficulties arise. It appears, that during embryogenesis there is no decrease, just the opposite: - there is an increase in the specific dissipation of heat; this would indirectly testify to departure from, rather than approach to, a final stationary state (Trinscher, 1967).

There are no experimental data to prove how changes in heat production proceed during plant ontogenesis so that any thesis about their possible departure from a steady state cannot be verified. Nevertheless, the well known facts relating to continuous differentiation of new structures and growth lasts over the whole life of a plant (Zelawski, 1981) indicates certain analogies with embryogenesis in animal organisms. In plant biology one speaks about embryogenesis in relation to embryos developing in seeds only, but from the point of view of the formation of all the essential structures of a plant, its whole life seems to be a prolonged phase of embryogenesis. Therefore the doubts raised by some biologists experimentating with animals as to the applicability of Prigogine's principle to embryogenesis seem to be valid for plant ontogenesis too.

The trend towards a final steady state, supposedly attainable during the development of an organism, would only be discernible if the biological systems were able to achieve it spontaneously. In the case of animal systems this would mean an attempt to achieve a state of full passivity. In the case of continusouly developing plant organisms this would lead towards cessation of growth. In both cases, however, the existence of a biological system is only conceivable if the non-equilibrium (non-stationary) state is continuously and actively maintained.

Chernavsky (1975) is probably right in saying that the hitherto existing concepts of thermodynamics do not suffice to describe, constructively, the evolution of the entire organism. For biological systems, and particularly for plant organisms, not only the distinct, non-equilibrium state of various parts but also the coexistence and the cooperation of all, living and dead elements of the whole structure are important. The dead elements usually make a framework of the entire construction on which a virtual structure, important for the function of the whole system, is based. Function in a plant organism is possible only because such a system is an open system in which accumulation of energy exceeds dissipative processes. Thus, the existence of a biological system is even much more expensive than the maintenance of a steady state; hence, the minimisation of this cost cannot be considered as a main principle of life.

In spite of opinions still held (note: e. g. 'Lindsay's thermodynamic imperative') the formation of a living organism does not proceed in contradiction to the Second Principle and is not related to entropy decrease of the system nor to negentropy uptake from the environment. The creation of a biological structure is not a consequence of ordering, in the sense of classical thermodynamics, because there is no analogy between this process and acrystallisation, nor even specific ordering in fluid crystals and dissipative structures that are also imposed by the action of external forces only.

Nor is the formation of a living structure a random selection of one, capable of life, from among a great number of possible configurations of elements. It is a process of organisation consistent with the action of constraints, resulting from genetic and environmental information. Despite the fact that the number of informational signals certainly plays a role in biological processes the increas of informational entropy in a biological system, if definable at all, cannot be considered as a measure of the thermodynamic entropy decrease; both notions should be seen as fundamentally divergent.

Development is not a process of achieving the final steady state by an open system, either. Some elementary processes which combine to produce this development may bear the character of non-equilibrium thermodynamic changes and be coupled in such a way that the possibility of local stationary states would appear. However, ontogenesis, as a whole, means the persistence of a biological system in a non-stationary state in which the course of spontaneous processes is affected by informational constraints.

The action of physical forces of an informational character which do not occur in balanced equations of classical, and also non-equilibrium, thermodynamics, means that thermodynamics does not yield a sufficient basis to describe the development of living organisms. In order to make possible a strict, plain usage of thermodynamic ideas in biology it is necessary to develop a theory that will quantitatively express the magnitude of constraints hindering the spontaneous course of events and determining the emergence of biological structures. Such a theory would simultaneously define a physical principle of the optimisation of life processes (Chernavsky, 1975). It seems, at the moment, that the minimisation of the energetic costs of information, in relation to the energetic action released in the system, might be the essence of the problem in this, main, principle of life.

Concluding Remarks

The present book opens with Leuschner's paper full of optimism and closes with Żelawski's and Galiński's article full of pessimism. To some extent this reflects the situation in the thermodynamics of biological processes. On the one hand, there are grounds for an optimistic view of the future: there appear more and more works in this field of biophysics (see Ivanitsky, 1982); the problems to be solved by a thermodynamics of biological processes have been outlined; results have been obtained which are of significance not only for theoretical biology and physics but for practical purposes too. On the other hand, the theoretical basis of the thermodynamics of biological processes is not reliable enough. The theory of thermodynamic systems far from equilibrium is not sufficiently elaborated. The thermodynamics of information processes is at the initial stage of development. At the same time all living organisms constitute systems far from equilibrium in which information processes play the decisive role.

As mentioned in previous monographs, the thermodynamics of biological processes ought to consider all the basic biological phenomena from a single standpoint, e. g. development, growth and aging of organisms; reaction of living systems to external and internal stimuli; adaptation of animals and plants to changing environmental conditions; historical development of individual animal groups and evolution of organisms as a whole. In other words the thermodynamics of biological processes must be the basis for theoretical biology. Not without reason did an outstanding Soviet scientist, E. S. Bauer, who made an attempt to construct a thermodynamics of biological processes on the basis of classical thermodynamics, entitled his book 'Theoretical Biology' (Bauer, 1935). The series, completed with the present monograph continues the development of theoretical biology founded by E. S. Bauer.

Perhaps, many of the theories and hypotheses presented in the monographs 'Thermodynamics of Biological Processes' (Berlin, 1978), 'Thermodynamics and Kinetics of Biological Processes' (Berlin, 1983) and in the present monograph 'Thermodynamics and Regulation of Biological Processes' (Berlin, 1985) will turn out to be erroneous and, will be rejected in the course of scientific development. But such is the mechanism of scientific progress: only those hypotheses and their extensions are selected that are close to reality, and, to the ideals that, at the moment, inspire the vast majority of scientific workers. Natural selection acts not only in nature, but in the sphere of scientific ideas, hypotheses and theories. And this difficult and sometimes cruel process gives rise to a beautiful, logically complete and, possibly, real edifice, called science. We hope that, in this series of books on the thermodynamics of biological processes there is enough material for natural selection in science.

List of Contributors

Alekseeva, T. A. - Institute of Developmental Biology, USSR Academy of Sciences (Moscow, USSR)

Bat', O. G. - Airplane Engineering Institute (Moscow, USSR)

Borckmans, P. - Université Libre de Bruxelles (Bruxelles, Belgium)

Chernavskaya, N. M. - Department of Scientific Information, Moscow State University (Moscow, USSR)

Chernavsky, D. S. - Physical Institute, USSR Academy of Sciences (Moscow, USSR)

Dewel, G. - Université Libre de Bruxelles (Bruxelles, Belgium)

Ebeling, W. - Section Physik der Humboldt-Universität zu Berlin (Berlin, DDR)

Eliseev, E. N. - Institute of Precambrian Geology and Geochronology, USSR Academy of Sciences (Leningrad, USSR)

Feistel, R. - Sektion Physik der Humboldt-Universität zu Berlin (Berlin, DDR)

Galiński, W. - Institute of Plant Biology, Warsaw Agricultural University (Warsaw, Poland)

Goodwin, B. C. - School of Biological Sciences, University Sussex (Sussex, England)

Haken, H. - Institut für Theoretische Physik, Universität Stuttgart (Stuttgart, FRG)

Ishida, K. - Department of Applied Chemistry, Ibaraki (Ibaraki, Japan)

Ivanov, K. P. - Institute of Physiology, USSR Academy of Sciences (Leningrad, USSR)

Jumarie, G. - Department of Mathematics, Université du Québec à Montréal (Montréal, Canada)

Keiper, W. - Institut für Theoretische Physik, Rheinisch-Westfälische Technische Hochschule Aachen (Aachen, FRG)

Keller, J. U. - Institut für Thermodynamik, Technische Universität Berlin (Berlin, FRG)

Khanin, M. A. - Airplane Engineering Institute (Moscow, USSR)

Lamprecht, I.- Institut für Biophysik, Freie Universität Berlin (Berlin, FRG)

Lapkin, V. V. - Institute of Inland Waters Biology, USSR Academy of Sciences (Borok, Yaroslavl Region, USSR)

Lerner, V. S. - Research Institute of Special Casting (Odessa, USSR)

Leuschner, D. - Institut für Obstforschung, Akademie der Landwirtschafts-wissenschaften (Dresden, GDR)

Lurié, D. - Facultad de Fisica, Universidad de Barcelona (Barcelona, Spain)

Mager, P. P. - Forschungsgruppe für Pharmakochemie, Bereich Medizin der Karl-Marx-Universität (Leipzig, GDR)

Nikolaev, L. A. - Medico-Biological Department, 2nd Moscow State Medical Institute (Moscow, USSR)

Pikovsky, A. S. - Institute of Applied Physics, USSR Academy of Sciences (Gorky, USSR)

Platonov, A. A. - Department of Cybernetics, Moscow Institute of Electronic Engineering (Moscow, USSR)

Poplavsky, R. P. - Physics Institute, USSR Academy of Sciences (Moscow, USSR)

Presnov, E. V. - Institute of Developmental Biology, USSR Academy of Sciences (Moscow, USSR)

Rabinovich, M. I. - Institute of Applied Physics, USSR Academy of Sciences (Gorky, USSR)

Rius, J. L. - Instituto de Fisica, UNAM (Mexico, Mexico)

Romanovsky, Yu. M. - Physics Department, Moscow State University (Moscow, USSR)

Rosen, R. - Department of Physiology and Biophysics, Dalhousie University (Halifax, Canada)

Rozonoer, L. I. - Institute of Control Problems, USSR Academy of Sciences (Moscow, USSR)

Schlögl, F. - Institut für Theoretische Physik, Rheinisch-Westfälische Technische Hochschule Aachen, FRG)

Schnakenberg, J. - Institut für Theoretische Physik, Rheinisch-Westfälische Technische Hochschule Aachen (Aachen, FRG)

Seese, A. - Forschungsgruppe für Pharmakochemie Bereich Medizin der Karl-Marx-Universität (Leipzig, GDR)

Stratonovich, R. L. - Physics Department, Moscow State University (Moscow, USSR)

Terentjeva, Z. A. - Chemical Department, Moscow State University (Moscow, USSR)

Vasiliev, V. A. - Physics Department, Moscow State University (Moscow, USSR)

Wagensberg, J. Facultad de Fisica Universidad de Barcelona (Barcelona, Spain)

Walgraef, D. Université Libre de Bruxelles (Bruxelles, Belgique)

Zakharov, V. M. Institute of Developmental Biology, USSR Academy of Sc Sciences (Moscow)

Zelawski, W. Institute of Plant Biology, Warsaw Agricultural University (Warsaw, Poland)

Zotin, A. I. Institute of Developmental Biology, USSR Academy of Sciences (Moscow)

References

Adams, R. S., Niswander, J. D. 1967. Developmental 'noise' and a cogenital malformation. Genet. Res. 10, N3, 313-317

Adolph, E. 1951. Some difference in response to low temperatures between warm-blooded and cold-blooded vertebrates. Amer. J. Physiol. 11, 92-101

Agladze, K. I., Krinski, V. I. 1982. Rotating spiral wave solutions of reaction-diffusion equations. Nature (London) 296, 424-426

Alekseev, V. M., Tikhomirov, V. M., Fomin, S. V. 1979. Optimal control. Moscow: Nauka (Russian)

Angrist, S. W., Hepler, L. G. 1967 a. Order and chaos - Laws of energy and entropy. New York: Basic Books, 237 p.

Angrist, S. W., Hepler, L. G. 1967 b. Demons, poetry and life - Laws of energy and entropy. Texas Quart. 10, 26-35

Ariens, E. J., Beld, A. J., De Miranda, R. J. F., Simonis, A. M. 1979. The Receptors. v. I. General Principles and Procedures. R. D. O. Brien (Ed.). New York: Plenum Press, 33

Arnold, V. I. 1971. Common differential equations. Moscow: Nauka, 240 p. (Russian)

Arnold, V. I. 1980. Chapitres supplémentaires de la Théorie des Equations Différentielles Ordinaires. Moscow: Mir

Ashby, W. R. 1956. Introduction to Cybernetics. London: Chapman & Hall

Asmussen, E., Chiodi, H. 1965. Muscular exercise. In: Handbook of Physiology. Sect. 3. Respiration. Vol. II. W. O. Fenn, H. Rann (eds.) Wa Washington: American Physiological Society

Astaurov, B. L. 1927. Study of hereditary disturbances of halters in *Drosophila melanogaster*. Schin. Zh. Eksp. Biol., Ser. A., 3, 1-62 (Russian)

Astaurov, B. L. 1974. Study of hereditary disturbances of bilateral symmetry in relation to variation of similar structures within the organism. In: B. L. Astaurov. Heredity and Development. Moscow: Nauka, 54-109 (Russian)

Atlan, H. 1972. L'organization biologique et la théorie de l'information. Paris: Hermann, 299 p.

Auerbach, F. 1902. Die Weltherrin und ihr Schatten. Jena: Fischer, 110 p.

Bader, R. S. 1965. Fluctuating asymmetry in the dentition of the house mouse. Growth 29, 291-300

Bailey, N. T. J. 1964. The elements of stocahstic processes. New York: Wiley, 249 p.

Balmukhanov, B. S., Ivashchenko, A. T., Malenkow, A. G. 1970. On the mechanism of regulation of biological homeostasis of the cell via the changes of the working regime of energy-supplying system. Izv. Acad. Nauk Kazach. SSR, 60-66 (Russian)

Baranowski, B. 1979. Termodynamika a życie. Wiadomosci Chemiczne 2(380), 67-74

Baranowski, B. 1981. Some thermodynamical comments to the origin and evolution of life. Viva Origino 9, 49-54

Bataille, J., Edelen, D. G. B., Kestin, J. 1982. On the structure of thermodynamic fluxes: a direct implementation of the dissipation inequality. In: Thermodynamics and Kinetics of Biological Processes (I. Lamprecht, A. I. Zotin, eds.), Berlin: Walter de Gruyter, 239-253

Bauer, E. S. 1931. Lebensdauer, Assimilationsgrenze, Rubnersche Konstante und Evolution. Biol. Zentralbl. 51, 74-80

Bauer, E. S. 1935. Theoretical Biology. Moscow, Leningrad: Izd. VIEM, 206 p. (Russian)

Beardmore, J. A. 1960. Developmental stability in constant and fluctuating temperatures. Heredity 14, 411-422

Beier, W. 1968. Biophysik. Leipzig: VEB Thieme. 420 p.

Beitinger, T. L., Fitzpatrick, L. C. 1979. Physiological and ecological correlates of preferred temperature in fish. Amer. Zool. 19, 319-329

Benzinger, T. 1969. Heat regulation: homeostasis of central temperatures in man. Physiol. Rev. 49, 671-759

Benedict, F. 1932. Carnegie Inst. Wash., Publ., N. 425, (ref. In: Hensel Temperatur und Leben. Berlin, 1955)

Bennett, A. F., Dawson, W. R. 1976. Metabolism. In: Biology of the Reptilia. New York: Academic Press, Vol. 5, 127-223

Bergé, P. 1981. In: Chaos and Order in Nature, Schloß Elmau (H. Haken, ed.), Berlin: Springer

Berry, F. A., Bollay, E., Norman, R., Beers, N. R. (eds.) 1945. Handbook of Meteorology. New York: McGraw Hill, 1068 p.

Bertalanffy, L. von 1977. Biophysik des Fließgleichgewichtes. Braunschweig: Vieweg

Bhandari, R. 1976. Entropy, information and Maxwell's demon after quantum mechanics. Pramana 6, 135-145

Blackshaw, S. E., Warner, A. E. 1976. Alterations in resting membrane properties during neural plate stages of development of nervous system. J. Physiol. 255, 231-247

Blake, C. C. F. 1978. Do genes-in-pieces imply protein-in-pieces. Nature (London) 273, 267-268

Bligh, J. 1973. Temperature Regulation in Mammals and other Vertebrates. New York: 437 p.

Blumenfeld, L. A. 1974. Problems of biological physics. Moscow: Nauka 355 p. (Russian)

Bogolubov, N. N. 1971. Lectures in Quantum Statistics. London: McDonald Vol. 2

Bogomolni, R. A., Baker, R. A., Lozier, R. H., Stoeckenius, W. 1980. Action spectrum and quantum efficiency for proton pumping in *Halobacterium halobium*. Biochemistry 19, 2152-2159.

Boiteux, A., Hess, B. 1980. Spatial dissipative structures in yeast extracts. Ber. Bunsenges. Phys. Chem. 84, 392-398.

Bongard, M. M. 1963. On the notion "useful information". In: Problems of cybernetics. Moscow: Fizmatgiz 9, 71-102 (Russian)

Borsellino, A., Fuortes, M. G. F. 1968. Responses to single photons in visual cells of Limulus. J. Physiol. 196, 507-539

Bowen, R. 1979. Methods of Symbolic Dynamics. Moscow: Mir 246 p. (Russian)

Bremermann, H. J. 1979. Theory of spontaneous cell fusion, Sexuality in cell populations as an evolutionary stable strategy. J. theoret. Biol. 76, 311- 334

Bremermann, H. J. 1980. Sex and polymorphism as strategies in host-pathogen interactions. J. theoret. Biol. 87, 671- 702

Bresler, S. E. 1981. On the origin of spontaneous mutations. In: Problems of modern history of evolutionary theory. Leningrad: Nauka 27-37 (Russian)

Brett, J. R. 1971. Energetic responses of salmon to temperature. A study of some thermal relations in the physiology and freshwater ecology of sockeye salmon (*Oncorhynchus nerka*). Amer. Zool. 11, 99-113

Bridgman, P. W. 1928. Note on the principle of detailed balancing. Phys. Rev. 31, 101-102

Bridgman, P. W. 1953. Reflections on thermodynamics. Proc. Amer. Arts Sci. 82, 301-309

Brillouin, L. 1951a. Maxwell's demon cannot operate: information and entropy. I . J. Appl. Phys. 22, 334-337

Brillouin, L. 1951b. Physical entropy and information. II. J. Appl. Phys. 22, 338-343

Brillouin, L. 1959. Information theory and its applications to fundamental problems in physics. Nature (London) 183, 501-502

Brillouin, L. 1962. Science and Information Theory. New York: Academic Press 351 p.

Brillouin, L. 1964. Scientific uncertainty and information. New York: Academic Press 164 p.

Brody, S. 1945. Bioenergetics and Growth. New York: Reinhold 1023 p.

Brostow, W. 1972. Between laws of thermodynamics and coding of information Science 178, 29-32

Brückner, D. 1976. The influence of genetic variability on wing asymmetry in honeybees (*Apis mellifera*). Evolution 30, 100-108

Burton, A. Edholm, O. 1955. Man in Cold Environment. London: 329 p.

Butler, J. A. V. 1946. Life and the second law of thermodynamics. Nature (London) 158, 153-154

Butler, T. 1971. Development of LINC methods. In: Proc. Second Intr. Conf. on Numerical Methods in Fluid Dynamics. M. Holt (ed.). Berlin: Springer 146-155

Callen, H. B. 1960. Thermodynamics. New York: Wiley.

Cannon, W. B. 1932. The Wisdom of the Body. New York: Norton, 312 p.

Carnot, S. 1924. Reflexions sur la puissance motrice du feu et les machine propres à développer cette puissance. Paris.

Cater, H. D. 1947. Henry Adams and his friends - a collection of his unpublished letters. Boston: Miftlin, 798 p.

Chambadal, P. 1963. Evolution et applications du concept d'entropie. Paris: Dunod 220 p.

Chatonnet, J., Cabanac, M. 1965. The perception of thermal comfort. Int. J. Biochim. Biometeorol. 9, 183-193

Chernasky, D. S., Dumaev, A. K., Savransky, V. V., Chernavskaya, N. M. Chizhov, I. V. 1982. Kinetics of photo-induced proton transport in bacteriorhodopsin: a physical model. Moscow: Physical Institute of the USSR Academy of Sciences. 1-24, Preprint.

Chernavsky, D. S. 1975. Thermodynamics and Biology. In: Theoretical and Experimental Biophysics. Kaliningrad 5, 3-27 (Russian)

Chernavsky, D. S. 1978. Principles of biological catalysis. In: Mathematical biology and medicine. Results of Science and Technology. Vol. 1 9-57 (Russian)

Chernavsky, D. S. 1980. Theoretical approach to the problem of life origin. Zh. Vsesoyuzn. khim. obsh. im. D. I. Mendeleeva 25, 404-411 (Russian)

Chernavsky, D. S., Korogodin, V. I. 1982. Surplus information and its role in evolution. In: Proceedings of Lectures and symposia, 1. Biophys. Congress. Moscow: Nauka 105-106 (Russian)

Chernavsky, D. S., Chernavskaya, N. M., Yaguzhinsky, L. S. 1982. Physical model of initial stages of energy transformation at oxidative phosphorilation. Biofizika 27, 114-119 (Russian)

Chernavskaya, N. M., Chernavsky, D. S. 1981. Physical aspects of bacterio-rhodopsin action model. Photosynthetica 15, 195-200

Chernousko, F. L. 1980. Some optimal branching structures in biomechanics. In: Mechanics of composite materials. Riga, 308-313 (Russian)

Chistyakov, I. G., Pikin, S. A. 1981. Liquid crystals. In: Modern crystallography. Moscow: Nauka 4, 425-472 (Russian)

Cohen, D. 1954. Optimal systems. I. The vascular system. Bull. math. Biophys. 16, 59-74

Cohen, D. 1955. Optimal systems. II. The vascular system. Bull. math. Biophys. 17, 219-227

Cohen, D. 1969. Optimizing reproduction in a randomly varying environment. J. theoret. Biol. 12, 119-129

Cohen, D. S., Neu, J. C., Rosales, R. R. 1978. Rotating spiral wave solutions of reaction-diffusion equations. SIAM J. Apl. Math. 35, 536-547

Colquhoun, D. 1973. The relation between classical and cooperative models for drug action. In: Drug Receptors (H. P. Rang ed.). London: Macmillan, 149 p.

Conrad, M. 1979. Bootstrapping on the adaptive landscape. Biosystems 11, 167-182

Cooper, M. S. 1979. Long-range dielectric aspects of the eukaryotic cell cycle. Physiol. Chem. Phy. 11, 435-442

Cooper, M. S. 1981. Coherent polarization waves in cell division and cancer. Collec. Phenom. 3, 273-288

Costa de Beauregard, O. 1960. Sur l'equivalence entre information et entropie dans la rapport I/k ln 2. C. rend. hebd. Acad. Sci. 251, 2898-2900

Coutent, C. C. 1977. Compilation of temperature preference data. J. Fish. Res. Board Canada 34, 739-745

Crescitelli, F. 1935. The respiratory metabolism of *Galleria mellonella* (bee moth) during pupal development at different constant temperatures. J. cell. comp. Physiol. 6, 351-368

Crick, F. H. C. 1968. The origin of the genetic code. J. mol. Biol. 38, 367-376

Croquette, V., Schosseler, F. 1982. Diffusive modes in Rayleigh-Bénard structures, J. Physique 43, 1183-1191

Crowther, J. G. 1979. James Clark Maxwell - demon of modern physics. From Edinburgh to first Cavendish Professor. New Scientist, 340-343

Csanyi, V. 1980. General theory of evolution. Acta Biol. Acad. Sci. Hung. 31, 409-434

Cziszar, I. 1963. Eine informationstheoretische Ungleichung und ihre Anwendung auf den Beweis der Ergodizität von Markoffschen Ketten. Magyar Tudomanyos Akad. Matemat. Kutato. Intezetenek Közlemenyei Budapest, 85-108

Daub, E. E. 1970/1971. Maxwell's demon. Studies in History and Philosophy of Sciences 1, 213-227

Dean, R. B. 1941. Theories of electrolyte equilibrium in muscle. Biol. Symp. Lancaster 3, 331-348

De Groot, S. R. 1951. Thermodynamics of Irreversible Processes. Amsterdam: North-Holland, 242 p.

De Groot, S. R., Mazur, P. 1963. Non-Equilibrium Thermodynamics. New York: North-Holland, 510 p.

Demers, P. 1944. Les démons de Maxwell et le second principe de la thermodynamique. Can. J. Res. 22, 27-51

Demers, P. 1945. Le second principe et la théorie des quanta. Can. J. Res. 23, 47-55

Denbigh, K. G. 1979. An Inventive Universe. London: Hutchinson, 250 p.

De Marinis, F. 1959. The nature of asymmetry and variability in the double Bar-eyeless[2] Drosophila. Genetics 44, 1101-1111

Deutsch, D. H. 1981. Can the second law of thermodynamics be circumvented? Amer. Lab. 13, 54-65

Devaney, R. 1976. Homoclinic orbits in Hamiltonian systems. J. different. Equat. 21, 431-438

Dewel, G., Borckmans, P., Walgraef, D. 1980. Concepts in the theory of nonequilibrium phase transitions. J. de Phys., CA-101

Dewel, G., Walgraef, D., Borckmans, P. 1981. Layered structures in two-dimensional nonequilibrium systems. J. Physique Lett. 42, L361-L364

Dimnikova, L. P., Ivanov, K. P. 1969. Thermosensitivity of hypothalamus. Fiziol. Zh. SSSR 55, 295-301 (Russian)

Dolnik, V. P. 1968. Energetic metabolism and evolution of animals. Usp. Sovr. Biol. 66, 276-293 (Russian)

Dolnik, V. R. 1978. Energetic metabolism and size of animals: physiological basis of relations between them. Zh. Obshch. Biol. 39, 805-816 (Russian)

Dontsova, G. V., Zotin, A. I. 1982. Relationship between maximal metabolism, body weight and standard metabolism of animals. In: Thermodynamics and kinetics of biological processes (I. Lamprecht, A. I. Zotin, eds.). Berlin: Walter de Gruyter, 369-386

Dreyfus, J. M., Guyon, E. 1980. Convective instabilities in nematues caused by an elliptical shear. J. Phys. 42, 283-292

Dreyer, F., Sterz, R. 1978. Determination of dose-response curves by quantitative ionophoresis at the frog neuromuscular junction. J. Physiol. 281, 395

Drischel, H. 1976. Einführung in die Biokybernetik. Berlin: Akademie-Verlag, 262 p.

Dubina, T. L., Razumovich, A. N. 1975. Introduction into experimental gerontology. Minsk: Nauka i Tekhnika. 168 p. (Russian)

Dubois, M., Bergé, P., Wesfreid, J. 1978. Non Boussinesq convective structures in water near 4 °C. J. Physique 39, 1253-1257

Dubois-Violette, E., Durand, G., Guyon, E., Manneville, P., Pieranski, P. 1978. Instabilities in nematic liquid crystals. In: Liquid Crystals (L. Liebert, ed.). Solid state Phys. Suppl. 14. New York: Academic Press, 147-208

Duschek, A. 1961. Vorlesungen über höhere Mathematik. Wien: Springer, Vol. 4, 67 p.

Ebeling, W. 1976. Strukturbildung bei irreversiblen Prozessen. Leipzig: Teubner, 262 p.

Ebeling, W., Feistel, R. 1976. Zur nicht-linearen Kinetik von homogenen Konkurrenzreaktionen. Z. phys. Chem., Leipzig 257, 705-720

Ebeling, W., Feistel, R. 1982. Physik der Selbstorganisation und Evolution. Berlin: Akademie-Verlag, 452 p.

Ebeling, W., Sonntag, I., Schimansky-Geier, L. 1981. On the evolution of biological macromolecules. II. Catalytic networks. Studia biophysica 84, 87-88

Eckert, N., Kramer, L. 1978. New proposal for the amplification mechanism in visual excitation. In: 6th Int. Biophys. Congr., Kyoto, Abstract VII-23-(510), 327

Eddington, A. S. 1955. The Nature of the Physical World. London: Dent, 345 p.

Ehrenberg, W. 1967. Maxwell's demon. Scient. Amer. 217, 103-110

Eigen, M. 1971. The selforganisation of matter and the evolution of biological macromolecules. Naturwissenschaften 58, 465-523

Eigen, M. 1976. Wie entsteht Information? Prinzipien der Selbstorganisation in der Biologie. Ber. Bunsenges. Phys. Chem. 80, 1059-1081

Eigen, M. 1978. How does information originate? Principles of biological self-organisation. Adv. Chem. Phys. 38, 211-262

Eigen, M. 1981. Darwin und die Molekularbiologie. Angew. Chemie 93, 221-229

Eigen, M., Schuster, P. 1977. The hypercycle. A principle of natural self-organization. Part A. Emergence of the hypercycle. Naturwissenschaften 64, 541-565

Eigen, M., Schuster, P. 1978 a. The hypercycle. A principle of natural self-organization. Part B: The abstract hypercycle. Naturwissenschaften 65, 7-41

Eigen, M., Schuster, P. 1978 b. Part C: The realistic hypercycle. Naturwissenschaften 65, 341-369

Eigen, M., Schuster, F. 1979. The abstract hypercycle erratum. Naturwissenschaften 66, 512

Eigen, M., Winkler, R. 1973/1974. Ludus vitalis. Mannheim: Mannheimer Forum, Studienreihe Boehringer, 53-139

Eliseev, N. A. 1967. Foundations of structural petrology. Leningrad: Nauka, 258 p. (Russian)

Eliseev, E. N. 1971. Variational physico-chemical analysis of crystallization processes of multicomponent systems. Leningrad: Nauka, 128 p. (Russian)

Eliseev, E. N. 1975. Physico-chemical modelling (exemplified by crystallization processes of multicomponent systems). Leningrad: Nauka, 156 p. (Russian)

Eliseev, E. N. 1980. Supermolecular regularities of selforganization of the developing physico-chemical systems. In: Regularities of development of complex systems (Evolution and supermolecular non-equilibrium phenomena). Leningrad: Nauka, 272-275, 283-300 (Russian)

Eliseev, E. N. 1982. Phases with inner degrees of freedom appearing under non-equilibrium conditions. In: Currents of ideas and regularities of natural history. Leningrad: Nauka, 208-239 (Russian)

Eroshenko, T. M. 1982. Problems of homeostating of basic physiological systems. In: Itogi nauki i tekhniki. Fiziologiya cheloveka i zhivotnikh. Moscow: Izd. VINITI 25, 19-111 (Russian)

Feistel, R. 1979. Selektion und nichtlineare Oszillationen in chemischen Modellreaktionen. Rostock: Dissertation.

Feistel, R., Ebeling, W. 1978. On the Eigen-Schuster concept of quasispecies in the theory of natural selforganization. Studia biophys. 71, 139

Feistel, R., Ebeling, W. Models of Darwin processes and evolution principles. Biosystems (in press).

Feistel, R., Romanovsky, Yu. M., Vasiliev, V. A. 1980. Evolution of Eigen hypercycles proceeding in coacervates. Biofizika 25, 882-887. (Russian)

Feynman, R. P., Leighton, R. B., Sands, M. 1963. The Feynman Lectures on Physics. Reading: Addison-Wesley, Vol. 1, 52.

Flagg-Newton, J. L., Loewenstein, W. R. 1980. Asymmetrical permeable membrane channels in cell junction. Science 207, 771-773.

Fleury, P. A. 1981. Phase transitions, critical phenomena, and instabilities. Science 211, 125.

Fisher, R. A. 1958. The genetical Theory of Natural Selection. Oxford, 1930, Dover N.4 291 p.

Fuortes, M. G. F., Hodgkin, A. L. 1964. Changes in time scale and sensitivity in the ommatidia of Limulus. J. Physiol. 172, 239-263.

Fritzsch, G. 1970. Zur Untersuchung des Wachstums von Einzelzellern mit Hilfe der Thermodynamik irreversibler Prozesse und offener Systeme. Biophysik 7, 33-

Fröhlich, H. 1968 a. Long-range coherence and energy storage in biological systems. Int. J. Quantum Chem. 2, 641-649

Fröhlich, H. 1968 b. Bose condensation of strongly excited longitudinal electric modes. Phys. Lett. 26 A, 402-403

Fröhlich, H. 1969. Quantum mechanical concepts in biology. In: Theoretical Physics and Biology (M. Marois ed.). Amsterdam: North-Holland, 13-22

Fröhlich, H. 1970. Long-range coherence and the action of enzymes. Nature (London) 228, 1093

Fröhlich, H. 1972. Selective long-range dispersion forces between large systems. Phys. Lett. 39 A, 153-154

Fröhlich, H. 1973 a. Collective behaviour of non-linearly coupled oscillating fields. With applications to biological systems. Collec. Phenom. 1, 101-109

Fröhlich, H. 1973 b. The connection between macro- and microphysics. Revista del Nuovo Cimento 3, 490-534

Fröhlich, H. 1975 a. Evidence for Bose condensation - like exitation of coherent models in biological systems. Phys. Lett. 51 A, 21-22

Fröhlich, H. 1975 b. The extraordinary dielectric properties of biological materials and the action of enzymes. Proc. Nat. Acad. Sci. USA 72, 4211-4215

Fröhlich, H. 1977 a. In: Synergetics. A Workshop (H. Haken ed.). Berlin: Springer, 241-246

Fröhlich, H. 1977 b. Dielectric theory and ion channels in nerve membranes. Bio. Sytems 8, 193-194

Fröhlich, H. 1977 c. Long-range coherence in biological systems. Revista del Nuovo Cimento 7, 399-418

Fröhlich, H. 1977 d. Neuroscience Res. Prog. Bull. 15, 67-72

Fröhlich, H. 1978. Coherent electric vibrations in biological systems and cancer problem. IEEE Tran. Microwave Theory and Techniques 26, 613-617

Fröhlich, H. 1980. The biological effects of microwaves and related questions. Adv. Electronics and Electron Phys. 53, 85-151.

Frolkis, V. V. Muradyan, Kh. K. 1982. Prolongation of life in experiment. In: Physiological mechanisms of aging. Leningrad: Nauka, 157-186 (Russian)

Fry, F. E. 1947. Effects of the environment on animal activity. Univ. Toronto Stud. Biol. Ser. 55, 1-62

Gabor, D. 1961. Light and information. Progress in Optics, 1, 109-153

Galanin, M. D. 1960. Resonance transfer of excitation energy in luminescing solutions. Trudy FIAN SSR 12, 3-53 (Russian)

Gantmacher, F. R. 1971. Matrizenrechnung. Vol. 2 Berlin: Verlag der Wissenschaften, 575 p.

Gatlin, L. L. 1972. Information Theory and Living Organism. New York: Columbia University Press, 300 p.

Geptner, V. G., Naumov, N. P., Yurgenson, P. B., Sludsky, A. A., Chirkova, A. F., Bannikov, A. G. 1967. Mammals of the Soviet Union Moscow: Visshaya Shk. 2, 1004 p. (Russian)

Gerber, G. E. Anderegg, R. J., Herlihych, P., Gray, K., Biemann, H. C., Khorana, H. C. 1979. Partial primary structure of BR: sequencing methods for membrane protein. Proc. Nat. Acad. Sci. USA 76, 227-231

Ghilarov, M. S. 1970. Regularities and trends of phylogenesis. Zh. Obshch. Biol. 31, 179-188 (Russian)

Ghilarov, M. S. 1976. Feedback and trends of evolutionary process. Vestn. An SSR, 68-76 (Russian)

Gilbert, W. 1978. Why genes in pieces ?. Nature (London) 271, 501.

Gillespie, T. D. 1976. A general method for numerically stimulating the stochastic time evolution of coupled chemical reactions in embryogenesis. J. theoret. Biol. 86, 757-770

Gillespie, T. D. 1978. Monte Carlo simulation of random walks with residence time dependent transition probability rates. J. comput. Phys. 28, 395-407.

Giró, A., Gonzalez, J. M., Padró, J. A., Torra, V. 1980. The structure of the liquid lead at 670° K through molecular dynamics. J. Chem. Phys. 73

Giró, A., Lurié, D., Padró, J. A., Valls, J., Wagensberg, J. 1982. Information theory, biomass diversity and Monte Carlo simulation: a new approach in biology. In: 2th World Conf. on math. at the service of man, Las Palmas de Gran Canaria, 1982

Glansdorff, P. 1960. On a non-linear law of the irreversible phenomena with stationary constraints. Mol. Phys. 3, 277-283

Glansdorff, P., Prigogine, I. 1954. Sur les propriétés différentielles de la production d'entropie. Physica (Amsterdam) 20, 773-780

Glansdorff, P. Prigogine, I. 1964. On a general criterion in macroscopic physics. Physica (Amsterdam) 30, 351-374

Glansdorff, P., Prigogine, I. 1970. Non-equilibrium stability theory. Physica (Amsterdam) 46, 344-366

Glansdorff, P., Prigogine, I. 1971. Thermodynamic Theory of Structure, Stability and Fluctuations. London: Wiley, 306 p.

Goldring, M. A., Lisman, J. E. Single photon transduction in Limulus photoreceptors and the Borselino-Fuortes model. Preprint 1982 (submitted to J. Physiol.)

Gollub, J. P., Steinman, J. F. 1981. Doppler imaging of the onset of turbulent convection. Phys. Rev. Lett. 47, 505-508

Gol'tsman, G. V. 1978. Problem of the thermodynamic cycles of muscle. Biophysics 23, 331-336

Goodwin, B. C. 1963. Temporal Organization in Cells. London: Academic Press, 163 p.

Goodwin, B. C. 1977. Mechanics, fields and statistical mechanics in developmental biology. Proc. Roy. Soc. London 199 B, 407-414

Goodwin, B, C., Trainor, L. 1980. A field description of the cleavage process.

Gorizontov, P. D.(Ed) 1981. Homeostasis. Moscow: Meditsina, 576 p. (Russian)

Gorshkov, K. A., Ostrovsky, L. A., Papko, V. V., Pikovsky, A. A. 1979. On the existence of stationary multisolitons. Phys. Lett. 74 A, 177-179

Górski, F. 1966. Plant Growth and Entropy Production. Krakow: Polish Academy Sciences, 102 p.

Graham, C. F., Morgan, R. W. 1966. Changes in the cell cycle during early amphibien development. Devel. Biol. 14, 439-460

Graham, R. 1974. Hydrodynamic fluctuations near the convection instability. Phys. Rev. 10 A, 1762-1784

Greco, F. A. 1983. On the 2nd law of thermodynamics: a summary of the replies to Deutsch's article. Int. Lab. 13, 38-47

Green, D., Goldberger, R. 1967. Molecular insights into living process. New York: Academic Press, 400 p.

Grekhova, M. T. (Ed.) 1981. Processes in the Systems with Diffusion. Gorky: Izd. IPF AN SSSR, 286 p. (Russian)

Grodins, F. S. 1963. Control Theory and Biological Systems. New York: Columbia University Press, 254 p.

Guazzeli, E. 1981. Thèse de 3e cycle Orsay (juin 1981).

Guazzelli, E., Guyon, E., Wesfreid, J. 1981. In: N. Boccara - Symmetry and Broken Symmetry in Condensed Matter Physics (IDSET, Paris).

Gulidov, M. V., Popova, K. S. 1979. Dynamics of hatching and morphological peculiarities of hatched embryos of *Rutilus rutilus* (L) depending on incubation temperature. Vorp. Ikhtiol. 19, 868-873 (Russian)

Gusak, P. P., Lerner, V. S. 1980. Dynamics of regenerator process and its optimal characteristics. In: Natural sciences in the service of public health. Novosibirsk 83-85 (Russian)

Gyarmati, I. 1965. Relations between the integral principles of thermodynamics and the Hamilton principle. Acta Chim. Acad. Sci. Hung. 47, 367-377

Gyarmati, I. 1970. Non-Equilibrium Thermodynamics. Berlin: Springer 184 p.

Hagan, P. S. 1981. Target patterns in reaction - diffusion systems. Adv. Appl. Math. 2, 400-416

Hagan, P. S. 1982. Spiral waves in reaction - diffusion equations. SIAM J. Appl. Math. 42, 762-786

Hagiwara, S., Jaffe, L. 1979. Electric properties of egg cell membranes. Ann. Rev. Biophys. Bioeng. 8, 385-416

Haken, H. 1978. Synergetics. Berlin: Springer 355 p.

Haken, H. 1983. Advanced Synergetics. Berlin: Springer

Halperin, B. 1979. In: Physics of Low Dimensional Systems. Proc. Kyoto Summer Institute (Y. Nagaoka, S. Hikami eds.).

Hämäläinen, R. P. 1978. Optimization concepts in models of physiological systems. In: Progress in Cybernetics and System Research (R. Trappl, G. J. Klir, L. Ricciardy eds.). New York: Academic Press, Vol. 3 539-553

Hammel, H. T. 1955. Thermal properties of fur. Amer. J. Physiol. 182. 369-376

Hart, J. S. 1956. Seasonal changes in insulation of fur. Canad. J. Zool. 34, 53-57

Hart, J. S. 1971. Rodents. In: Comparative Physiology of Thermoregulation (G. C. Whittow ed.). New York: McGraw, Vol. 2, 1-149

Harvey, A. F. 1965. Techniques of superhigh frequences. Moscow: Sovetskoye radio 2, 774 p. (Russian)

Hassenstein, B. 1967. Biologische Kybernetik. Jena: VEB Fischer, 144 p.

Hayward, S. B., Stroud, R. M. 1981. Projection structure of purple membrane determined to 3-7 Å resolution by low temperature electron microscopy. J. mol. Biol. 151, 491-517

Heimann, P. M. 1970/1971. Molecular forces, statistical representation and Maxwell's demon. Studies in History and Philosophy of Science 1, 189-211

Heinrich, R., Sonntag, I. 1981. Analysis of the selection equations for multivariable population model. J. Theoret. Biol. 93, 326-361

Hellstrom, B. Hammel, H. 1967. Some characteristics of temperature regulation in the unanaesthetized dog. Amer. J. Physiol. 213. 547-560

Hendes, R. V., Duarte, J. T. 1981. Decomposition of vector fields and mixed dynamics. J. math. Phys. 22, 1420-1422

Hensel, H. 1973. Neural processes in thermoregulation. Physiol. Rev. 53, 948-1017

Hemmingsen, A. H. 1960. Energy metabolism as related to body size and respiratory surface, and its evolution. Rep. Steno Memor. Hospital Nordisk Insulinlabor 9, 7-110

Hentschel, H. C. E. 1978. Application of the dynamic renormilization group to non-equilibrium instabilities: the k = 0 hard mode transition. Z. Physik 31 B, 401-406

Hess, W. R. 1914. Das Prinzip des Kleinsten Kraftverbrauches im Dienste Hämodynamischer Forschung. Arch. Anat. Physiol. (Physiol. Abteil) 1, 1-62

Hill, A. V. 1924. Thermodynamics in physiology. Nature (London) 113, 859-862

Hochachka, P. W., Somero, J. N. 1973. Strategies of Biochemical Adaptation. New York: Saunders, 390 p.

Hofbauer, J., Schuster, P., Wolff, R. 1980. Dynamical systems under constant organization. II. Homogeneous growth function of degree p = 2. SIAM J. Appl. Math. 38, 282-304

Horstfield, K., Cummings, G. 1967. Angles of branching and diameters of branches in the n - H human bronchian tree. Bull. math. Biophys. 20, 245-259

Houston, A. H., Rupert, R. 1976. Immediate response of the hemoglobin system of the goldfish *Carassius auratus*, to temperature change. Canad. J. Zool. 54, 1734-1741

Huneeus, F. 1976. Una fundamentación cientifico-cultural del concepto de Guestalt. Acta psiquiàt. psicol. Amérlat. 22, 88-97

Insarov, G. E., Semenov, S. M. 1982. Mathematical Modelling of Biocenosis on the Basis of Holdane-Semevsky Principle. In: Thermodynamics and Kinetics of Biological Processes (I. Lamprecht, A. I. Zotin eds.). Berlin: de Gruyter, 315-324

Irving, L., Krog, J., Monson, M. 1955. The metabolism of some Alaskan animals in winter and summer. Physiol. Zool. 28, 173-185

Ishida, K. 1975. Non-equilibrium thermodynamics of temporally oscillating chemical reaction. J. theoret. Biol. 52, 343-363

Ishida, K. 1981 a. Non-equilibrium thermodynamics of hypercycles. J. Non-Equilib. Thermodyn. 6, 229-240

Ishida, K. 1981 b. Non-equilibrium thermodynamics of the selection of biological macromolecules. J. theoret. Biol. 88, 257-273

Ishida, K. 1982. Self-amplification of fluctuations in the competition of two classes of short hypercycles J. Non-Equilib. Thermodyn. 7, 293-308

Ito, S., Hori, N. 1966. Electrical properties of Triturus egg cells during cleavage. J. Gen. Physiol. 49, 1019-1027

Ivanitsky, G. R. 1982. Pulsating process of science development. Priroda, 14-21 (Russian)

Ivanitsky, G. R., Krinsky, V. I., Sel'kov, E. E. 1978. Mathematical biophysics of the cell. Moscow: Nauka, 310 p. (Russian)

Ivanitsky, G. R., Krinsky, V. I., Zaikin, A. N., Zhabotinsky, A. M. 1981. Soviet Scientific Reviews. In: Biology Reviews, Sec. D. 2, 279

Ivanov, K. P. 1965. Muscle system and chemical thermoregulation. Moscow/ Leningrad: Nauka, 126 p. (Russian)

Ivanov, K. P. 1972. Bioenergetics and temperature homeostasis. Leningrad: Nauka, 170 p. (Russian)

Ivanov, K. P. 1980. The muscle heat production after adaptation to cold. Med. Biol. 58, 76-81

Ivanov, K. P., Dymnikova, L. Khalilov, E. 1981. The convergence onto posterior hypothalamic neurons of signals thermosensors. J. therm. Biol. 6, 37-41

Ivanov, K. P., Kalinina, M. K. 1972. Oxygen consumption and its "critical" tension for the brain cortex. Fiziol. Zh. SSSR 58, 1469-1475 (Russian)

Ivanov, K. P., Kislyakov, Yu. Ya. 1979. Energetic requirements and oxygen supply of the brain. Leningrad: Nauka, 212 p. (Russian)

Ivanov, K. P., Slepchuk, N. A. 1971. On the elaboration of controlling signal in the organism thermoregulation system. Fiziol. Zh. SSSR 57, 103-110 (Russian)

Ivlev, V. S. 1959. Experience of estimating evolutionary significance of the levels of energetic metabolism. Zh. Obshch. Biol. 20, 94-103 (Russian)

Ivlev, V. S. 1963. On the parameters characterizing the levels of energetic metabolism of animals. In: Application of biological methods in biology. Leningrad: Izd. LGU 2, 146-151 (Russian)

Ivlev, V. S., Leyzerovich, Kh. A. 1960. Ecological analysis of distribution of animals in gradient temperature conditions. Trudy Murman. Morsk. Biol. In-ta 1, 3-27 (Russian)

Jaynes, E. T. 1957 a. Information theory and statistical mechanics. I. Phys. Rev. 106, 620-630

Jaynes, E. T. 1957 b. Information theory and statistical mechanics. II. Phys. Rev. 108, 171-190

Johnson, H. A., Knudsen, K. D. 1965. Renal efficiency and information theory. Nature (London) 206, 930-931

Jones, M. N. 1979. The scope of thermodynamics in biochemistry. In: Biochemical Thermodynamics (M. N. Jones ed.). Amsterdam: Elsevier 1-14

Jumarie, G. 1978. Théorie relativiste de l'information et télécommunication, perspectives. Ann. Télé-communications, Paris 33, 13-27

Jumarie, G. 1979 a. Théorie relativiste de l'information. II. Information de Shannon, entropie de Renyi, ensembles flous relativistes. Ann. Télécommunications, Paris 34, 491-507

Jumarie, G. 1979 b . Théorie relativiste de l'information. III. Sur la signification et l'utilisation effective de l'entropie de Renyi dans les problèmes de codages. Ann. Télé-communications, Paris 34, 521-530

Jumarie, G. 1980 a. An approach to encoding in the presence of subjectivity. J. Inform. Optimiz. Sci. India 1, 166-186

Jumarie, G. 1980 b. Théorie relativiste de l'information. IV. Sur l'introduction de facteurs subjectifs dans les processus de communication. Ann. Télé-communications, Paris 35, 281-296

Jumarie, G. 1981. A general paradigm for subjectivity in communication, subjective transinformation and subjective transuncertainty. J. Inform. Optimiz. Sci. India 2, 273-296

Jumarie, G. 1982 a. Théorie relativiste de l'information. V. Transfer d'information, transfer d'incertitude, bilan informationel. Ann. Télé-communications, Paris 37, 201-212

Jumarie, G. 1982 b. Subjective information and limits in learning without teacher. J. Inform. Optimiz. Sci. India 3, 133-148

Jumarie, G. 1983. Subjectivity, Information, Systems. Introduction to a Theory of Relativistic Cybernetics. New York: Gordon and Breach

Junge, H. D. (Ed.) 1975. Automatisierung. Leipzig: VEB Brockhaus 738 p.

Kalabukhov, N. I. 1946. Preservation of energetic balance of the organism as the basis of adaptation process. Zh. Obshch. Biol. 7, 56-74 (Russian)

Kalabukhov, N. I. 1950. Technique of experimental studies on ecology of terrestrial vertebrates. Moscow: Sovetskaya Nauka, 177 p. (Russian)

Keiper, W. 1981. Stochastische Untersuchungen zum Latenzzeitverhalten chemischer Reaktionen. Aachen: Diploma Thesis RWTH.

Keister, M., Buck, J. 1964. Respiratory rate in relation to temperature. In: The Physiology of Insecta (M. Rockstein ed.). New York: Academic Press Vol. 3, 630-639

Keller, J. 1975. Über den Zweiten Hauptsatz in der Thermodynamik Irreversibler Prozesse. Berlin: Walter de Gruyter.

Keller, J. U. 1976. The fundamental inequality for thermodynamic systems with heat and mass transfer. J. Non-Equilib. Thermodyn. 1, 67

Keller, J. U. 1977. Thermodynamik der irreversiblen Prozesse mit Aufgaben, Rechenweg und Lösungen. Teil I. Thermostatik und Grundbegriffe. Berlin: de Gruyter, 570 p.

Keller, J. U. 1984. Evolution of biothermodynamic systems. In: Springer Texts on Synergetics (E. Frehland ed.). Berlin: Springer.

Khanin, M. A., Bukharov, I. B. 1980. A mathematical model of the functional state of the oxygen system. Bull. math. Biol. 42, 627-645

Khanin, M. A., Dorfman, N. L. 1982. Natural selection and optimal principles. In: Thermodynamics and Kinetics of Biological Processes. (I. Lamprecht, A. I. Zotin eds.). Berlin: Walter de Gruyter, 325-336

Khanin, M. A., Dorfman, N. L., Bukharov, I. B., Levadniy, V. G. 1976. Extremal Principles in Biology and Physiology. Moscow: Nauka, 256 p. (Russian)

Kharkevich, A. A. 1963. On information value. In: Problems of cybernetics. Moscow: Fizmatgiz 9, 71-102 (Russian)

Khinchin, A. Ya. 1953. Notion of entropy and probability theory. Usp. mat. nauk. 8, 3-20 (Russian)

Khurgin, Yu. I., Chernavsky, D. S., Shnol, S. E. 1967. Molecule of protein-enzyme as a mechanical system. In: Oscillation processes in biological and chemical systems. Moscow: Nauka 42-50 (Russian)

Kimura, M. 1968. Evolutionary of the molecular level. Nature (London) 127, 624-626

Kimura, M., Ohta, T. 1971. Protein polymorphysm as a phase of molecular evolution. Nature (London) 229, 467-469

King, J. C. 1967. Inbreeding, heterosis and information theory. Amer. Nat. 57, 403-410

Kittel, C. 1973. Physik der Wärme. München: Oldenbourg, Wiley, 462 p.

Kleiber, M. 1961. The Fire of Life. An Introduction to Animal Energetics. New York, Wiley 454 p.

Klein, M. J. 1970. Maxwell, his demon and the second law of thermodynamics. Amer. Scientist 58, 84-97

Klimontovich, Yu. L. 1982. Statistical physics. Moscow: Nauka, 310 p. (Russian)

Kobozev, N. I., Kobozev, N. I. 1948. Elements of total theory of vector-Brown processes and the laws of biological kinematics. Bull. MOIP, Ser. biol. 53, 3-29 (Russian)

Kobozev, N. I. 1971. Studies in the field of thermodynamics of information and thinking processes. Moscow: Izd. MGU, 195 p. (Russian)

Kogan, A. B. 1977. Biological cybernetics. Moscow, Visshaya shkola, 305 p. (Russian)

Kokurewicz, B. 1970. The effect of temperature on embryonic development of *Tinca tinca* (L.) and *Rutilus rutilus* (L.). Zool. Polon. 20, 317-337

Konoplev, V. A., Zotin, A. I. 1975. Orderliness criterion and its application to some evolutionary problems. In: Theoretical and experimental biophysics. Kaliningrad 5, 98-108 (Russian)

Konoplev, V. A., Sokolov, V. E., Zotin, A. I. 1975. Classification of animals according to energetic metabolism. Zool. Zh. 54, 1429-1441 (Russian)

Konyukova, L. G., Orlova, V. V., Shver, Ts. A. 1971. Climatic characteristics of the USSR by months. Leningrad: Gidrometeotsentr, 144 p. (Russian)

Kopell, N., Howard, L. N. 1981. Target patterns and Horseshoes from a perturbed central-force problem: some temporal periodic solution to reaction-diffusion equations. Stud. Appl. Math. 64, 1-56

Kosloff, R. 1981. Thermodynamic aspects of the quantum-mechanical measuring process. Adv. Chem. Phys. 46, 153-193

Kosterlitz, J. M., Thouless, D. J. 1973. Ordering, metastability and phase transitions in two-dimensional systems. J. Phys. 6C, 1181-1203

Krasovsky, A. A. 1963. Dynamics of continuous selftuning systems. Moscow: Fizmatgiz, 468 p. (Russian)

Krasovsky, A. A. 1974. Phase space and statistical theory of dynamic systems. Moscow: Nauka, 232 p. (Russian)

Krebs, H., Kornberg, H. 1957. Energy Transformations in Living Matter. Berlin: Springer, 86 p.

Kuhn, H., Waser, J. 1981. Molekulare Selbstorganisation und Ursprung des Lebens. Angew. Chemie 93, 495-515

Kullback, S. 1951. Information Theory and Statistics. New York: Wiley.

Kullback, S. 1953. A note on information theory. Lett. J. Appl. Phys. 24, 106-107

Kullback, S. 1959. Information Theory and Statistics. New York: Wiley, 395 p.

Kuramoto, Y., Tsuzuki, T. 1974. Reductive pertubation approach to chemical reactions. Prog. theor. Physics 56, 1399-1401

Kuramoto, Y., Tsuzuki, T. 1975. On the formation of dissipative structures in reaction-diffusion systems - reductive pertubation approach. Prog. theor. Physics 54, 687

Kuramoto, Y., Yamada, T. 1976 a. Turbulent state in chemical reactions. Prog. theor. Physics 56, 679-681

Kuramoto, Y., Yamada, T. 1976 b. Pattern formation in oscillatory chemical instabilities. Prog. theor. Physics 52, 1399-1401

Kurdyumov, S. P. 1982. Proper functions of burning of non-linear medium and constructive laws of its organization. In: Modern problems of calculational mathematics and physics. Moscow, Nauka 217-243 (Russian)

Kuschmitz, D., Engelhard, M., Hess, B. 1980. Interaction between the retinal chromophore and amino acid residues in bacteriorhodopsin. Biophys. Struct. Mech. 6, suppl. 122

Kushner, D. J. (ed.) 1978. Microbial Life in Extreme Environments. London: Academic Press, 465 p.

Laing, R. 1974. Maxwell's demon and computation. Philos. Sci. 41, 171-178

Laskowski, W., Pohlit, W. 1974. Biophysik. Stuttgart: Thieme, 508 p.

Lamprecht, I. 1978. Application of the concepts of classical thermodynamics. In: Thermodynamics of Biological Processes (I. Lamprecht, A. I. Zotin eds.). Berlin: Walter de Gruyter 5-18

Lamprecht, I. 1978. Review of the theory of dissipative structures. In: Thermodynamics of Biological Processes (I. Lamprecht, A. I. Zotin eds.). Berlin: de Gruyter, 261-276

Lamprecht, I., Zotin, A. I. (eds.) 1978. Thermodynamics of Biological Processes. Berlin: de Gruyter, 428 p.

Lamprecht, I., Zotin, A. I. (eds.) 1982. Thermodynamics and Kinetics of Biological Processes. Berlin: de Gruyter, 536 p.

Landau, L., Lifshitz, E. M. 1959. Fluid Mechanics. Mass.: Addison-Wesley.

Landau, L., Lifshitz, E. M. 1969. Statistical Physics. Mass.: Addison-Wesley.

Lange, O. L., Kappen, L., Schulze, E. D. (eds.) 1976. Water and Plant Life. Problems and modern Approaches. Berlin: Springer, 536 p.

Lansberg, P. T. 1978. Thermodynamics and Statistical Mechanics. London: Oxford University Press.

Lapkin, V. V., Svirsky, A. M., Golovanov, V. K. 1981. Age dynamics of selected and letal temperatures of fishes. Zool. Zh. 60, 1792-1801 (Russian)

Lapkin, V. V., Svirsky, A. M., Sopov, Yu. N. 1979. Selected temperature and temperature of fish acclimation. Zool. Zh. 58, 1659-1670 (Russian)

La Salle, J. A., Lefschetz, S. 1961. Stability by Liapunov's Direct Method with Applications. New York: Academic Press, 134 p.

Lehninger, A. 1965. Bioenergetics. Amsterdam: Benjamin 258 p.

Lenk, R. (Ed.) 1972. Physik. Leipzig: VEB Brockhaus, 1783 p.

Lerner, A. J. 1970. Grundzüge der Kybernetik. Berlin: VEB Verlag Technik, 344 p.

Lerner, V. S. 1969. Application of physical approach to some control problems. Kishinev: Kartia Moldavii (Russian)

Lerner, V. S. 1972. Optimal control over imposed macroprocesses on the basis of physical approach. Izv. Vissh. Uchebn. Zaved., Radiofizika 15, 1608-1626 (Russian)

Lerner, V. S. 1980. Macromodels of distributed systems and their identification. In: Control systems. Tomsk: Izd. TGU, 61-74 (Russian)

Lerner, V. S. 1981. Some inverse problems of complex system theory. Differ. uravneniya 17, 1498-1501 (Russian)

Lerner, V. S., Gusak, P. P. 1978. Macromodels of probability distributions of microstructures. In: Statistical properties of microstructures. Moscow: Nauka 63-65 (Russian)

Lerner, V. S., Gusak, P. P., Prekup, A. V. (1978). Dynamic macromodel of regenerator process. In: 3rd All-State Conf. on Biolog. and Med. Cybernetics. Moscow: Nauka 2, 71-74 (Russian)

Leuschner, D. 1968. Kristall- und Magnetstrukturuntersuchungen an intermetallischen Verbindungen mit experimentellen Ergebnissen an $ZrCO_2$. Dresden: Dissertation TU, 169 p.

Leuschner, D. 1974 a. Einführung in die numerische Taxonomie. Jena: Fischer, 139 p.

Leuschner, D. 1974 b. Zur mathematischen Behandlung taxonomischer Merkmale. Biom. J. 16, 213-216

Leuschner, D. 1975. Mathematisch-phänomenologische Betrachtung zum Sokalschen Experiment zur genetisch geregelten Lebensdauer. Biom. J. 17, 233-239

Leuschner, D. 1979 a. Grundbegriffe der Thermodynamik. Berlin: Akademie-Verlag, 186 p.

Leuschner, D. 1979 b. Evolutions- und Stabilitätskriterien: Ein thermodynamisches Konzept - auch bei Konvektion. Vorträge Heft II der Tagung "Transportprozesse in turbulenten Strömungen", Eisenach, 20.-24. 11. 1978. Akad. Wissensch. DDR, Zentralinstitut für Mathematik und Mechanik, Berlin

Leuschner, D. 1981 a. Biomathematics on the basis of thermodynamics. Biom. J. 23, 87-104

Leuschner, D. 1981 b. Biomathematics in embryogenesis and the conception of the phoronomics of the entropy. Biom. J. 23, 823-829

Leuschner, D. 1981 c. A remark on the mathematical motivation of the second law of thermodynamics from a biological point of view. Biom. J. 23, 311-314

Leuschner, D. 1981 d. A specification of numerical taxonomy: thermodynamical taxonomy. Biom. J. 23, 611-620

Leuschner, D. 1981 e. Laminare und turbulente Strömung unter thermodynamischem Aspekt. Wissensch. Z. Hochschule Verkehrswesen "Friedrich List" in Dresden 28, 575-582

Leuschner, D., Walden, F., Hackschmidt, M. 1981. Paradigma zur Anwendung der irreversiblen Thermodynamik auf die Turbulenz, insbesondere auf die ebene Kanalströmung. Techn. Mechanik 2, 82-87

Lewin, R. 1980. Evolutionary theory under fire. Science 210, 883-887

Lomov, I. A. 1975. Relative power and basic metabolism. In: Quantitative aspects of the growth of organism. Moscow: Nauka, 244-248 (Russian)

Lorenz, E. N. 1963. Deterministic nonperiodic flow. J. Atmospheric Sci. 20, 130-141

Loschmidt, J. 1869. Der zweite Satz der mechanischen Wärmetheorie. - Sitzungsber. Akad. Wissensch. Wien, Math. Nat. Klasse, 2. Abt., 59, 395-418

Lundström, A. 1967. Genetic aspects of variation in tooth width based on asymmetry and twin studies. Hereditas 57, 403-410

Lurié, D., Wagensberg, J. 1982. On biomass diversity in ecology. Bull. math. Biol. (in press)

MacArthur, R. 1955. Fluctuations of animal populations and measure of community stability. Ecology 36, 533-536

MacPherson, E. 1981. "Informe Benguela II". Dirección General de Pesca. Barcelona

Malenkow, A. G. 1976. Ionic homeostasis and autonomous behaviour of tumors. Moscow: Nauka, 171 p. (Russian)

Manneville, P. 1978. Non-linearities and fluctuations at the threshold of a hydrodynamic instability in nematic liquid crystals. J. Phys. 39, 911-925

Manriquez, M., Rucabado, J. 1976. "Datos informativos n$^{\underline{o}}$ I". Instituto de Investigaciones Pesqueras, Barcelona

Mantelman, I. I. 1958. On distribution of young fishes of some species under thermogradient conditions. Izv. Vsesoyuzn.n/i in-ta ozern. rechn. khoz. 47, 1-62 (Russian)

Margalef, R. 1958. Information theory in ecology. Gen. Syst. 3, 36-71

Margalef, R. 1972. Homage to Evelyn Hutchinson, or why there is an upper limit to diversity. Trans. Conn. Acad. Arts Sci. 44, 211-236

Mather, K. 1953. Genetical control of stability in development. Heredity 7, 297-336

Maxwell, J. C. 1872. The Theory of Heat. London: Longmans, Green, 308 p.

Mayr, E. 1967. Artbegriff und Evolution. Berlin: Hamburg, 617 p.

McClare, C. W. F. 1971. Chemical machines, Maxwell's demon and living organisms. J. theoret. Biol. 30, 1-34

McLennan, J. A. 1964. The formal statistical theory of transport processes. Adv. Chem. Phys. 5, 261-317

Meixner, J. 1968. Entropie im Nichtgleichgewicht. Rheol. Acta 7, 8

Meixner, J. 1969. Thermodynamik der Vorgänge und die Charakterisierung der klassischen Thermodynamik irreversibler Prozesse. Z. Physik. 219, 79

Meixner, J., Reik, H. G. 1959. Thermodynamik der irreversiblen Prozesse. In: Handbook of Physics (S. Flügge, ed.). Berlin: Springer, Vol. III/2, 413-523

Metrópolis, N., Rosenbluth, A. W., Rosenbluth, M. N., Teller, A. H. 1953. Equation of state calculations by fast computing machines. J. Chem. Phys. 21, 1087

Meyer-Eppler, W. 1959. Grundlagen und Anwendungen der Informationstheorie. Berlin: Springer, 446 p.

Milsum, J. H. 1968. Optimization aspects in biological control theory. In: Advances in Biomedical Engineering and Medical Physics (S. N. Levine, ed.). New York: Interscience Publishers, Vol. I, 243-278

Minut-Sorokhtina, O. P. 1972. Physiology of thermoreception. Moscow: Meditsina, 210 p. (Russian)

Mishchenko, A. S., Fomenko, A. T. 1980. Course of differential geometry and topology. Moscow: Izd. MGU, 439 p. (Russian)

Monod, J. 1974. Reduction and related problems. New York: Macmillan

Mori, H. 1969. Correlation function method for transport phenomena. Phys. Rev. 115, 298-300

Morowitz, H. J. 1960. Some consequences of the application of the second law of thermodynamics to cellular systems. Biochem. Biophys. Acta 40, 340

Morrison, P., Ryser, F. 1952. Weight and body temperature in mammals. Science 116, 231-240

Muradyan, H. K., Timchenko, A. N. 1982. Temperature relationship of the life span, intensity of RNA biosynthesis and liberation of carbonic acid in Drosophila. Zh. Obshch. Biol. 43, 560-565 (Russian)

Murray, C. D. 1926. The physiological principle of minimum work applied to the angle of branching of arteries. J. Gen. Physiol. 9, 835-841

Nagle, J. F., Mille, M. 1980. Theory of hydrogen bonded in bioenergetics. J. Chem. Phys. 72, 3959-3971

Nagle, J. F., Morowitz, H. J. 1978. Molecular mechanism for proton transport in membranes. Proc. Nat. Acad. Sci. USA 75, 298-302

Neu, J. 1979. Chemical waves and the diffusive coupling of limit cycle oscillators. SIAM J. Appl. Math. 36, 509-515

Nicolis, G., Prigogine, I. 1977. Self-Organization in Non-equilibrium Systems. New York: Wiley, 306 p.

Niewkoop, P. D. 1977. Origin and establishment of embryonic polar axes in amphibian development. Current Topics in Developmental Biology, Vol. 11, 115-132

Nikitin, V. N. 1982. Experimental approaches to the prolongation of life. In: Biology of aging. Leningrad: Nauka, 548-568 (Russian)

Nikolaev, L. A. 1976. Foundations of physical chemistry of biological processes. Moscow: Visshaya shkola, 260 p. (Russian)

Nikolaev, L. A. 1982. Evolution of chemical systems and coding principle. Zh. Fiz. Khimii 58, 787-798 (Russian)

Nitzan, A. 1977. Chemical instabilities as critical phenomena. In: Synergetics - A Workshop (H. Haken, ed.). Berlin: Springer, 122-132

Novoseltsev, V. N. 1978. Control theory and biosystems. Analysis of conservation properties. Moscow: Nauka, 320 p. (Russian)

Olnyanskaya, R. P., Slonim, A. D. 1947. On the adaptation of organisms to very low temperatures of the environment. Izv. AN SSSR, Ser. Biol., 245-250 (Russian)

Onsager, L. 1931 a. Reciprocal relations in irreversible processes. I. Phys. Rev. 37, 405-426

Onsager, L. 1931 b. Reciprocal relations in irreversible processes. II. Phys. Rev. 38, 2265-2279

Oparin, A. I. 1964. The Chemical Origin of Life. Chicago: Thomas

Oparin, A. I. 1968. Life, its nature, origin and development. Moscow: Nauka, 256 p. (Russian)

Ortoleva, P. 1978. Dynamic Padé approximation in the theory of periodic and chaotic chemical center waves. J. Chem. Phys. 69, 300-307

Osgood, D. W. 1978. Effects of temperature on the development of meristic characters in *Natrix fasciata*. Copeia, 33-47

Ottolenghi, M. 1980. The photochemistry of rhodopsins. Adv. Photochem. 12, 99-200

Ovchinnikov, Yu. A., Abdulaev, N. G., Feygina, M. G., Kiselev, A. V., Lobanov, N. A., Nazimov, N. V. 1978. On the problem of primary structure of bacteriorhodopsin. Bioorgan. Khimiya 4, 979-981 (Russian)

Ovsyannikov, L. V. 1981. Lectures on the bases of gas dynamics. Moscow: Nauka, 368 p. (Russian)

Ozernyuk, N. D., Zotin, A. I., Konoplev, V. A., Zotin, A. A. 1978. Possible mechanism of the origin of bacteria. In: Thermodynamics of Biological Processes. (I. Lamprecht, A. I. Zotin, eds.). Berlin: Walter de Gruyter, 329-340

Pacault, A. 1971. Einführung in eine phänomenologische Untersuchung der Evolution von Systemen. Vortrag Nr. 215, RW Akad. Wissensch., Westdeutscher Verlag, Köln und Opladen

Palmer, J. F., Slack, C. H. 1970. Some bio-electric parameters of early Xenopus embryos. J. Embryol. Exp. Morph. 24, 535-553

Päsler, M. 1975. Phänomenologische Thermodynamik mit einer Einführung in die Thermodynamik der irreversiblen Prozesse von J. U. Keller. Berlin: Walter de Gruyter

Pantaloni, J., Cerisier, P., Bailleux, R., Gerbaud, C. 1981. Convection de Bénard-Maranqoni; un pendule de Foucault? J. Phys. Lett. 42, L147-L150

Patashinskii, A. Z., Pokrovskii, V. L. 1979. Fluctuation Theory of Phase Transitions. Oxford: Pergamon Press

Patlak, C. S. 1956. Contributions to the theory of active transport. Bull. math. Biophys. 18, 271-315

Patlak, C. S. 1957. Contributions to the theory of active transport. II. The gate type non-carrier mechanism and generalizations concerning tracer flow, efficiency, and measurement of energy expenditure. Bull. math. Biophys. 19, 209-235

Pavlov, D. S., Sbikin, Yu. N., Uspensky, D. S. 1967. Effect of temperature on some functional properties of fish vision. In: Behaviour and reception of fishes. Moscow: Nauka, 86-89 (Russian)

Parkhomenko, I. M. 1968. Kinetics of biological reactions. In: Biophysics. Moscow: Visshaya shkola, 48-74 (Russian)

Petukhov, B. S. (ed.) 1971. Theory of heat exchange, terminology. Collection of recommended terms. Moscow: Nauka, N83, 81 p. (Russian)

Pierce, J. R. 1961. Symbols, Signals and Noise: The Nature and Process of Communication. New York: Harper & Row, 305 p.

Platonov, A. A. 1979. About the time-reversibility of processes in Markovian dynamic systems. In: Abstracts of VI. Int. Conf. on Systems Science. Worclaw, Poland, 149

Platonov, A. A. 1981. Equations of Markov theory taking into account the final values of the process. In: Abstr. of the 11th Int. Conf. "Functional differential systems and related topics". Warszawa, 42-43 (Russian)

Platonov, A. A. 1982. Thermodynamical Approach to the Analysis of Statistical Properties of Non-linear Systems. In: Thermodynamics and Kinetics of Biological Processes. (I. Lamprecht, A. I. Zotin, eds.) Berlin: Walter de Gruyter, 19-33

Poluektov, R. A. (ed.) 1974. Dynamic theory of biological populations. Moscow: Nauka, 455 p. (Russian)

Pomeau, P., Manneville, P. 1979. Stability and fluctuations of a spatially periodic convective flow. J. Phys. Lett. 40, L609-L612

Pontryagin, L. S., Boltyansky, V. G., Gamkrelidze, R. V., Mishchenko, E. F. 1969. Mathematical theory of optimal processes. Moscow: Nauka (Russian)

Poplavsky, R. P. 1975. Thermodynamic models of information processes. Usp. Fiz. nauk 115, 465-502 (Russian)

Poplavsky, R. P. 1979. Maxwell demon and relationships between information and entropy. Usp. Fiz. nauk 128, 165-176 (Russian)

Poplavsky, R. P. 1981. Thermodynamics of information processes. Moscow: Nauka, 255 p. (Russian)

Presnow, E. V. 1978. Formalism of non-equilibrium phenomenological thermodynamics. In: Thermodynamics of Biological Processes (I. Lamprecht, A. I. Zotin, eds.) Berlin: Walter de Gruyter, 31-52

Presnov, E. V. 1978. Strengthened evolution criterion in developmental biology. In: Thermodynamics of Biological Processes (I. Lamprecht, A. I. Zotin, eds.). Berlin: Walter de Gruyter, 105-109

Presnov, E. V., Malyghin, S. N. 1982. Axiomatic models of phenomenological thermodynamics. In: Thermodynamics and Kinetics of Biological Processes (I. Lamprecht, A. I. Zotin, eds.) Berlin: Walter de Gruyter, 255-264

Prigogine, I. 1947. Etude thermodynamique des phénomeènes irrèversibles. Liège: Desoer

Prigogine, I. 1955. Introduction to Thermodynamics of Irreversible Processes. Chicago: Thournes, 115 p.

Prigogine, I. 1961. Introduction to Thermodynamics of Irreversible Processes. New York: Wiley, 119 p.

Prigogine, I. 1967. Introduction to Thermodynamics of Irreversible Processes. New York, Wiley, 147 p.

Prigogine, I. 1972. La thermodynamique de la vie. La Recherche 3, 547-562

Prigogine, I. Wiame, J. M. 1946. Biologie et thermodynamique des phénomènes irréversibles. Experientia 2, 451-453

Primavesi, G. R. 1979. Maxwell's demon - an analyst's query on the second law. Amer. Lab. 11, 105-108

Pritchard, D. H. 1977. Doppler velocity selective studies of excited state collisions: Maxwell's demon realized. Bull. Amer. Phys. Soc 22, 32

Prosser, L. (ed.). 1973. Comparative Animal Physiology. Philadelphia: Saunders

Ptitsyn, O. B., Finkelstein, A. V. 1980. Similarities of protein topologies, evolutionary divergence, functional convergence or principles of folding. Quart. Rev. Biophys. 13, 339-386

Quastler, H. 1964. The Emergence of Biological Organization. New Haven: Yale Univ. Press

Rabinovich, E. Z. 1982. Kinetics of transition processes in living systems - load, adaptation, stability. In: Thermodynamics and Kinetics of Biological Processes (I. Lamprecht, A. I. Zotin, eds.). Berlin: Walter de Gruyter, 387-396

Rabinovich, M. I. 1978. Stochastic autooscillations and turbulence. Usp. Fiz. Nauka 125, 123-168 (Russian)

Rashevsky, N. 1960. Mathematical Biophysics. Physico-mathematical Foundations of Biology. New York: Dover, Vol. I, 488 p.

Rasmuson, M. 1951. Variation in inbred lines of *Drosophila melanogaster*. Hereditas 37, 561

Rasmuson, M. 1960. Frequency of morphological deviants as a criterion of developmental stability. Hereditas 46, 511-535

Rastogi, P. P., Zundel, G. 1981. Proton translocation in hydrogen bonds with large proton polarizability formed between a Schiff base and phenols. Biochem. Biophys. Res. Commun. 99, 804-812

Ratiu, T. 1977. Appendix to lecture 1: Bifurcation, semiflows and Navier-Stokes equations. In: "Turbulence Seminar" (P. Bernard, T. Ratiu, eds.). Lecture Notes in Mathematics, Berlin: Springer, 615

Reeve, E. C. R. 1960. Some genetic tests on asymmetry of sternopleural chaeta number in Drosophila. Genet. Res. 1, 151-172

Reeve, E. C. R., Robertson, F. W. 1953. Analysis of environmental variability in quantitative inheritance. Nature (London) 151, 874-875

Rényi, A. 1966. Wharscheinlichkeitsrechnung. Berlin: Deutscher Verlag der Wissenschaften, 547 p.

Rényi, A. 1970. Probability Theory. Amsterdam: North-Holland, 560 p.

Reznichenko, P. N., Kotlyarevskaya, N. V., Gulidov, M. V. 1962. Effect of constant incubation temperature on viability of rutilus roe. Trudy In-ta morf. Zhivotn., AN SSSR, N49, 247-253 (Russian)

Rhim, W. K., Pines, A., Waugh, J. S. 1971. Time-reversal experiments in dipolar-coupled spin systems. Phys. Rev. 3B, 684-696

Robertons, F. W. 1957. Studies in quantitative inheritance. X. Genetic variation of ovary size in Drosophila. J. Genet. 55, 410-427

Robertson, F. W., Reeve, E. C. R. 1952. Heterocygosity, environmental variation and heterosis. Nature (London) 170, 286

Rokitsky, P. F. 1974. Introduction into statistical genetics. Minsk, Vysheyshaya shkola, 447 p. (Russian)

Romanovsky, Yu. M. 1981. Selforganization processes in physics, chemistry and biology. Moscow: Znanio, 47 p. (Russian)

Romanovsky, Yu. M. 1982. Two classes of selforganization processes. Pushino: Inst. Biophys. (Preprint) (Russian)

Romanovsky, Yu. M., Stepanova, N. V., Chernavsky, D. S. 1974. Kinetische Modelle in der Biologie. Jena: VEB Fischer, 164 p.

Romanovsky, Yu. M., Stepanova, N. V., Chernavsky, D. S. 1975. Mathematical modelling in biophysics. Moscow: Nauka, 343 p. (Russian)

Rosen, R. 1967. Optimality Principles in Biology. London: Butterworths, 198 p.

Rosen, R. 1978 a. Feedforwards and global system failure: a general mechanism for senescence. J. theoret. Biol. 74, 579-590

Rosen, R. 1978 b. Fundamentals of Measurement and the Representation of Natural Systems. New York: North-Holland, 221 p.

Rosenblat-Rot, M. 1957. Entropy of stochastic processes. Dokl. ANSSSR 112, 16-19 (Russian)

Rozonoer, L. I. 1973. Metabolism and distribution of resources (generalized thermodynamic approach). Automatica i telemekhanika, 82-103 (Russian)

Rozonoer, L. I. 1982. Irreversible Thermodynamics far from Equilibrium. In: Thermodynamics and Kinetics of Biological Processes (I. Lamprecht, A. I. Zotin, eds.). Berlin: Walter de Gruyter, 219-238

Rozonoer, L. I., Orlov, V. N. 1981. Variational principle for the equations of macroscopic dynamics and its applications in chemical kinetics. Zh. Vychisl. matem. i fiziki 21, 1192-1205 (Russian)

Rozonoer, L. I., Tsirlin, A. M. 1983. Optimal control over thermodynamic processes. Avtomatika i telemekhanika 23, 70-79, 50-64, 88-101 (Russian)

Rubin, A. B. 1976. Thermodynamics of biological processes. Moscow: Izd. MGU, 240 p. (Russian)

Rubner, M. 1908. Das Problem der Lebensdauer und seine Beziehungen zu Wachstum und Ernährung. München

Ruelle, D. 1979. Sensitive dependence on initial condition and turbulent behavior of dynamical systems. Ann. N. Y. Acad. Sci. 316, 408-416

Ruelle, D. 1981. Differentiable dynamical systems and the problem of turbulence. Bull. Amer. Math. Soc. 5, 29-42

Ruelle, D., Takens, F. 1971. On the nature of turbulence. Commun. Math. Phys. 20, 167-192

Ryan, J. P. 1980. Information - entropy interfaces and different levels of biological organization. J. theoret. Biol. 84, 31-48

Sacher, G. A. 1976. Evolution of the entropy and information terms governing mammalian longevity. In: Interdiscipl. Topics in Gerontology. Basel: Karger, Vol. 9, 69-82

Schaarschmidt, B., Lamprecht, I. 1978. Heat production in life processes. In: Thermodynamics of Biological Processes (I. Lamprecht, A. I. Zotin, eds.). Berlin: Walter de Gruyter, 167-179

Scharf, H. J. 1977. Wachstum. Verh. Anat. Ges. 71, 29-58

Scharf, H. J. 1979. Der Schritt von der heuristischen Wachstumskurve zum biophysikalischen Modell. Gegenbaurs morph. Jahrb. 125, 585

Schlögl., F. 1967. On the statistical foundation of the thermodynamic evolution criterion of Glansdorff and Prigogine. Ann. Phys. 45, 155-163

Schlögl, F. 1971 a. On stability of steady states. Z. Physik 243, 303-310

Schlögl, F. 1971 b. On thermodynamics near a steady state. Z. Physik 248, 446-458

Schlögl, F. 1978. A characteristic class of quantities in non-equilibrium thermodynamics and a statistical justification of the local equilibrium approximation. Z. Physik B. 20, 177-184

Schlögl, F. 1980. Stochastic measures in non-equilibrium thermodynamics. Phys. Reports (Rew. Sect. Phys. Lett.) 62, 267-380

Schmalhausen, I. I. 1968 a. Cybernetic Problems of Biology. Novosibirsk: Nauka, 223 p. (Russian)

Schmalhausen, I. I. 1968 b. Factors of Evolution. Moscow: Nauka, 452 p. (Russian)

Schmalhausen, I. I. 1969. Problems of Darwinism. Moscow: Nauka, 493 p. (Russian)

Schmidt-Nielsen, K. 1972. How Animals Work. Cambridge: Univ. Press, 139 p.

Schnakenberg, J. 1981. Thermodynamic Network Analysis of Biological Systems. Berlin: Springer, 95 p.

Schnakenberg, J., Tiedge, J. 1979. Bond graph representation of a photoreception model. J. Franklin Inst. 308, 327-341

Scholander, P. F., Hock, R., Walters, W. 1950 a. Heart regulation in some arctic and tropical mammals and birds. Biol. Bull. 90, 237-249

Scholander, P. F., Walters, W., Hock, R., Irving, L. 1950 b. Body insulation of some arctic and tropical mammals and birds. Biol. Bull. 99, 225-258

Schottky, W., Ulrich, H., Wagner, C. 1973. Thermodynamik. Berlin: Springer (Reprint of the 1929 edition)

Schrödinger, E. 1944. What is Life? Cambridge: University Press, 131 p.

Schuster, P., Sigmund, K. 1980. Self-organization of biological macromolecules and evolutionary stable strategies. In: Dynamics of Synergetics Systems (H. Haken, ed.). Berlin: Springer, 156-169

Schuster, P., Sigmund, K., Wolff, R. 1979. Dynamical systems under constant organization. III. Cooperative and competitive behavior of hypercycles. J. diff. Equations 32, 357-368

Schuster, P., Sigmund, K., Wolff, R. 1980. Mass action kinetics of self-replication in flow reactors. J. Math. Anal. Appl. 78, 88-112

Schwartz, L. S. 1963. Principles of coding, filtering and information theory. Baltimore: Cleaver, 253 p.

Selfridge, O. G. 1958/1959. Pandemonium: a paradigma for learning. In: Proc. Symp. Nat. Phys. Lab. 10, Mechanization of Thought Processes, 511-531

Semenchenko, V. K. 1950. Gosia Williard Gibbs, his life and principle studies. In: G. W. Gibbs. Thermodynamic studies. Moscow. Leningrad: Gostekhteorlit, 11-26 (Russian)

Semevskaya, V. A., Semevsky, F. N. 1977. About stability of ecological systems of the type "parasite-host", "predator-victim" in terms of optimality principle. Zool. Zh. 56, 5-9 (Russian)

Sendov, B. Kh. 1976. Mathematical models of division and differentiation of the cells. Moscow: Izd. MGU, 58 p. (Russian)

Severtsov, A. N. 1934. Main trends of evolutionary process. Moscow, Leningrad: Izd. AN SSSR, 150 p. (Russian)

Shakhparonov, M. I. 1965. On resonance migration of intramolecular oscillations. Zh. fiz. khimii 39, 2370-2375 (Russian)

Shannon, C. E. 1948. A mathematical theory of communication. Bell Syst. Techn. J. 27, 379-423

Shannon, C. E. 1949. Communication in the presence of noise. PIRE 37, 10-21

Shannon, C. E. 1951. Prediction and entropy of printed English. Bell Syst. Techn. J. 30, 50-64

Shannon, C. E., Weaver, M. 1949. The Mathematical Theory of Communication. Urbana: University Illinois Press, 117 p.

Shcherbakov, A. P. 1935 a. Metabolism intensity and life span of Drosophila. I. Introductory notes and survey of literature. Arkhiv biol. nauk 38, 639-650 (Russian)

Shcherbakov, A. P. 1935 b. Metabolism intensity and life span of Drosophila. II. Life span and intensity of metabolism in *Drosophila melanogaster* at various population density. Arkhiv biol. nauk. 38, 651-655 (Russian)

Shcherbakov, A. P. 1937. Metabolism intensity and life span of Drosophila. IV. Intensity of metabolism and life span of *Drosophila melanogaster* at different temperatures. Arkhiv biol. nauk. 45, 73-86 (Russian)

Shidlovsky, V. A. 1982. Modern theoretical notions on homeostasis. In: Results of Science and Technology. Physiology of man and animals. Moscow: VINITI 25, 3-18 (Russian)

Shield, J. 1972. Accumulation and energy metabolism of the dingo, *Canis dingo* and the coyots, *Canis latrans*. J. Zool. 168, 483-501

Shishkin, M. A. 1981. Regularities of ontogenesis evolution. Zh. Obshch. Biol. 42, 38-54 (Russian)

Shnol, S. E. 1979. Physico-chemical factors of biological evolution. Moscow: Nauka, 262 p. (Russian)

Shotter, J. 1966. The existence of the crossroads policeman. Nature (London) 211, 343-345

Showalter, K. 1980. Pattern formation in a ferrion-bromate system. J. Chem. Phys. 73, 3735-3742

Shub, M. 1974. Dynamical systems, filtrations and entropy. Bull. Amer. Mathem. Soc. 80, 27-41

Siegel, M. I., Doyle, W. J. 1975 a. The differential effects of prenatal and postnatal audiogenic stress on fluctuating dental asymmetry. J. exp. Zool. 191, 211-214

Siegel, M. I., Doyle, W. J. 1975 b. The effects of cold stress on fluctuating asymmetry in the dentition of the mouse. J. exp. Zool. 193, 385-389

Slonim, L. D. 1961. Principles of general ecological physiology of mammals. Moscow-Leningrad: AN SSSR, 432 p. (Russian)

Slonim, L. D. 1982. Adaptation to cold and high latitudes. In: Ecological physiology of animals. Leningrad: Nauka, 67-80 (Russian)

Smoes, M. L. 1980. Chemical waves in the oscillatory Zhabotinskii system. A transition from temporal to spatio-temporal organization. In: Dynamics of Synergetic Systems (H. Haken, ed.). Berlin: Springer, 80-96

Smoluchowski, M. von 1912. Experimentell nachweisbare, der üblichen Thermodynamik widersprechende Molekularphänomene. Physik Zeitschr. 13, 1069-1080

Sneath, P. H. A., Sokal, R. R. 1973. Numerical Taxonomy. The Principles and Practice of Numerical Classification. San Francisco: Freeman, 573 p.

Sokal, R. R., Sneath, P. H. A. 1963. Numerical Taxonomy. San Francisco: Freeman, 359 p.

Solyanik, G. I., Chernavsky, D. S. 1980. Mathematical models of morphogenesis. Moscow: FIAN SSSR, Preprint N 8. 16 p. (Russian)

Sonntag, I., Feistel, R., Ebeling, W. 1981. Random networks of catalytic biochemical reactions. Biometr. J. 23, 501-515

Soulé, M. 1967. Phenetics of natural populations. II. Asymmetry and evolution in a lizard. Amer. Nat. 101, 141-160

Soulé, M., Baker, B. 1968. Phenetics of natural populations. IV. The population asymmetry parameter in the butterfly *Coenonempha tullia*. Heredity 23, 611-614

Spray, D. C., Harris, A. L., Bennett, M. 1979. Voltage dependence of functional conductance in early amphibian embryos. Science 204, 432-434

Stengel, R. F. 1976. "Maxwellian demon" improves reactor efficiency. Design News 32, 40-41

Stieve, H. 1982. Photorezeption und ihre molekularen Grundlagen. In: Biophysik (W. Hoppe, W. Lohmann, H. Markl, H. Ziegler, eds). Berlin: Springer, 734-755

Stoeckenius, W., Bogomolni, R. H. 1979. Bacterorhodopsin and purple membrane of Halobacteria. Biochem. Biophys. Acta 505, 215-278

Stong, C. L. 1973. Machines that work like muscles, and how Maxwell's demon was captured in a bottle. Scient. Amer. 228, 112-117

Stratonovich, R. L. 1965. On information value. Izv. AN SSSR, Ser. Tekhnich. kibernetika, 3-12 (Russian)

Stratonovich, R. L. 1967. On important relationships of thermodynamics of irreversible processes. Vestnik MGU, Ser. 3, Fizika, Astronomia, 84-89 (Russian)

Stratonovich, R. L. 1969. On linear thermodynamics of non-equilibrium processes. Vestnik MGU, Ser. 3, Fizika, Astronomia, 40-46 (Russian)

Stratonovich, R. L. 1970. Fluctuation-dissipative thermodynamics with temporal-even and temporal-uneven variables. Vestnik MGU, Ser. 3 Fizika, Astronomia, 479-486 (Russian)

Stratonovich, R. L. 1971. Application of linear and non-linear fluctuation-dissipation thermodynamics to mechanics. Vestnik MGU, Ser. Fizika, Astronomia, 3-10 (Russian)

Stratonovich, R. L. 1975. Information theory. Moscow: Sovetskoye radio 423 p. (Russian)

Stratonovich, R. L. 1980. Phase transitions in non-equilibrium radiophysical systems. Izv. vyssh. uchebn. zav., Radiofizika 23, 942-955 (Russian)

Stugren, B. 1974. Grundlagen der allgemeinen Ökologie. Jena: VEB Fischer, 223 p.

Sugawara, M. 1977. Public nuisance and entropy increase. Ecotoxicol. Environm. Safety 1, 361-364

Sussman, M. V. 1974. Seeing entropy - the incomplete thermodynamics of the Maxwell demon bottle. Chem. eng. Educ. 8, 149-156

Sussman, M. V. Demonstration device. Patent: US 32 89 321, Date 661206

Suzuki, M. 1976. Scaling theory of non-equilibrium systems near the instability point. II. Anomalous fluctuation theorems in the extensive region. Prog. theor. Phys. 56, 477-493

Suzuki, M. 1977. Anomalous fluctuation and relaxation in unstable systems. J. stat. Phys. 16, 477-504

Suzuki, M. 1981. Passage from an initial unstable state to a final stable state. Adv. Chem. Phys. 46, 195-278

Swift, J., Hohenberg, P. C. 1977. Hydrodynamic fluctuations at the convective instability. Phys. Rev. 15A, 319-328

Swinney, H. L., Gollub, J. P. 1981. Hydrodynamic Instabilities and the Transition to Turbulence. Berlin: Springer

Szilard, L. 1925. Über die Audehnung der phänomenologischen Thermodynamik auf die Schwankungserscheinungen. Z. Physik 32, 753-788

Szilard, L. 1929. Über die Entropieverminderung in einem thermodynamischen System bei Eingriffen intelligenter Wesen. Z. Physik 53, 840-856

Symposium. 1979. Thermoregulation in ectotherms. Amer. Zool. 19, 193-384

Tatarko, K. I. 1977. Carp anomalies and the role of temperature factor in their development. In: Biological regime of reservoirs-coolers of heat and power plants and temperature effect on hydrobionts. Moscow: Nauka, 157-196 (Russian)

Tebb, G., Thoday, J. M. 1954 a. Stability in development and relational balance of X chromosomes in *Drosophila melanogaster*. Nature (London) 174, 1109-1110

Tebb, G., Thoday, J. M. 1954 b. Genetic effects of diurnal temperature change in laboratory populations of *Drosophila melanogaster*. Proc. IX Int. Cong. Genet. Caryologia, 789-791

Terentjeva, Z. A., Kobozev, N. I. 1976. On possible relations between entropy of information theory and thermodynamic entropy. Zh. fiz. khimii 50, 877-881 (Russian)

Thoday, J. M. Homeostasis in a selection experiment. Heredity, 1958, 12, 401-415

Thom, R. 1975. Structural Stability and Morphogenesis. New York: Benjamin, 362 p.

Thomson, W. 1853. On the dynamical theory of heat. Trans. Roy. Soc. Edinburgh 20, 261-288

Tiedge, H. 1981. Deterministische und stochastische Modelltheorien zur Photorezeption des Pfeilschwanzkrebses *Limulus polyphemus*. Aachen: Dissertation, RWTH

Tikhonov, A. N. 1952. Systems of differential equations containing small parameters as derivatives. Matematichesky sbornik 31(73), 575-586 (Russian)

Timofeeff-Ressovsky, N. V., Voronzov, N. N., Jablokov, A. V. 1975. Kurzer Grundriß der Evolutionstheorie. Jena: Fischer, 299 p.

Tomito, K., Daido, H. 1980. Possibility of chaotic behaviour and multi-basins in forced glycolytic oscillations. Phys. Lett. 79A, 133-137

Tonnelat, J. 1979. Qu'est-ce qu'un etre vivant? La Recherche 10, 614-622

Tribus, M. 1961 a. Thermostatics and Thermodynamics. Princeton: Van Nostrand, 649 p.

Tribus, M. 1961 b. Information theory as the basis for thermostatics and thermodynamics. J. Appl. Mech. 28, Ser. E., 1-8

Tribus, M., McIrvine, E. C. 1971. Energy and information. Scient. Amer. 59, 179-188

Tricomi, F. G. 1954. Lezioni sulle Equationi a Derivate Parziali. Torino: Gheroni

Trinscher, K. 1967. Biologie und Information. Eine Diskussion über Probleme der biologischen Thermodynamik. Leipzig: Teubner

Trinscher, K. S. 1975. Information and biological thermodynamics. In: Entropy and Information in Science and Philosophy (K. Zeman Ed.). Amsterdam: Elsevier, 105-123

Trukhaev, R. I., Lerner, V. S. 1974. Dynamic models decisions taking. Kishenev, Stiinza (Russian)

Tupikova, N. V. 1949. Nutrition and pattern of daily activity of shrews. Zool. zh. 28, 402-409 (Russian)

Tupper, J., Saunders, J., Edwards, C. 1970. The onset of electrical communication between cells in the developing starfish embryo. J. Cell Biol. 46, 187-191

Turing, A. 1952. On the chemical basis of the morphogenesis. Phil. Trans. Roy. Soc. London 237 B, 37-72

Turner, J. S., Roux, J. C., McCormick, W. D., Swinney, H. L. 1981. Alternating periodic and chaotic regimes in a chemical reaction - experiment and theory. Phys. Lett. 85 A, 9-12

Valentine, D. W., Soulé, M. E., Samollow, P. 1973. Asymmetry analysis in fishes: a possible indicator of environmental stress. Fishery Bull. 71, 357-370

van Kampen, N. G. 1976. The expansion of the matter equation. Adv. Chem. Phys. 34, 245-309 van Kampen, N. G. 1981. Stochastic Processes in Physics and Chemistry. Amsterdam: North-Holland 180 p.

van Valen, L. 1962. A study of fluctuating assymetry. Evolution 16, 125-142

Vasiliev, V. A., Romanovsky, Yu. M., Chernavsky, D. S. 1982. Elements of dissipative structure theory: Relation to the problems of structure formation. In: Mathematical Biology of Development. Moscow: Nauka 82-101 (Russian)

Vasiliev, V. A., Romanovsky, Yu. M., Yakhno, V. G. 1979. Autowave processes in distributed kinetic systems. Usp. fiz. nauk 128, 625-641 (Russian)

Vidal, C., Pacault, A. 1982. Spatial chemical structures, chemical waves. In: Proc. Schloss Elmau Conf., Synergetics (H. Haken ed.)

Volkenstein, M. V., Chernavsky, D. S. 1978. Information and biology. J. Soc. Biol. Struct. 95-108

Volkenstein. M. V. 1978. General biophysics. Moscow: Nauka 592 p. (Russian)

Volkenstein, M. V. 1980. Physics and biology. Moscow: Nauka 152 p. (Russian)

Volkenstein, M. V. 1981. Biophysics. Moscow: Nauka 575 p. (Russian)

Volkenstein, M. V., Chernavsky, D. S. 1979. Physical aspects of application of information theory in biology. Izv. AN SSSR Ser. biol., 531-547 (Russian)

Völz, H. 1982. Information genauer betrachtet. Wissensch. Fortschr. 32, 234-237

Voroshilov, N. V. 1969. Variability during inbreeding. Genetika 5, 169-175 (Russian)

Voytenko, V. P. 1982. Biological age. In: Physiological Mechanisms of aging. Leningrad: Nauka 144-156 (Russian)

Waddington, C. H. 1957. The Strategy of the Genes. London: Allen, Unwin, 262 p.

Waddington, C. H. 1962. New Patterns in Genetics and Development. New York: Columbia University Press 271 p.

Waddington, C. H. 1968. Basic biological conceptions. In: Towards a Theoretical Biology. Vol. 1 Prolegomena (C. H. Waddington ed.). Edinburgh: University Press 11-38

Walgraef, D., Dewel, G., Borckmans, P. 1980. Fluctuations near nonequilibrium phase transitions to nonuniform states. Phys. Rev. 21 A, 397-404

Walgraef, D., Dewel, G., Borckmans, P. 1981. Stochastic nonlinear systems in physics, chemistry and biology. (L. Arnold, R. Lefever eds.). Berlin: Springer 72-83

Walgraef, D., Dewel, G., Borckmans, P. 1982 a. Nonequilibrium phase transitions and chemical instabilities. Adv. Chem. Phys. 49, 311-355

Walgraef, D., Dewel, G., Borckmans, P. 1982 b. Melting of 2D nonequilibrium structures at low and high constraints. Z. Phys. 48 B, 167-173

Walker, I. 1971. Biological memory. Acta biotheoret. 21, 203-235

Walker, I. 1976. Maxwell's demon in biological systems. Acta biotheoret. 25, 103-110

Wang, Yuh-Ying L., Wang Wei-Kung. 1976. Is the active transport in contradiction to the second law of thermodynamics ? Chin. J. Phys. 14, 89-92

Wang, Yuh-Ying L., Wang Wei-Kung. 1979. Membrane potential and active transport - an information theory approach. Physiol. Chem. Phys. II, 77-82

Walter, R., Lamprecht, I. 1978. Modern theories concerning the growth equation. In: Thermodynamics of Biological Processes (I. Lamprecht, A. I. Zotin eds.). Berlin: Walter de Gruyter 143-162

Waever, W., Shannon, C. E. 1975. The Mathematical Theory of Communication. New York: Benjamin 362 p.

Webb, S. J. 1980. Laser-Raman spectroscopy of living cells. Phys. Rep. 60, 201-224

Weizsäcker, C. F. von. 1976. Evolution und Entropiewachstum. Regensburg: Vortrag Jahr. Deutsche Gesellschaft für Biophysik.

Wentzel, J. S. 1964. Elemente der Spieltheorie. Leipzig: Teubner 66 p.

Wiener, N. 1961. Cybernetics. New York: M. I. T. Press 212 p.

Wieser, W., Gnaiger, E. 1980. Über die Thermodynamik von Lebensprozessen. Biologie in unserer Zeit 10, 104-110

Wilson, K., Kogut, J. 1974. Thesis (Princeton University). Phys. Rept. 12 C, 75

Winberg, G. G. 1935. Intensity of metabolism and life span of Drosophila . III. Respiration intensity, size and life span of *Drosophila simulans*. Arkhiv biol. nauk 38, 657-663 (Russian)

Winberg, G. G., Shcherbakov, A. P. 1937. Intensity of metabolism and life span of three species of Drosophila. Zool. Zh. 16, 311-323 (Russian)

Winfree, A. T., 1978. Stably rotating patterns of reactions and diffusion. In: Theoretical Chemistry. Vol. 4, 2-51 (D. Henderson, H. Eyring eds.). Academic 1978

Winfree, A. T. 1980. The Geometry of Biological Time. New York: Springer 530 p.

Winfree, A. T. 1981. Multi-armed vortices in an active chemical medium. In: Chemical Dynamics (C. Vidal, P. Pacault eds.). Berlin: Springer 156 p.

Wong, F. 1977. Mechanisms of the phototransduction process in invertebrate photoreceptors. New York: Dissertation, The Rockfeller University

Wong, F. 1978. Nature of light-induced conductance changes in ventral photoreceptors of Limulus. Nature (London) 276, 76-78

Yablokov, A. V. 1968. On different forms of progressive development in organic nature. Problemy evolutsii 1, 98-115 (Russian)

Zaikin, A. N., Zhabotinskii, A. M. 1970. Concentration wave propagation in two-dimensional liquid phase self-oscillating systems. Nature (London) 225, 535-537

Zakharov, V. M. 1981 a. Criterium of estimation of stability of development in natural populations of animals. Dokl. AN SSSR 258, 254-256 (Russian)

Zakharov, V. M. 1981 b. Effect of incubation temperature on phenotypical variability of *Lacerts agilis*. In: Voprosy gerpetologii. Leningrad: Nauka 56-57 (Russian)

Zakharov, V. M. 1981 c. Asymmetry of morphological structures of animals as an index of insignificant changes of the state of the environment. In: Problems of Ecological monitoring and modelling of ecosystems. Leningrad: Gidrometioizdat 4, 59-66 (Russian)

Zakharov, V. M. 1982. Phenotypical aspect in the studies of natural populations. In: Population Phenetics. Moscow: Nauka 45-55 (Russian)

Zakharov, V. M., Baranov, A. S., Valetsky, A. V. 1982. Effect of incubation temperature on the time of development of *Lacerta agilis* (Sguamata, Lacertidae). Zool. Zh. 61, 883-889 (Russian)

Zakharov, V. M., Zyuganov, V. V. 1980. On the asymmetry of bilateral characters as populational characteristics. Ecology 10-16 (Russian)

Zar, J. H. 1968. Standard metabolism comparisons between orders of birds. Condor 70, 278-279

Zavadsky, K. M. 1971. To the studies of moving forces of arogenesis. Zh. Obshch. Biol. 32, 515-529 (Russian)

Zaytsev, G. F. 1975. Theory of automatic control and regulation. Kiev: Vishcha Shkola 422 p. (Russian)

Zelawski, W. 1981. Modelling of the dry matter accumulation in plants by means of asymptotic (logistic) and exponential function. Studia Forestalia Suecia 160, 31-38

Zeldovich, Ya. B., Malomed, B. A. 1982. Complex wave regimes in distributed dynamic systems. Izv. Vyssh. Uchebn. Zaved., Radiofizika, 1982, 25, 591-618 (Russian)

Ziegler, H. 1977. Der zweite Hauptsatz als Folge des Prinzips vom Maximum der Entropieproduktion. Z. Angew. Math. Phys. 28, 331-337

Ziesche, P. 1968. About a new introduction of entropy in statistical mechanics due to Macke. In: Sitzungsberichte vom Symposium über Informationstheorie (A. Renyi ed.). Budapest 515-518

Zimmermann, U., Vienken, J., Pilwat, G. 1981. Rotation of cells in an alternating electric field: the occurence of a resonance frequency. Z. Naturforsch. 36 C, 173-177

Zotin, A. I. 1974. Thermodynamic approach to the problems of development. growth and aging. Moscow: Nauka 184 p. (Russian)

Zotin, A. I. 1978. The second law, negentropy, thermodynamics of linear irreversible processes. In: Thermodynamics of Biological Processes (I. Lamprecht, A. I. Zotin eds.). Berlin: Walter de Gruyter 19-30

Zotin, A. I. 1979 a. Bound dissipation function - divergence between the data of direct and indirect calorimetry in living organisms. Zh. Obshch. Biol. 40, 229-237 (Russian)

Zotin, A. I. 1979 b. Classification of birds according to energetic metabolism. In: Aktuàlne Problemêmy Aviàrnej Genetiky. III. Celostàne Symposium Medzinarodou Ulastou. Smolenice 33-55 (Russian)

Zotin, A. I., Konoplev, V. A. 1978. Direction of the evolutionary progress of organisms. In: Thermodynamics of Biological Processes (I. Lamprecht, A. I. Zotin eds.). Berlin: Walter de Gruyter 341-348

Zotin, A. I., Konoplev, V. A., Presnov, E. V. 1975. Physical meaning and statistical interpetation of evolution criterion. Zh. Fiz. Khimii 49, 2901-2905 (Russian)

Zotin, A. I., Konoplev, V. A., Presnov, E. V. 1978 Non-linear phenomenological equations. In: Thermodynamics of Biological Processes (I. Lamprecht, A. I. Zotin eds.). Berlin: Walter de Gruyter 115-120

Zotin, A. I., Krivolutsky, D. A. 1982. Rate and direction of evolutionary progress of organisms. Zh. Obshch. Biol. 43, 3-13 (Russian)

Zotin, A. I., Lamprecht, I. H. D. 1982. Calorimetry and psiumetry. J. Non-Equilibr. Thermodyn. 7, 323-338

Zotin, A. I., Ozernyuk, N. D. 1966. Effect of temperature on respiration and ATP level during cleavage of loach eggs. Dokl. AN SSSR 171, 1002-1004 (Russian)

Zotin, A. I., Ozernyuk, N. D., Zotin, A. A., Konoplev, V. A. 1975. Possible way of procaryota origination. Zh. Obshch. Biol. 36, 163-172 (Russian)

Zotin, A. I., Zotina, R. S. 1977. External dissipation function of energy and evolution of an open system to the stationary state. In: Theoretical and Experimental Biophysics. Kaliningrad 7, 10-18 (Russian)

Zotin, A. I., Zotina, R. S., Konoplev, V. A. 1978. Theoretical basis for a qualitative phenomenological theory of development. In: Thermodynamics of Biological Processes (I. Lamprecht, A. I. Zotin eds.). Berlin: Walter de Gruyter 85-98

Zotina, R. S., Zotin, A. I. 1982. Kinetics of constructive processes during development and growth of organisms. In: Thermodynamics and Kinetics of Biological Processes (I. Lamprecht, A. I. Zotin eds.). Berlin: Walter de Gruyter 423-436

Zubarev, D. N. 1965. Locally equilibrium Gibbs ensemble and its relation to the theory of fluctuations and transfer phenomena. Dokl. AN SSSR 162, 532-535 (Russian)

Subject Index

Bei Fragen zur Produktsicherheit wenden Sie sich bitte an:
If you have any questions regarding product safety,
please contact:

Walter de Gruyter GmbH
Genthiner Straße 13
10785 Berlin
productsafety@degruyterbrill.com